ADULT SPECIAL EDUCATION DEPT.
V. C. C. - KING EDWARD CAMPUS
BOX No. 24620, STA. 'C'
1155 EAST BROADWAY
VANCOUVER, B.C. V5T 4N3

Holt
General Science

William L. Ramsey

Lucretia A. Gabriel

James F. McGuirk

Clifford R. Phillips

Frank M. Watenpaugh

HOLT, RINEHART AND WINSTON, PUBLISHERS
New York • Toronto • London • Sydney • Mexico City • Tokyo

THE AUTHORS

William L. Ramsey
Former Head of the Science Department
Helix High School
La Mesa, California

Lucretia A. Gabriel
Science Consultant
Guilderland Central Schools
Guilderland, New York

James F. McGuirk
Head of Science Department
South High Community School
Worcester, Massachusetts

Clifford R. Phillips
Head of the Science Department
Mount Miguel High School
Spring Valley, California

Frank M. Watenpaugh
Head of the Science Department
Helix High School
La Mesa, California

About the Cover. The theme of this cover is flight. In life science we observe several examples of flying organisms, such as the sparrow pictured on the cover. The fossil of the reptile of the Jurassic period is an example of an ancient flying organism uncovered by earth scientists. The principles of physical science have allowed humans to build flying machines such as the space shuttle, also pictured.

Photo credits appear on pages 583–584
Cover design by Caliber Design Planning, Inc.
Cover photos by: (left Stephen
Dalton/Photo Researchers, (center)
E. R. Degginger/Bruce Coleman, (right) Hank Morgan.
Copyright © 1983, 1979 by Holt, Rinehart and Winston, Publishers
All Rights Reserved
Printed in the United States of America

ISBN: 0-03-059951-2

3 4 5 6 071 9 8 7 6 5 4 3

PREFACE TO THE STUDENT

In this text you will explore the four largest branches of science: chemistry, physics, earth science, and life science. When you have finished reading this text, you will have learned many interesting facts about each of these branches of science. Some of the things you will learn about are:

The basic properties of matter that allow us to be able to launch a space shuttle.

Various types of waves and how some of these can be useful and others can be very destructive.

How the decayed bodies of animals that died a very long time ago are still playing an important part in our everyday lives.

How the systems of your body work and what you should do to keep them functioning properly.

How To Use This Book

This book is divided into six main parts. The first part is called "What Does A Scientist Do?" This part of the text will help you to learn some of the basic skills used by scientists. These are skills that you will also use as you continue to study science. Following "What Does A Scientist Do?" are three units covering chemistry and physics. Unit 4 covers earth science and Unit 5 deals with life science.

Each unit is divided into chapters. At the end of each chapter there are questions to help you review what you have learned. Each chapter is also divided into sections. Each section begins with several statements called objectives that are set off by the symbol ●. These objectives tell you what you will learn in the section. Each section also has at least one activity. In the activities you can explore, and thereby better understand, the ideas presented in the section. The activity objective is set off by the symbol ▲. There are also many photographs and diagrams that will also help you understand each section. Special vocabulary words are printed in **boldface** type. The definitions of these words are found in the margin. Each section ends with a summary that will help you review what you learned. Questions follow each summary. These give you a chance to test yourself. If you can answer these questions, you have achieved the objectives. For more information on how to use this text, see the appendix.

In certain sections of this text, such as the material on the history of the earth and on natural selection, scientific data has been used to present the material as theory rather than fact. The information presented allows for the widest possible interpretation.

Throughout this text special features on careers in science will tell you about various jobs that you may find interesting.

Safety

In your study of science, you will learn many fascinating things about the natural world. However, the study of science can also involve many potential dangers. Science classrooms and laboratories contain equipment and chemicals that can be dangerous if not handled properly. You should always follow the directions and cautions in the book when doing an activity.

In addition, your teacher will outline the precautions and rules that must be followed to insure the safety of you and your classmates. Carefully following these rules and demonstrating a positive attitude toward laboratory safety is your responsibility.

Acknowledgements

Special thanks are due the consultants who contributed many helpful suggestions and criticisms in the manuscript state of this program:
Dr. Jerry Faughn, East Kentucky State University, Richmond, Kentucky; Dr. William Frase, University of Cincinnati, Cincinnati, Ohio; Lynette Christian McRae, Science Teacher, Franklin D. Roosevelt High School, Brooklyn, New York; J. Peter O'Neil, Science Teacher and Science Coordinator, Waunakee Public Schools, Waunakee, Wisconsin; Karen E. O'Neil, Energy Consultant for D.O.E. and Science Teacher, Monument Mountain High School, Great Barrington, Massachusetts; Dr. Robert W. Ridky, University of Maryland, College Park, Maryland; Dr. Marylu Shore Simon, Science Consultant, South Brunswick Public Schools, South Brunswick, New Jersey; Gerald N. Slutzky, Safety Specialist and Coordinator of Science and Health Education, Hauppauge Public Schools, Hauppauge, New York.

CONTENTS

To the Student	iii
Acknowledgements	iv

WHAT DOES A SCIENTIST DO?	**1**
1 Problem Solving	1
2 Measuring Length and Temperature	8
3 Mass, Volume, and Density	

UNIT 1 — WHAT MAKES UP OUR WORLD?   17

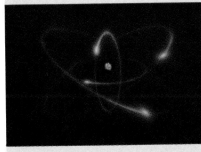

Chapter 1 / MATTER	**18**
1 The Properties of Matter	18
2 Special Properties of Matter	22
2 The Makeup of Matter	24
3 Changing Matter	29
END OF CHAPTER REVIEW	34

Chapter 2 / ATOMS	**36**
1 The Not-So-Solid Atom	36
2 The Nucleus	41
3 The Electron Cloud	46
4 Nuclear Activity	51
5 Atomic Models	56
END OF CHAPTER REVIEW	59

Chapter 3 / ENERGY	**61**
1 What Is Energy?	61
2 Conservation of Energy	65
3 Uses of Energy	71
END OF CHAPTER REVIEW	76

Chapter 4 / CHEMICAL CHANGES	**78**
1 Types of Atoms	78
2 Chemical Activity	82
3 Chemical Families	88
4 Chemical Bonding	93
5 Chemical Reactions	99
END OF CHAPTER REVIEW	104
CAREERS IN CHEMISTRY	106

UNIT 2 / HOW DOES ENERGY AFFECT MATTER? 109

Chapter 5 / HEAT AND TEMPERATURE — **110**
1. Heat Energy — 110
2. Heat Transfer — 115
3. Temperature and Heat — 121
4. Behavior of Gases — 127
5. Liquids and Solids — 133
   CAREERS IN HEATING — 140
   END OF CHAPTER REVIEW — 142

Chapter 6 / FORCES AND MOTION — **144**
1. Speed and Accelerated Motion — 144
2. Forces — 148
3. Work and Power — 154
4. Using Simple Machines — 158
   END OF CHAPTER REVIEW — 165

Chapter 7 / WAVES AND SOUND — **167**
1. Energy and Waves — 167
2. Wave Motion — 173
3. Sound Waves — 177
   END OF CHAPTER REVIEW — 183

Chapter 8 / LIGHT — **185**
1. Behavior of Light — 185
2. Movement of Light Waves — 189
3. Color — 194
4. Bending Light Rays — 201
   END OF CHAPTER REVIEW — 209

UNIT 3 / HOW DO WE MAKE AND USE ENERGY? 211

Chapter 9 / ELECTRICITY AND MAGNETISM — **212**
1. Electric and Magnetic Forces — 212
2. Electric Circuits — 221
3. Measuring Electricity — 227
   END OF CHAPTER REVIEW — 235
   CAREERS IN ELECTRICITY — 236

CHAPTER 10 / ENERGY FOR EVERYDAY USE — **239**
1. Electric Power — 239
2. Making Electricity — 244
3. Home Heating — 251
   HOME HEATING SYSTEMS — 254
4. Energy for Everyday Use — 257
   END OF CHAPTER REVIEW — 266

Chapter 11 / SOURCES OF ENERGY	**268**
1 Fossil Fuels	268
2 Energy Today and Tomorrow	274
3 Energy Alternatives	281
END OF CHAPTER REVIEW	287

UNIT 4

HOW IS OUR PLANET CHANGING?   289

Chapter 12 / CHANGES HAPPENING NOW	**290**
1 Earthquakes	290
2 Volcanoes	298
3 Moving Plates	304
4 Building Forces	311
5 Wearing Away the Land	317
END OF CHAPTER REVIEW	322
MINERALS	324
Chapter 13 / CHANGES IN THE ROCKS	**328**
1 Magma and Igneous Rocks	328
2 Sediments	334
3 Changed Rocks	340
END OF CHAPTER REVIEW	345
Chapter 14 / CHANGES THROUGH TIME	**347**
1 Fossils	347
2 Reading Earth's Diary	354
3 The Earth Through Time	360
END OF CHAPTER REVIEW	367
Chapter 15 / CHANGES IN THE ATMOSPHERE	**369**
1 Weather Factors: Heat and Pressure	369
2 Weather Factors: Moisture	376
3 Weather Changes	381
END OF CHAPTER REVIEW	390
CAREERS: WEATHER WATCH	392
Chapter 16 / OUR CHANGING FRONTIERS	**393**
1 The Oceans	393
2 The Earth in Space	401
3 The Moon	408
4 The Solar System	416
END OF CHAPTER REVIEW	425

UNIT 5 / WHAT MAKES UP OUR LIVING WORLD? 427

Chapter 17 / LIFE ON EARTH — 428
1. Earth: A Special Place — 428
2. Nature's Recycling Business — 434
3. The Activities of Life — 440
   END OF CHAPTER REVIEW — 445

Chapter 18 / ALL LIVING THINGS ARE SIMILAR — 447
1. Animal and Plant Cells — 447
2. Classification: Key to Understanding — 456
3. Classifying: Plants and Animals — 462
   END OF CHAPTER REVIEW — 468
   CAREERS IN LIFE SCIENCE — 470

Chapter 19 / COMMUNITY RELATIONSHIPS — 471
1. Ecosystems — 471
2. The Chain of Life — 476
3. The Living Community — 484
4. Changes in Populations — 490
   END OF CHAPTER REVIEW — 498
   CAREERS IN LIFE SCIENCE — 500

Chapter 20 / CONTINUING THE SPECIES — 501
1. Life Goes On — 501
2. Patterns of Inheritance — 506
3. Human Inheritance — 511
4. Adaptation and Survival — 516
   END OF CHAPTER REVIEW — 522

Chapter 21 / THE HUMAN ORGANISM — 524
1. Support and Movement — 524
2. Digestion — 531
3. Energy Release and Transport — 536
4. Wastes and Excretion — 540
5. Control Systems — 542
   END OF CHAPTER REVIEW — 550

Chapter 22 / THE QUALITY OF OUR ENVIRONMENT — 552
1. Pollution: Our Problem — 552
2. The Size of the Problem — 557
3. Hanging in the Balance — 562
   END OF CHAPTER REVIEW — 568
   APPENDIX — 569
   GLOSSARY
   PHOTO CREDITS
   INDEX

WHAT DOES A SCIENTIST DO?

1 PROBLEM SOLVING

Which of these four people is a scientist? You probably recognize Dr. Albert Einstein in the lower left photo. Dr. Einstein was a famous *physicist*. His research contributed to our understanding of gravity, motion, and atomic energy. When you finish this section, you will be able to:

- Describe the steps in solving problems scientifically.

- Explain the importance of accurate *observations*.

● Explain why scientists use *controlled experiments*.

▲ Use a microscope properly.

What about the people in the other photos? Actually, they are all scientists. Dr. Marion Parks (top left) is an *industrial chemist*. Her job is to develop new chemicals and materials for products used every day in our homes. Jim Morgan (upper right) is a *field geologist*. His job includes identifying and mapping deposits of minerals and rocks for a mining company. Dolores Lugo (lower right) is a *medical technician* in the public health department of a large city. Among her responsibilities are carrying out tests to identify the causes of certain diseases.

What makes these people scientists? They are all involved in solving problems or answering questions in particular ways. Scientists sometimes work alone. Often the answers to scientific questions come from the combined efforts of many people.

What are some of the things scientists do when solving problems scientifically? Making careful **observations** is the first step. An *observation* is any information that is gathered through our senses. Everything we see, hear, taste, touch, or smell is an observation. Scientists make careful observations because they want to learn everything they can about the problems they are working on. They also do a lot of reading to learn what others have already found out about the problem. The following story is one example of how scientists solve problems.

One summer day a large manufacturing company treated its employees and their families to a company picnic. The company hired a food service to provide the picnic meal. Everyone at the picnic enjoyed themselves, but the next day many of them became very ill. Several had to be hospitalized. The public health department of the city needed to find the cause of the illness. It had a problem to solve.

The investigators began by making observations and gathering information. They noted that all of the people

Observation: Any information that comes to us through our senses.

1. *What could cause several of these people to become ill?*

who became ill were at the same picnic. From this observation, several possible reasons for the illness were suggested. First, there could have been something wrong with the food. Second, there could have been something wrong with the water. Third, someone at the picnic could have had a disease that was passed to others.

These suggestions are called **hypotheses.** Forming a *hypothesis* is the second step of scientific thinking. From the observations, measurements, readings, or other information, scientists propose a possible answer or explanation to a problem. This possible answer or explanation is called a hypothesis. Can you think of other hypotheses that might explain the illness?

Hypothesis: A prediction or "educated guess" based on patterns in observations.

The next step for the investigators was to determine which of these hypotheses was the reason for the illness. They had to gather more information and make more observations to test each hypothesis. The investigators questioned many people, including some who did not become ill. They asked what each person did and what they had to eat and drink during the picnic. All of the people who became ill had consumed the food and water supplied by the food service. Those who did not eat or drink at the picnic did not become ill. Therefore, the investigators suspected the food or water rather than a contagious disease. Samples were taken to the health department laboratory for testing. Doctors and technicians like Dolores Lugo ran several tests to see if anything was wrong with the food or water. Several kinds of scientific tools were used for these tests.

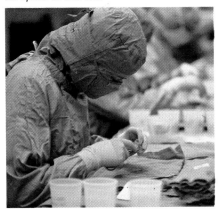

2. *Dolores Lugo examining samples in the lab.*

Tools are needed because the senses are not sensitive enough to provide accurate observations. The tools of a scientist may include microscopes, lasers, computers, thermometers, balances, and even rulers.

These instruments help scientists collect the information they need to form hypotheses. In trying to find the cause of the illness at the picnic, many samples of the food and water were examined under microscopes. If the investigators did not have microscopes, they would not have been able to find the cause of the illness. They would not have known how to treat the ill people. In the Skill Building Activity on pages 4 and 5, you will learn how to use a microscope just as a scientist does.

SKILL BUILDING ACTIVITY

Materials
Compound microscope
Lens paper
Prepared slide
Glass slide
Cover slip
Tweezers
Water
Newspaper page
Scissors
Medicine dropper
Paper
Pencil

I. THE MICROSCOPE

Below is a diagram of a compound microscope. Identify each part on your own microscope and learn what it does. When using a microscope, keep the following rules in mind:

1. Always carry a microscope with both hands. Hold the arm with one hand and place the other hand under the base.

2. Always keep the microscope away from the edge of the work table so that it cannot be knocked off and broken.

3. Always use lens paper to clean the lenses. Other types of paper or tissue may scratch the lenses.

A. Obtain the materials listed in the margin.

B. Locate the objectives and eyepiece on your microscope.
  1. What power is the low power objective?
  2. What power is the eyepiece?
  3. Multiply these numbers to determine the magnification.
  4. What magnification does the high power objective give?

C. Turn the nosepiece so that the low power objective is in position under the tube. Be sure it clicks into position.

D. If your microscope does not have its own light source, turn it so it faces a light or a window. Do not place it in bright sunlight.

3. *The compound microscope.*

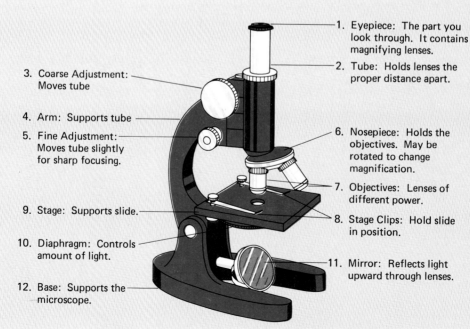

1. Eyepiece: The part you look through. It contains magnifying lenses.
2. Tube: Holds lenses the proper distance apart.
3. Coarse Adjustment: Moves tube
4. Arm: Supports tube
5. Fine Adjustment: Moves tube slightly for sharp focusing.
6. Nosepiece: Holds the objectives. May be rotated to change magnification.
7. Objectives: Lenses of different power.
8. Stage Clips: Hold slide in position.
9. Stage: Supports slide.
10. Diaphragm: Controls amount of light.
11. Mirror: Reflects light upward through lenses.
12. Base: Supports the microscope.

What Does a Scientist Do?

E. While looking through the eyepiece, adjust the mirror so that light reflects upward.

5. Where is the diaphragm located?

F. Open the diaphragm to adjust the light.

G. Place the prepared slide on the stage. Center it over the hole in the stage. Use the stage clips to hold the slide.

H. Look at the microscope from the side. Use the coarse adjustment to lower the low power objective until it *almost* touches the slide.

6. Which way do you turn the knob to lower the objective?

I. Look through the eyepiece and use the coarse adjustment to slowly raise the tube until the slide is in focus.

J. Use the fine adjustment to sharpen the focus.

7. Describe what you see.

8. Can you focus on more than one layer of material?

K. To switch to high power, turn the coarse adjustment to raise the tube. Turn the nosepiece until the high power lens clicks into place.

L. Looking from the side, lower the objective until it *almost* touches the slide. To focus, use the fine adjustment.

9. Describe what you see.

II. PREPARING A WET MOUNT

A. Cut a letter "e" from a newspaper and place it on a slide.

B. Using a medicine dropper, place a drop of water on the "e."

C. Hold a cover slip with forceps as shown in Fig. 4. Draw the cover slip toward the drop until they touch.

D. Gently lower the cover slip onto the slide.

E. Place the slide on the stage so that "e" is facing you.

F. Examine under low power.

1. Is the letter rightside up or upside down?

2. Move the slide to the right. Which way did the "e" move?

G. Examine under high power.

3. Does this look different from low power? How?

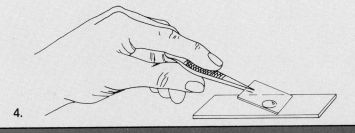

4.

Controlled experiment: Two experimental tests in which all factors are the same except the one being tested.

Experimental factor: The aspect that varies in an experiment to test a hypothesis.

5. *Steps in solving problems scientifically.*

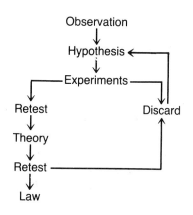

Each of the samples of food and water was run through several tests. These types of activities are called *experiments*. Scientists design experiments to test hypotheses. By comparing the samples of food and water from the picnic to pure food and water, each of the hypotheses could be checked. These are examples of **controlled experiments**. By comparing pure water with water that may be causing a disease, you can determine if there is anything unusual in the suspected water. The pure water is called the *control*. Everything about it is known to be all right. The suspected water is called the *experiment*. If there is anything different about it when compared with the pure water, the difference may be the cause of the disease.

Controlled experiments are very important to scientists. They provide a way to test the effect of one **experimental factor** at a time. Any difference between the results of a control and an experiment must be due to the factor being tested.

The water from the picnic grounds proved to be pure. Thus the hypothesis that this might have been the cause of the illness had to be discarded. This often happens in scientific experiments. The information gained from experimenting may not support the hypothesis being tested. It is then necessary to either discard the hypothesis or to change it.

Most of the food also proved to be pure. But when the investigators tested the potato salad, they found a bacteria growing in it. This type of bacteria grows and develops when foods are not properly refrigerated, especially during warm weather. The bacteria was shown to be the cause of the disease.

Throughout the world, scientists are asking questions and identifying problems. In order to find answers to these questions and problems, scientists follow processes that involve making observations and forming hypotheses. These hypotheses are then tested. If a hypothesis does not stand up to the tests, it must be changed or discarded completely. Even if experiments seem to support a hypothesis, it must be tested again and again. Only when a hypothesis stands up to many tests can it be accepted as correct.

However, the process does not stop there. Scientists are constantly testing and checking. They may find new evidence that indicates a hypothesis does not work in

Theory: A hypothesis that has withstood repeated testing.

Scientific law: A scientific theory that has continuously been upheld by experiments.

all situations. Then it must be changed again. If the hypothesis continues to be supported by evidence from many tests or experiments, it is called a **theory**. However, even *theories* are constantly rechecked. If a theory appears to be correct in all cases, it may become accepted as a **scientific law**. But even a *law* may be proven wrong if new information is uncovered.

SUMMARY & QUESTIONS

Scientific thinking means solving problems in an orderly way. The process includes making observations, forming hypotheses, and testing the hypotheses by experimenting. This method can also be used to solve problems in other than scientific areas.

1. Briefly state the steps of scientific problem solving.
2. Why are accurate observations necessary in scientific problem solving?
3. Explain the term hypothesis.
4. What is the purpose of an experiment in scientific problem solving?
5. Define the terms control and experimental factor.
6. What does a scientist do if an experiment proves a hypothesis to be correct? If it proves the hypothesis to be incorrect?
7. Why are hypotheses and theories that seem correct constantly rechecked?

2 MEASURING LENGTH AND TEMPERATURE

Have you ever seen the Thanksgiving Day parade in New York City? One of the biggest attractions in the parade are the giant balloons of cartoon and movie characters. These balloons are made from plastic-coated nylon and other synthetic fibers. They are examples of the thousands of products created by the chemical industry. When you finish this section, you will be able to:

● Describe the metric system of measurement for length and temperature.

● Explain the importance of accurate measurements.

▲ Make measurements of length and temperature.

Meter (m): The standard unit of length in the metric system.

6. *Accurate measurements are important parts of all scientific work.*

In her job as an industrial chemist, Dr. Parks makes many kinds of observations. Some of these observations are measurements. Dr. Parks often has to find the length or thickness of a sample. She may also have to measure the distance an object is propelled by a certain type of fuel. These measurements must be very accurate. Accurate measurements are very important in scientific experiments. If measurements are not accurate, the results of an experiment may be meaningless.

The simplest tool used for measuring length is the metric ruler. The basic unit for measuring distance in the metric system is the **meter (m)**. To measure a smaller distance, the *meter* may be divided into smaller parts called *decimeters* (dm). There are 10 decimeters in one meter. Each decimeter may also be divided into 10 equal parts called *centimeters* (cm). Therefore, there are 10 × 10 or 100 centimeters in a meter. Each centimeter can be divided into 10 equal parts called *millimeters* (mm). There are 10 × 10 × 10 or a total of 1,000 mm in a meter.

What Does a Scientist Do?

Sometimes it is also necessary to measure distances much larger than a meter. For example, the distance a car can run on a tankful of gas. This type of measurement is given in *kilometers* (*km*). A kilometer is equal to 1,000 meters. To convert from meters to kilometers, divide the number of meters by 1,000. Thus, the 1,500 meter run could also be called the 1.5 kilometer run.

You have seen that it is important for scientists to make accurate measurements. It is also important for you to make accurate measurements as you study science and in your daily activities. For example, you may wish to measure a window for curtains. You may need to find the area of a floor in order to buy new carpeting. You may need to measure a stereo set to see if it will fit into your room.

You can practice some simple measurements in the following Skill Building Activity.

SKILL BUILDING ACTIVITY

Materials
Metric ruler (30 cm)
Pencil
Paper

MEASURING LENGTH

A. Obtain the materials listed in the margin.

B. The numbered lines on the ruler are centimeters (cm).

  1. How many full cm long is your ruler?

C. Line up the first numbered line of your ruler with the top edge of this page. Write the number of the last numbered line that touches the bottom edge of the page.

D. The small lines between the cm marks are millimeters. This means each small line is 1/10 (0.1) of a cm. Count the number of these lines from the last full cm to the lower edge of this page. Multiply this number by 0.1. Add this number to your full cm reading.

E. Subtract the value of the first numbered line where you lined up the edge of the page.

  2. What is the actual length of this page?

  3. Why should you begin measuring at a numbered line instead of at the edge of the ruler?

F. Practice this skill by measuring the following distances to the nearest mm.

7. *The metric ruler.*

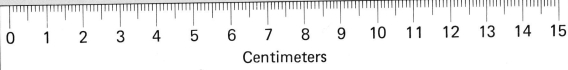

4. The length of your pencil.
5. The length and width of a piece of paper.
6. The length and width of your desktop.
7. The length of your index finger.
8. The length of your thumb.

G. Measure several other objects that are available around the classroom. Be as accurate as possible.

H. For the next part of this activity, work in groups of five. Select one object such as a pencil. Have each member of your group measure this object.

9. Did everyone get the same measurement? If not, what reasons can you give to explain this?

Experiments in chemistry often involve changes in temperature. As an industrial chemist, Dr. Parks needs to know what happens to various substances at different temperatures. She needs to know at what temperatures certain substances freeze, boil, or even explode. Why do you think she must know this? Suppose Dr. Parks was asked to develop a new type of antifreeze for automobiles. She could not make this from a substance that would freeze at normal winter temperatures. She also would not want to use a substance that would explode at the normal running temperature of the automobile engine.

Every day you and the people around you rely on temperature readings for different things. For example, you check the temperature to decide how to dress before going outdoors. Various oven temperatures are used to cook different types of food, and so on.

Scientists use a Celsius (**sell**-see-us) thermometer. On the Celsius scale the boiling point of water is 100°C and the freezing point is 0°C.

The Skill Building Activity on page 11 will teach you how to make temperature measurements by using a laboratory thermometer. Before you begin this or any other activity, there is one thing you should keep in mind. For your safety and the safety of those around you, be sure you understand all directions before doing any activity. Pay attention to all notes of caution and follow all safety procedures. If there is something you do not understand, ask your teacher for help.

SKILL BUILDING ACTIVITY

MEASURING TEMPERATURE

A. Obtain the materials listed in the margin. **CAUTION:** THERMOMETERS ARE GLASS, BE CAREFUL NOT TO BREAK THEM. IF A THERMOMETER DOES BREAK, DO NOT TOUCH IT. CALL YOUR TEACHER.

B. Notice the markings on your thermometer.
  1. How many degrees does each line marked on your thermometer represent?

C. Place several ice cubes in a beaker. Then place the bulb end of your thermometer into the beaker of ice. Wait until the reading stops changing.
  2. What is the temperature of the ice?

D. Fill a beaker about ½ full with room temperature water. Place the thermometer into the beaker. Wait until the temperature reading stops changing.
  3. What is the temperature?

E. Place an ice cube and the thermometer into the beaker of room temperature water. Record the temperature of the water every minute until it stops changing. Stir *gently* with the stirrer. **CAUTION:** DO NOT USE THE THERMOMETER TO STIR.
  4. How many degrees did the ice cube lower the temperature?

Materials
Chemical thermometer
  ($-10°C$ to $110°C$)
100 mL beaker
Water (room temperature)
Stirrer
Ice cubes
Paper
Pencil

SUMMARY & QUESTIONS

Accurate measurements are very important in science and in our everyday lives. Safety is also important to scientists and in the science class.

1. Explain how the metric units for measuring length are related to each other.
2. Why is it important for chemists to know how temperature affects substances?
3. At what temperature does water boil on the Celsius scale? At what temperature does it freeze?

What Does a Scientist Do?–2 Measuring Length and Temperature

3 MASS, VOLUME, AND DENSITY

One of the newer tools used by scientists who study the earth is the space satellite. Every day these satellites send back hundreds of pictures of the earth's surface. You have probably seen some of these photographs on television weather reports. One type of photo has made it possible to map remote parts of the earth. These maps are used by field geologists in their search for new deposits of minerals. When you finish this section, you will be able to:

- List reasons for organizing *data* into tables.
- Make measurements of *mass* and *volume*.
- Explain how to calculate *density*.
- ▲ Use a laboratory balance and a graduate cylinder properly.

Mass: The measure of the amount of matter contained in an object.

Data: Information collected from observations.

8. *Accurate measurements are an important part of all scientific work.*

As a field geologist, Jim Morgan uses measurements constantly. The **mass** of an object is one measurement he often makes. *Mass* is the amount of matter an object contains. When working in the field, it is often necessary to measure the mass of several objects. Much information can be gathered this way. Scientists often organize this information, or **data,** into tables. This makes working with this information easier.

You have just learned that mass is a measure of the amount of matter an object contains. This property is sometimes called weight, but this term is not scientifically correct. To a scientist, weight means the pull of gravity on an object.

For example, look carefully at the pictures of the space shuttle in Fig. 9. Does the shuttle have the same amount of matter in it in each picture? Of course. Does it weigh

9. *The shuttle has both mass and weight on the earth. Does it have mass when it is in space? Weight?*

TABLE 1

1,000 grams (g)	= 1 kilogram (kg)
1 gram (g)	= 10 decigrams (dg)
1 gram (g)	= 100 centigrams (cg)
1 gram (g)	= 1,000 milligrams (mg)

Gram (g): A small unit of mass in the metric system.

the same in each picture? No. The pictures demonstrate the difference between the terms mass and weight. For measurements taken on earth, the two terms are often used as if they had the same meaning. On earth the measurement systems for both mass and weight are the same. In space, where there is no gravity, there is no weight. However, there is still mass. So you see, mass and weight do not mean the same thing.

The basic unit for measuring mass in the metric system is the **gram.** This unit is about equal to the mass of a paper clip. Just like the meter, the *gram* can be divided into smaller units or grouped into larger units. These units are listed in Table 1.

An instrument called a *balance* is used to determine an object's mass. There are several types of balances. The two types most commonly used in schools are the equal arm and the triple beam balances. The Skill Building Activity below will teach you how to use a balance. You can also practice organizing your data into tables.

MEASURING MASS

The triple beam balance has a pan where the unknown mass is placed. There are three beams along which riders are moved in order to balance the unknown mass. The rear beam is marked off in 10 g units. The center beam is marked in 100 g units. The front beam is marked in 1 g units with the lines between equal to 0.1 g.

Another type of balance is the pan balance. This works by balancing an object of unknown mass with objects of known mass. See Fig. 10. If you have a pan balance, your teacher will assist you in using it.

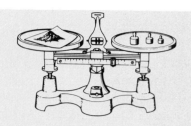

10. *The equal arm balance.*

Materials
Triple beam balance
2 solid objects
Pencil
Paper

TABLE 2

	Objects' mass (g/cm³)	
	My Data	Partner's Data
Object #1		
Object #2		
Object #3		

A. Obtain the materials listed in the margin. In this activity, you will work with a partner.

B. A balance must first be checked without a mass on it. One person should place the balance on a level surface. Move each rider to the zero position on its beam. See Fig. 11.

C. Adjust the balancing screw until the indicator reads zero. The balance should always swing a little when it is used. When the pointer moves equally on both sides of the scale, the instrument is "in balance." See Fig. 11.

D. Place the pencil on the pan. Move the 100 g rider until the pointer goes below the center of the scale. Now move the rider back one number.

E. Move the 10 g rider until the pointer goes below the center line. Now move the rider back one number.

F. Move the 1 g rider until the pointer swings equally on both sides of the center line.

G. The mass of the pencil is equal to the total gram readings of the riders.

  1. What is the mass of the pencil?

H. Now repeat this procedure using the other objects.

I. Have your partner measure the mass of each of the objects you used.

  2. Record your data in a table like Table 2.

  3. Add your partner's data to your table. Compare the results.

  4. What reasons can you think of for the differences between your results and those of your partner?

11. *The triple beam balance.*

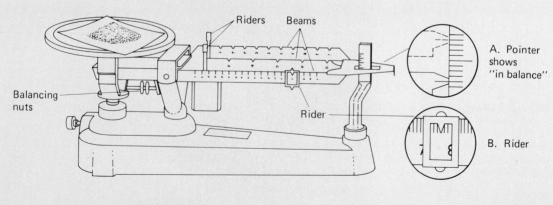

Volume: The amount of space an object takes up.

Liter (L): The standard unit of volume in the metric system.

Density: The amount of matter in a given unit of volume.

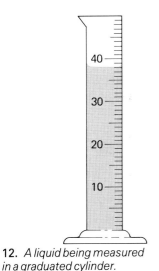

12. *A liquid being measured in a graduated cylinder.*

Another measurement Dr. Morgan uses in his work as a field geologist is **volume**. This measurement indicates how much space an object takes up. The basic unit for measuring the *volume* of a liquid in the metric system is the **liter**. The *liter* can be divided into smaller units in the same way as the meter. Therefore, you can have deciliters, milliliters, and centiliters. What do each of these prefixes mean?

The tool most often used by scientists to measure the volume of liquids is a *graduated cylinder*. You will use this tool in the Skill Building Activity on page 16.

You can also measure the volume of a solid. This measurement is not given in liters. It is given in cubic centimeters (cm^3). To find the volume of a cube or a rectangular solid, measure the length, the width, and the height. Then multiply the length times the width times the height. Written as a formula, this is $V = l \times w \times h$. The Skill Building Activity on page 16 reviews this measurement. The method for finding the volume of an irregular solid will be discussed in Chapter 1.

One of the ways a field geologist identifies a mineral is by calculating its **density**. The *density* tells you how much matter is in a given unit of volume. Therefore, in order to calculate the density, you need to know the volume and mass of an object. Dividing the mass by the volume gives the density.

If you had a brick and a sponge that were exactly the same size, they would each have the same volume. Can you explain why? However, the brick would have more mass. Therefore, the brick is said to be more *dense* than the sponge.

Density is an important measurement in many areas of science. It is part of the reason why there are air currents and ocean currents. It is also involved in many of the properties of matter you will learn about in Chapter 1. Density causes some objects to float and some to sink. For example, a brick is more dense than water. A sponge is less dense than water. So, if you put a brick in a bucket of water, it will sink. If you put a sponge in a bucket of water, it will float.

Since density is really a combination of mass and volume, the units usually used for density are grams per cubic centimeter (g/cm^3). In the Skill Building Activity on page 16, you will learn how to calculate density.

SKILL BUILDING ACTIVITY

Materials
Graduated cylinder
Source of water
3 containers of varying size
5 cylinders for demonstration
3 regularly shaped solids
Balance
Paper
Pencil

I. MEASURING VOLUME

The Volume of a Liquid

A. Obtain the materials listed in the margin.

B. Place the graduated cylinder on a level surface.

1. How many mL are represented by each small line on your graduated cylinder?

C. The surface of a liquid in a glass graduated cylinder is curved. Read the mark closest to the bottom of the curve.

D. Your teacher has set up five cylinders containing water.

2. List the five volumes of water set up by your teacher.

E. You will be given 3 containers. Determine the total liquid volume for each container.

3. List the volumes for each container.

The Volume of a Solid

F. Your teacher will give you 3 regularly shaped solids.

4. Set up a data table to list the length, width, height and volume of each solid.

II. MEASURING DENSITY

A. Add the following information to the data table you made in part I, question 4.

1. What is the mass of each of the three solids?
2. What is the density of each of the three solids?

B. The density of water is 1 gm/cm^3.

3. Will each of the three solids you are working with sink or float in water?

SUMMARY & QUESTIONS

Scientists use data tables to organize information. Density is a property of matter that is related to many other properties. To calculate density, the mass and volume of an object must be known.

1. Define the term density. How is it calculated?
2. How do scientists organize observations?
3. How are mass and weight different?
4. How do you find the volume of a rectangular object?

What Does a Scientist Do?

UNIT 1 WHAT MAKES UP OUR WORLD?

CHAPTER 1

MATTER

1 THE PROPERTIES OF MATTER

The fiery launch of the space shuttle *Columbia* is a fantastic sight. The eighteen-story high combination of spacecraft and rockets seems to leap from the launch pad. For miles around the ground is shaken by the force of the rocket engines as they propel 2 million kilograms of mass toward space. The shuttle appears to perch on top of columns of orange and white flames and smoke as it is hurled skyward. This event is a spectacular example of some of the basic principles of the universe. When you finish this section, you will be able to:

- Define the terms *matter* and *energy*.
- List examples of the *general properties* of matter.
- List examples of the *special properties* of matter.
- Demonstrate that all three *states of matter* take up space.

Matter: Anything that takes up space and has mass.

Energy: The property of something that makes it able to do work.

Each time the shuttle *Columbia* is launched, millions of kilograms of **matter** are changed into **energy**. The *matter* is the fuel used by the rocket engines. The *energy* is the force that lifts the spacecraft into orbit. The launch is a good example of a basic scientific law: Matter and energy cannot be created or destroyed. However, it is possible for one to be changed into the other.

To a scientist, matter is anything that takes up space and has mass. The space it occupies is called its volume. Rocket fuels are examples of two of the forms in which matter can exist. These forms are also called *states of matter*. The fuel in the large central tank consists of oxygen and hydrogen in the *liquid state*. The two smaller rockets contain a *solid* rocket fuel. When these fuels are burned, they produce a third state of matter —*gases*. These escaping gases provide the force that lifts the shuttle skyward.

Under the right conditions, matter can exist in any of these three states. The most familiar example is water. When the temperature is low enough, water exists in a solid state and is called ice. A solid is defined as having a definite shape and a definite volume. An ice cube is an example of a solid. Can you give three more examples of the solid state of matter?

At warmer temperatures, water exists as a liquid. Can you name three more examples of the liquid state? Like solids, liquids have definite volumes. These volumes can be measured with a graduated cylinder. However, liquids do not have definite shapes. Think of all the different shapes of bottles and glasses in your home. When a liquid is poured from one container to another, it takes the shape of the new container but its volume does not change.

1–1. *Liquids take the shape of the container they are in.*

Chapter 1–1 The Properties of Matter

1-2. *How many different states of matter can you identify here?*

Displacement: When one object is moved out of a space by a second object.

General properties: Characteristics possessed by all matter.

Special properties: Characteristics that make one type of matter different from another.

The third state in which matter can exist is as a gas. When water is in this state, it is an invisible vapor. When a substance becomes a gas, it has neither a definite shape nor a definite volume. It spreads throughout the container it is in. For example, if you were to spray an air freshener in one corner of a room, the odor could soon be detected in all parts of the room. The particles of spray move between the air particles and spread throughout the room.

Since all matter has mass and volume, it also has density. Remember, density is found by dividing an object's mass by its volume. As we will see later, density is the basis for many properties of matter.

There is still another property common to all matter. You demonstrate this property each time you put ice cubes into a glass that is already filled with a liquid. Since two pieces of matter cannot occupy the same space at the same time, the ice cubes will push some liquid out of the space the ice occupies. This is called **displacement.** The volume of the ice pushes out or *displaces* an equal volume of liquid.

There are many other **general properties** that most types of matter possess. Color, odor, taste, and hardness are examples. Other properties include *texture*, which is the way a material feels, and *luster*, which is the way it reflects light. These *general properties* are often used to tell the different types of matter apart.

There are other properties that are not common to all matter. These **special properties** make certain materials useful to us. For example, metals are good conductors of electricity so they are used for electrical wires. Other materials are useful because they do not have certain properties. Rubber does not conduct electricity, therefore, it is used to cover these wires. Scientists are always searching for new ways to use the *special properties* of matter. Some of these special properties are described on pages 22 and 23.

Materials
2 graduated cylinders
Sand (30 mL)
Water
Small stone
Pencil
Paper

A. Measure 30 mL of dry sand into one graduated cylinder.

B. Measure 50 mL of water into the other graduate.

   1. If the sand were added to the water, what would you expect the total volume to be?

C. Pour the water into the graduate containing the sand.

   2. What is the total volume reading in the graduate now?

   3. Is this less than what you expected it to be?

D. Dry sand contains air spaces between the sand particles.

   4. Why is the volume of the combined sand and water **less** than the volume of the sand alone plus the volume of the water alone?

   5. Did the air in the sand take up space? Explain.

   6. Which substance in this activity is a solid? A gas? A liquid?

   7. Do all of these take up space? Explain.

E. Place 30 mL of water into an empty graduated cylinder.

F. Gently place a stone into the water.

   8. What is the reading of the water level now?

   9. What is the volume of the stone? Explain.

   10. Why is this an example of displacement?

The universe is made of matter and energy. We have seen that all matter has certain general properties in common. In addition, there are many special properties of matter that have proven to be very useful to our society.

1. Define matter. Describe and give an example of each of the three common states of matter.
2. List several properties of matter that can be used to tell the different types of matter apart.
3. What is meant by the property of texture? Give several examples of different textures.
4. List three special properties of metals. Give an example of one way each property is used.

Chapter 1–1   The Properties of Matter

Special Properties of Matter

MALLEABILITY

Most metals can be hammered or rolled into thin sheets without breaking. This property is called *malleability* (mal-ee-uh-**bill**-uh-tee). A good example of this is aluminum foil. Metals can also be formed into almost any necessary shape because of this property.

DUCTILITY

Ductility (duk-**till**-uh-tee) is the ability of a material to be drawn out into very thin wires without breaking. Materials with this property are used for making wire for electric cables, jewelry, pipes, and chains. Platinum and copper are very *ductile* materials.

CONDUCTIVITY

Conductivity (con-duk-**tiv**-uh-tee) is the ability of a material to allow either heat or electricity to flow through it. Metals are the best *conductors* of both. For this reason, we use metal pots, pans, and radiators to conduct heat. We also use metal wires to conduct electricity. Materials that will not conduct electricity can be used as *insulators.*

ELASTICITY

Some materials will return to their original shapes after being stretched. This is called *elasticity* (ee-las-**tiss**-uh-tee). Rubber and coiled metal have this property. Rubber strands can be woven into cloth to make it elastic. This material is used in stretch fabrics. Most materials are *nonelastic*.

SOLUBILITY

When a teaspoon of sugar is stirred into a glass of water, it spreads throughout the water, or *dissolves.* We say the sugar is *soluble* in water. Many materials are soluble in water or in other liquids. The ability to be dissolved is called *solubility* (sol-u-**bill**-uh-tee). Substances which will not dissolve are called *insoluble.*

OPTICAL PROPERTIES

Clear glass allows light to pass through it undisturbed. This type of glass is called *transparent* (trans-**pair**-ent). Stained glass disturbs the light as it passes through and objects cannot be seen clearly. Stained glass is *translucent* (trans-**loo**-cent). Some substances, such as brick, do not allow light to pass through at all. These are called *opaque* (oe-**pake**).

MELTING AND BOILING POINTS

Many materials have a characteristic *melting point* at which they change from a solid to a liquid. Ice melts at 0°C. Many materials also have a *boiling point* at which they become a gas. Water boils at 100°C and becomes water vapor. Different substances boil at different temperatures.

Ability to React with Other Materials

Under the right conditions there are very few materials that will not react with other materials to form new substances. These changes are very important in the development of industrial products, such as synthetic fibers. Chemical reactions are also essential to life.

2 THE MAKEUP OF MATTER

Chemistry has come a long way since the primitive workshop shown in the drawing at the right. The laboratories of today's scientists are quite different in their equipment. New methods are also being used. But it may surprise you to learn that the basic purpose is still the same: to understand matter.

Another difference is how scientists use the knowledge they have gained. Today's chemists are trying to develop new materials for our modern society. The man in the picture at right is trying to make gold. When you finish this section, you will be able to:

● Describe early attempts to discover the nature of matter.

● Compare and contrast *elements* and *atoms*.

● Compare and contrast *compounds* and *molecules*.

▲ Use the library as a tool for research.

Unit 1 What Makes Up Our World?

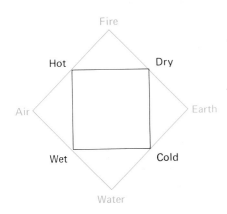

1-3. *The four forms of matter of Aristotle's theory and the shared properties. All materials were combinations of these four.*

1-4. *This alchemist has just discovered a new "glowing" element: phosphorus.*

As you look around, it is easy to see that there are many different types of matter in the world. Some are solids. Others are liquids. Still others are gases. But people have long wondered, "What makes up matter? Are all these different types of matter made of simpler things? If matter is made of simpler things, is it possible to find out what? Can we then put these simple things together to make the kinds of matter we want?"

Among the first persons to seriously wonder about the nature of matter were the ancient Greeks. They developed two major theories. One stated that there were only four different basic forms of matter: Air, Water, Fire, and Earth. Each of these forms possessed two basic properties. Earth was dry and cold. Water was cold and wet. Air was wet and hot. Fire was hot and dry. All other types of matter were combinations of these four basic forms.

The second theory was proposed by Democritus (Duh-**mock**-ree-tus). He stated that all matter was made of particles so tiny they could not be cut any smaller. He also said the particles were always moving and that they combined with each other in different ways. Democritus believed that these particles differed from each other only in shape and arrangement.

However, Aristotle (**Ar**-uh-staht-'l), one of the best known teachers of that time, spoke out against these ideas. He supported the four forms of matter theory. Because Aristotle was so famous, the ideas of Democritus were soon forgotten.

Aristotle's beliefs influenced scientists for 2,000 years. People believed that all matter was only a combination of the four basic forms. Therefore, they thought it should be possible to change one type of matter into another. So, the search began for the right combination to make gold.

For the next 2,000 years, scientists called *alchemists* (**al**-kuh-mists) searched for a way to make gold. They examined just about every material known at the time. They never succeeded, but they did learn a lot about the nature of different materials. They also developed many of the basic procedures used in chemistry today. The alchemists discovered several different substances that they could not change into simpler substances.

Element: A simple form of matter that cannot be changed into any simpler form of matter by ordinary means.

In the late 1600's these substances were given the name **elements.**

At that time a procedure for identifying *elements* was developed. The procedure involved trying every method possible to break a substance into something simpler. If the substance could not be changed into a simpler form, it was an element. Of course, new methods changed some of the old ideas. Scientists soon realized that many of the materials that had been used for centuries were elements. Gold, silver, tin, and iron were examples. Within 100 years, the number of identified elements had reached 23. The four element theory of Aristotle had finally been proven wrong.

Since that time, chemists have found a total of 90 different elements in nature. Other scientists have developed a way to change some of the elements into new ones. Using these methods, scientists have "made" over a dozen new elements. These elements will be discussed in chapter 2. Thus, instead of the simple, four element universe of Aristotle, we now know of over 100 "building blocks of matter."

Around the year 1800, John Dalton, an English school teacher, rediscovered the ideas of Democritus. He stated again that all elements are made of tiny particles which cannot be divided further. He used the term **atoms** for these particles. This term comes from a Greek word which means "indivisible." Dalton said that *atoms* of one element were all alike, but they differed in size, shape, and weight from the atoms of other elements. The atoms were always moving and

Atom: The smallest particle of an element that has all the properties of that element.

1–5. *Examples of pure elements. Many of these had been used by people for ages without knowing what they really were. (left) sulfur, (middle) copper, (right) gold.*

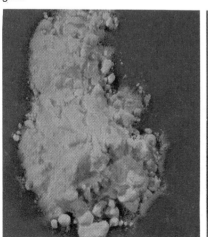

1–6. *Examples of natural and manufactured compounds. Many of the materials we use and need are compounds.*

Compound: A substance made by the joining of atoms of two or more elements.

Molecule: The smallest particle of a compound that has the properties of the compound.

they combined with each other in different ways.

What about substances that could be changed? What about wood which burned, giving off gases and smoke and leaving a black material? What about sugar which gave off gases and left a black material when it was mixed with acid? If these were not elements, what were they? It became clear that these materials must be made of two or more different elements. The atoms of these elements must be joined in some special way. Substances such as these were called **compounds.**

In time, chemists came to realize that you could take a *compound*, such as common salt, and break it into smaller and smaller pieces. Eventually you would come to the smallest particle that was still salt. This smallest particle would still have the properties of salt. If you broke this particle any further, you would no longer have salt, but some other, simpler materials. This particle is now called a **molecule.** The same is true of all compounds. The smallest particle of each compound that has all the properties of that compound is called a *molecule* of the compound.

So, after 2,000 years, science came full circle. The four element theory had been proven wrong. It could not stand up to new evidence. Continuous testing and experimenting had resulted in new observations which disproved this old theory. It had to be discarded. A new theory, based on the early hypotheses of Democritus, explained the observations better. In the last 200 years, even this theory has been changed because of new observations and experiments. Only time will tell what additional changes might be necessary.

SKILL BUILDING ACTIVITY

Materials
Pencil
Paper

A. Obtain the materials listed in the margin.

B. Use your school or local library to write a report (about 200 words) on one of the following topics:

   1. Use encyclopedias to report on Aristotle.
   2. Use the card catalog to locate two or more books for a report on *alchemists.* Include the call number of the books in your bibliography.
   3. Use *The Reader's Guide to Periodic Literature* to find 3 articles for a report on "Modern Chemistry and You."
   4. Summarize these: Putnam, John J., "Quicksilver and Slow Death," *National Geographic,* October, 1972. Asimov, Isaac, "Thinking Up Names," *SciQuest,* October, 1980. Include the magazine's volume and number in your bibliography.

C. If you are not familiar with these library tools, ask a librarian or teacher for help.

SUMMARY & QUESTIONS

The ancient alchemist and the modern chemist share a common purpose: to understand the nature of matter. The alchemist sought to use that knowledge to make gold. The chemist seeks to create a wide range of new compounds to improve the world we live in. Scientific research has shown that matter is composed of basic particles called atoms. Elements are made of like atoms. Unlike atoms join to form molecules of compounds.

1. Describe Aristotle's ideas on the nature of matter.
2. Compare Aristotle's ideas with those of John Dalton.
3. Explain briefly how Aristotle might come to the idea that some matter contained fire and earth while other matter contained fire and air.
4. How are the terms element and atom related?
5. How are the terms compound and molecule related?
6. How are the terms molecule and atom related?
7. Name two contributions of ancient alchemists to modern chemistry.

3 CHANGING MATTER

Fire. It has been a tool of humans for centuries. It has also been a dangerous enemy. Fire can be useful as a source of heat for your home or to cook your food. Fire can also be a disaster. It can damage or destroy that same home. It can injure or kill people, animals, or whole forests. Yet whether a fire is useful or harmful, one thing is certain. It changes whatever it comes in contact with in some way. When you finish this section, you will be able to:

- Define and list examples of *physical changes*.
- Define and list examples of *chemical changes*.
- Explain the differences between a compound, a *mixture*, and a *solution*.
- ▲ Observe examples of changes in matter and decide if they are physical or chemical changes.

1-7. *Fire causes both physical and chemical changes.*

Physical change: A change in matter which does not change the individual molecules.

1-8. *Change of state in a metal from solid to liquid and back to solid.*

Have you ever walked through a field or woods after a fire? Very little of what was in the path of the fire remains unchanged. Even things that were not burned were changed by the heat. You kick up clouds of black or gray ash, cinders, and soot. The wood crumbles into black powder. Pieces of metal seem to have softened and sagged out of shape. Even the stones seem to have been baked. Many different changes have taken place. These changes can all be placed into two groups.

The first of these are called **physical changes**. A *physical change* is one in which the appearance of an object changes but its chemical makeup does not. Nothing new is produced. For example, a piece of paper can be crumpled into a ball. It can be torn into little shreds. But, it is still only paper. The molecules have not changed. These are examples of physical changes.

Think of all the things you could do to a piece of

1-9. *What is happening to the powder? (Is this a physical or a chemical change?)*

Solution: A type of mixture where the molecules of one substance fit between the molecules of another. One of the substances is usually a liquid or a gas.

Mixture: A substance made of two or more elements or compounds which have not combined with each other.

metal that would cause a physical change. Did you think of heating it until it melted? This is a physical change because even as a liquid, the chemical makeup is the same. What happens to metal when it melts is called a change of state. It changes from a solid to a liquid. If it is cooled down, it will change to a solid again. The changing of ice from a solid to a liquid state is also a physical change. So is the changing of water into water vapor, which is a gas. In all three states the chemical makeup of water is the same. Therefore, these are physical changes.

What could you do to a sugar cube that would be a physical change? You probably thought of crushing it into smaller pieces. Did you also think of stirring it into water? When sugar is stirred into water, it seems to disappear. But you can tell that the sugar hasn't changed its makeup because the water tastes sweet. Actually, the sugar molecules are mixed between the water molecules. When this happens, we say the sugar dissolved in the water.

Sugar dissolved in water forms a **solution**. A *solution* is formed when the molecules of one material mix between the molecules of another. Salt dissolved in water is another example. So is the air you breathe. Air is a solution of gases.

A solution is actually a special type of **mixture**. Remember, a compound is made of two or more elements whose atoms have joined together. A *mixture* is made of two or more elements or compounds that are mixed together, but their atoms have not joined with each other. For example, a bowl of mixed nuts, a bowl of milk and cereal, a tossed salad, and a pizza with everything on it are all mixtures. So are concrete, sand, and soil.

1-10. *Mixtures are made of several materials that have not joined chemically.*

1–11. *(top)* Burning is a chemical change.

1–12. *(bottom)* Rusting is a chemical change. The dents are physical changes.

Chemical change: A change in matter in which new molecules are formed.

The materials that make up a mixture can usually be separated by a simple physical change. If you do not like the pepper and onions on the pizza, you can pick them off. It would be difficult to separate the cheese from the tomato sauce, but not impossible. Even the salt in the water can be removed by a physical change. If the solution is heated, the water can be boiled away, leaving the salt behind. These physical changes would not change the makeup of the molecules of each substance.

Changes in the second group are different. In these cases, the chemical makeups of the substances change. The molecules are different. They may break apart. New molecules are formed. New substances are produced. These are called **chemical changes.** Think of what happened to the wood in the fire. The ashes and soot are no longer wood. The molecules of wood have been changed into new substances. Some of these were given off as gases as the wood burned. Others were left as ash and soot.

Burning gasoline in the engine of a car is a *chemical change*. The gasoline molecules are broken down and new molecules are formed. Energy is released. Think of all the different materials we use that are not natural. Most were produced by chemical changes.

Some chemical changes need heat to occur. Others, such as rusting of metal, do not need heat. A piece of iron left outdoors will rust. The iron atoms join with oxygen atoms in the air to make a new compound.

Often both chemical and physical changes occur together. The fire we discussed earlier is one example. So is the peeling of paint on a house. Some of the changes are chemical, some are physical. The next time you see an old car, look for examples of both types of changes.

A. Obtain the materials listed in the margin.

B. Break one wood splint in half. Break these pieces into smaller pieces. Break the wood into the smallest pieces you can.

1. Are the smallest pieces you have still wood?

2. Is this a physical or a chemical change? Why?

C. Attach the large candle to the glass plate with a few drops of melted wax.
CAUTION: WEAR SAFETY GOGGLES AND TIE BACK LONG HAIR.

D. Place some pieces of the

Unit 1 What Makes Up Our World?

Materials
2 wood splints
Small (birthday) candle
Metal spoon
Candle (15 cm)
Metal jar lid
Matches
Glass plate
Clothespin (spring type)
Goggles

small candle in the bowl of a metal spoon. Hold the handle of the spoon with a spring type clothespin. Heat the spoon over the candle flame.

3. Describe what happens to the wax in the spoon.

4. Is the material in the spoon still wax?

5. Is this a physical or a chemical change? Why?

E. Remove the spoon from the flame and allow it to cool.

6. What happens to the wax?

7. Is this a physical or a chemical change? Why?

8. Examine the burning candle. Is a physical or chemical change taking place? Explain.

F. Place one or two broken pieces of wood into the jar lid. Keep the lid away from anything else that will burn. Light another splint from the candle. Use this to light the wood pieces in the lid. Add the splint to the lid.

9. Describe what happens to the wood pieces.

10. Wait until the wood stops burning. Is the material in the lid still wood? How do you know?

11. Describe the appearance and feel of the material in the lid.
CAUTION: DO NOT TOUCH IT UNTIL IT HAS STOPPED BURNING AND HAS COOLED.

12. Is this a physical or a chemical change? Why?

13. Are the wood and wax elements or compounds? Why?

Matter is constantly being changed. These changes can be grouped as either physical or chemical changes. During a physical change, the makeup of the substance does not change. It is a change in appearance only. In a chemical change, new molecules are formed.

1. Explain the difference between a physical and a chemical change.
2. Identify each of the following as either physical or chemical changes: a puddle of water drying up, sanding a piece of wood, melting ice cream, rusting metal.
3. Explain the difference between a compound and a mixture.

Chapter 1–3 Changing Matter

VOCABULARY REVIEW

Match the description in Column II with the term it describes in Column I.

Column I
1. energy
2. element
3. mixture
4. matter
5. molecule
6. physical change
7. atom
8. chemical change
9. compound
10. solution

Column II
a. A substance made by the joining of atoms of two or more elements.
b. A substance made of two or more materials whose atoms have not joined with each other.
c. The smallest particle of an element that still has the properties of the element.
d. A mixture formed when molecules of one substance fit between the molecules of another.
e. The property of something that makes it able to do work.
f. A change in matter in which new molecules are formed.
g. Anything that takes up space and has mass.
h. A change in appearance but not chemical make-up.
i. A form of matter that cannot be changed into simpler forms by ordinary means.
j. The smallest particle of a compound that has the properties of the compound.

REVIEW QUESTIONS

Choose the letter of the phrase that best completes each of the following statements.

1. The smallest particle of sugar that can be identified as sugar is a(n) (a) element (b) electron (c) solution (d) molecule.
2. Which of the following is a physical change? (a) melting ice cubes (b) burning paper (c) rusting iron (d) burning gasoline.
3. The three common states or phases of matter are (a) solid, water, gas (b) ice, water, gas (c) solid, liquid, gas (d) solid, liquid, air.
4. Matter that contains more than one kind of material is called a(n) (a) element (b) compound (c) atom (d) pure substance.
5. The idea that all matter was composed of only four "elements" was believed by (a) Einstein (b) Dalton (c) Aristotle (d) all of these.

6. Matter and energy (a) can be destroyed (b) can change from one to the other (c) cannot be changed (d) are always the same.
7. The general properties of matter are possessed by (a) solids but not by liquids or gases (b) all matter (c) only certain types of matter (d) all matter except gases.
8. Since sugar can be changed into simpler substances by chemical changes, sugar must be a(n) (a) compound (b) element (c) atom (d) mixture.
9. Which of the following is an example of a solution? (a) water (b) gold (c) salt (d) orange soda.
10. Which of the following is an example of a special property of matter? (a) density (b) color (c) ductility (d) mass.

REVIEW EXERCISES

Give brief but complete answers to each of the following exercises.

1. How do the general properties of matter differ from the special properties? How are the special properties useful?
2. How are the following pairs of terms related? Element and atom; compound and molecule; mixture and solution.
3. Compare and contrast physical changes with chemical changes.
4. Briefly describe the early atomic theory of Democritus and John Dalton.
5. Explain some of the alchemists' contributions to modern chemistry.
6. A chemist discovers that when an orange powder is heated in a test tube, it changes. It gives off a gas identified as oxygen. It also leaves a silvery material called mercury on the inside of the tube. Further tests show that neither the oxygen nor the mercury can be broken into simpler substances. Which of these materials are elements? Which are compounds? Explain why.
7. Is the example in question 6 a physical or a chemical change? Explain why.
8. How are the following pairs of terms different? Element and compound, atom and molecule.
9. Compare and contrast the properties of solids, liquids, and gases.
10. An artist decides to create a metal sculpture. Starting with a single piece, she heats it until it is soft enough to bend and twist. She draws a part out into a thin wire and cuts off pieces. Then she files the edges smooth. Next she hammers the metal into shape, drills holes, and polishes it. Then she paints some of the parts. Finally, she uses an acid to burn her name in a corner. Which of these changes are physical? Which are chemical? Why?

CHAPTER 2 ATOMS

1 THE NOT-SO-SOLID ATOM

In Mexico, one of the customs at Christmas is breaking the piñata (pin-**yaht**-uh). A decorated clay pot or papier-maché doll is hung. The children are blindfolded and given sticks. They then try to hit and break the doll. Sometimes the children miss completely. Sometimes they hit it a glancing blow that does no damage. Finally, someone hits the piñata straight on and breaks it. Out tumble surprise gifts such as candy, fruits, and coins. Breaking the piñata is not unlike probing the secrets of the atom. When you finish this section, you will be able to:

- Describe some early ideas of the structure of the atom.
- Explain the purpose of using a *model* in science.
- Name and describe two particles that make up atoms.
- Determine the shape of an object you cannot see.

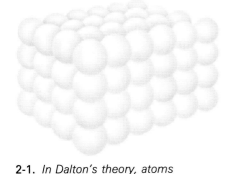

2-1. *In Dalton's theory, atoms were solid, indivisible little balls.*

Electron: A very light, negatively charged atomic particle.

Like all theories, John Dalton's ideas about atoms were subjected to many tests and experiments. Dalton believed that atoms were the tiniest particles of matter that could exist. To him they were solid little balls. All the atoms of one kind of element were alike. The atoms of different elements were not alike. His ideas held for almost one hundred years. Then about 1900, scientists in England did a series of experiments that showed that some parts of Dalton's theory had to be changed.

In one of these experiments, a strong electrical current was sent through a glass tube. The tube was partly filled with gas. A ray of light became visible. This ray could be bent by magnetic or electrical forces. From this and other information, it was concluded that the ray was made of tiny particles that had a negative electrical charge. These particles were smaller than atoms. They were called **electrons**.

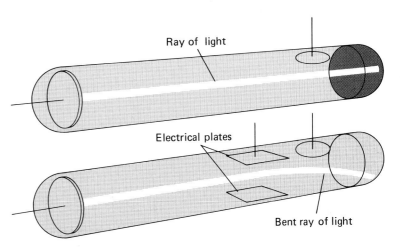

2-2. *A beam of negatively charged particles can be made to bend by strong magnetic forces.*

Chapter 2–1 The Not-So-Solid Atom

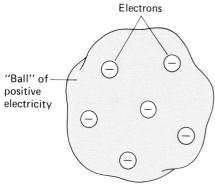

2-3. One early atomic model had the electrons scattered through a ball of positive electricity.

Scientific Model: A way of representing something that cannot be observed, based on properties that have been observed.

It was already known that atoms are *neutral*. This means that they do not have an electrical charge. Therefore, the idea was suggested that there must be an equal positive charge in the atom that cancelled out the negative charge of the *electron*. A new hypothesis was suggested about the structure of the atom. It said the atom was not solid, but a ball of some kind of positive electricity with negative electrons scattered through it—something like chocolate chips scattered through ice cream.

Comparing the structure of the atom to chocolate chip ice cream may seem rather strange to you. But remember, no one has ever seen an atom. They are too small. So to explain what one might look like, scientists use **models**. *Models* are a great help in understanding ideas or objects that cannot be seen directly. Every time you read a novel, you develop a "mental model" of what the people and places in the story look like. When chocolate chip ice cream was mentioned, you were able to form a mental picture of what an atom looks like.

But models are not always exactly like the objects they seek to explain. Often they must be changed to account for new information. So it was with the model of the atom.

Other scientists had discovered different kinds of energy rays coming from some elements such as radium and uranium. One of these rays was found to have a positive electrical charge. When a beam of these positive particles was aimed at a very thin piece

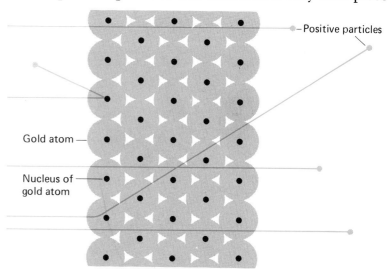

2-4. The scattering of positive particles lead to the discovery of the nucleus.

Unit 1 What Makes Up Our World?

of gold foil, a strange thing happened. Like children batting at the piñata, scientists were aiming at atoms they couldn't see. When they did hit something, they were very surprised and excited at the results.

The scientists expected their particles to go right through the foil. Most of them did. However, a few particles really surprised the scientists. They came almost straight back at the source of the beam. As one of the scientists said, "It was like shooting bullets at a piece of tissue paper and having one bounce back at you."

It was already known that two similar electrical charges will push away from each other when they are brought together. Since most of the positive particles in the beam went right through the foil, they couldn't have bumped into any positive charges on the way. However, a few of the beam particles did bounce back. Therefore, each atom of gold had to have positive charges concentrated in a very small space. If positive charges were scattered throughout the atom, more of the beams would have bounced back.

Later it was shown that the positive charges in the atoms were also particles. These particles were labeled **protons.** Each *proton* has an electrical charge that is opposite but equal in strength to the charge on an electron. If the number of protons in the center of an atom was equal to the number of electrons around it, their charges would cancel each other. The atom would have no overall electrical charge.

So the model of the atom was changed. In a brief period of about twenty-five years, it had been shown that the atom, small as it is, is mostly empty space. There were positively charged protons concentrated in the center or **nucleus** of the atom. An equal number of negatively charged electrons were moving around the *nucleus*. The idea of atoms as solid little balls had changed completely.

Scientists often have to describe things they cannot see directly. The ocean bottom, the structure of molecules, the interior of the earth, and the structure of atoms are all examples. They do this by gathering as much indirect evidence as possible. From this information they "hypothesize" the structure. Of course, as new information is discovered, the model must be changed.

Proton: A positively charged atomic particle found in the center of an atom.

Nucleus: The small central core of an atom.

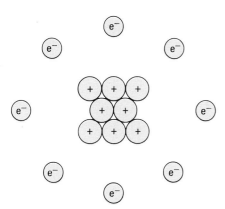

2-5. *The revised atom model after the discovery of the nucleus.*

Materials
"Hidden objects" (at least 4)
Marble
Paper
Pencil

In this activity, you will try to describe an object you cannot see. This is similar to finding the structure of an atom.

A. Obtain the materials listed in the margin.

B. Do not look under the cardboard until you have finished the activity.

C. Hidden under the cardboard is an object with a simple shape. You are to describe the shape and size from what happens to the marble as you roll it under the cover. First answer these questions.

  1. If a rubber ball is dropped straight down onto a flat surface, which way does it bounce?

  2. If the ball hits a wall at an angle, which way does it bounce?

D. You have 10 trials to gather information about the shape of the object beneath the cardboard. Roll the marble under the cardboard and observe the direction of its bounce. You may roll the ball in from any side.

  3. What is the shape of the object beneath the cardboard?

E. Repeat the process with the other hidden objects.

  4. Describe the shapes beneath each of the cardboards.

  5. Describe the indirect evidence you used to form these hypotheses.

F. When the class has finished with all the set-ups, check your answers.

  6. How is this activity similar to the way protons were discovered?

Early ideas about atoms were that they were tiny solid spheres. Experiments performed around 1900 showed this theory to be wrong. New evidence showed that the atom contained tiny, positively charged protons in the nucleus and negatively charged electrons around it.

1. Summarize Dalton's ideas about the nature of the atom.
2. Compare the electron and the proton in terms of their electrical charges.
3. What is the nucleus of the atom?
4. What evidence indicated that there were positive charges in the nucleus of an atom?
5. Why are models useful to scientists?
6. Briefly discuss three different models of atoms discussed in this lesson.

2 THE NUCLEUS

How small are atoms? That is very hard to answer. Hydrogen atoms are so small that it would take more than two billion to stretch across this page. This paper is only about one million atoms thick. Yet the atom is mostly empty space.

Imagine that you are in the top row of seats in a large stadium. With powerful binoculars you look at a baseball lying on the pitcher's mound. Suddenly you see an ant on the ball. If an atom of hydrogen were the same size as the stadium, you would be sitting in the path of the electron, looking at the ant-sized nucleus. When you finish this section, you will be able to:

● Define the terms *atomic number* and *atomic mass*.

● Describe the *neutron*.

● Draw the structure of the atom when given its *atomic number* and *mass*.

▲ Use a simple model to show the location of *subatomic particles* within an atom.

Atomic mass unit: A special unit used to express the mass of atomic particles and atoms.

Small as they are, atoms are still matter. Therefore, they must have mass. However, grams, the usual units for measuring mass, are too big for this job. Rather than use small fractions of grams, scientists developed a new mass unit for the particles that make up atoms. It is called the **atomic mass unit,** or *amu* for short. The proton has a mass of one *atomic mass unit*. When the mass of an electron was compared to that of a proton, it was found that it took 1,836 electrons to equal one proton. So, the electron has a mass of 1/1,836 amu.

All protons, no matter what type of atom they come from, are exactly alike. The same is true for all electrons. If these atomic particles that form atoms are all alike, how can there be different kinds of atoms? The answer lies in how many of these particles the atom contains. Different atoms have different numbers of protons in the nucleus. Hydrogen is the simplest atom. It has one proton in the nucleus. However, hydrogen, like most atoms, has no charge. It is neutral. Therefore, it must also have one electron circling the nucleus. The negative charge of the electron cancels the positive charge of the proton. Carbon is different from hydrogen. Carbon has six protons and six electrons in each atom. These different numbers of protons and electrons give each atom its different properties.

Atomic number: The number of protons in the nucleus of an atom.

Eventually, it was shown that each type of atom had a certain number of protons. The number was the same for all atoms of that element. We now call this the atom's **atomic number.** This number tells us how many protons are in every atom of that element. Since most atoms are neutral, the *atomic number* also tells us how many electrons are in that atom.

It was also noted that the known elements could be

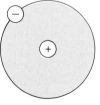

Hydrogen
atomic number = 1

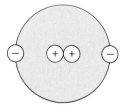

Helium
atomic number = 2

2-6. *The atomic number is the number of protons in the nucleus.*

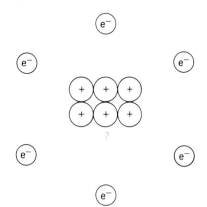

2-7. *Carbon's mass is greater than its atomic number. The difference was identified as the neutrons.*

Atomic mass: The sum expressed in amu of the mass of protons and neutrons in an atom.

Sub-atomic particles: The tiniest particles of matter that make up atoms.

Neutron: An atomic particle with no charge that is found in the nucleus.

2-8. *Large machines like this are used to study the atom.*

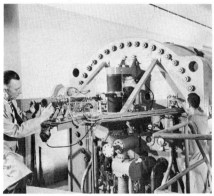

arranged in a sequence according to their atomic numbers. Hydrogen's atomic number is 1. Helium's is 2, Lithium's is 3, all the way up to uranium with an atomic number of 92. This means that uranium has 92 protons in its nucleus and 92 electrons circling that nucleus.

But there were gaps in the sequence. Some numbers didn't have elements matched to them. This caused scientists to predict that some elements hadn't been discovered yet. In time, elements with the proper atomic numbers were discovered to fill the gaps. Scientists also found that elements with higher atomic numbers can exist.

Arranging atoms in order of atomic numbers also helped point up another problem. For example, hydrogen's atomic number is 1. The one proton in its nucleus has an amu of 1. The mass of an electron is so small, it is not usually included. So the **atomic mass** of hydrogen is also 1. Carbon's atomic number is 6. It has six protons in its nucleus. Carbon should have an *atomic mass* of 6, but it doesn't. The atomic mass of carbon is 12. This was found to be the case with all atoms except hydrogen. Their masses were greater than the number of protons in their nuclei. What caused this difference? It was suggested that there was another **sub-atomic particle** in the nucleus that had the same mass as a proton but did not have an electrical charge. Although its existence was suggested about 1914, the particle wasn't identified until 1932. Because this particle has no electrical charge, it is called a **neutron**.

Have you ever seen the trail of ice crystals a jet plane sometimes leaves high in the sky? From this trail you can tell a lot about the size, speed, and direction of a jet even though you can't see it. Scientists use an instrument called a cloud chamber to see similar trails left by *subatomic particles*. As they streak through a chamber filled with cool, moist air, the particles leave a trail of droplets. The tracks of electrons and protons can be easily identified because they can be bent by magnets. It was in such a chamber that the track of a *neutron* was discovered. Its path could not be made to curve since it has no charge.

An atom's atomic mass, then, is the total of the protons and neutrons in its nucleus. If you know the atomic mass and the atomic number of an atom, you can find the number of neutrons in the nucleus by subtracting.

Chapter 2–2 The Nucleus

For example, the atomic number of nitrogen is 7. Its atomic mass is 14. Therefore:

atomic mass − atomic number = number of
(number of protons − (number of neutrons
and neutrons) protons)
14 − 7 = 7

Thus, an atom of nitrogen has 7 neutrons in the nucleus.

Iron has an atomic number of 26 and an atomic mass of 56. Therefore, it has 30 neutrons in its nucleus. Notice too that the number of neutrons does not have to equal the number of protons. In fact, the higher the atomic number is, the greater the difference between the protons and neutrons.

2-9. *Atoms with the same number of protons but different numbers of neutrons are called isotopes.*

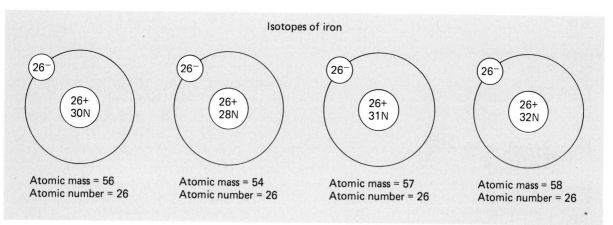

Isotopes: Atoms of the same element whose nuclei contain the same number of protons but different numbers of neutrons.

Later it was discovered that not all atoms of an element had the same mass. Most iron atoms have a mass of 56, but some were found with masses of 54, 57, or 58. Since all of them had 26 protons in the nucleus, they were each iron atoms. Therefore, the difference must be in the number of neutrons. Most iron atoms have 30 neutrons, but by subtracting, we can see that some have 28, and others 31 or 32 neutrons. The term **isotope** is used to identify atoms that have the same number of protons but different numbers of neutrons. There are 21 atoms that do not have natural *isotopes*. All the other elements exist as two or more isotopes. There are two natural isotopes of hydrogen. Oxygen has three. Tin has the most with ten. But remember, all ten are tin because they all have the same number of protons in the nucleus. The only difference is the number of neutrons.

Materials
Small circles of blue, red, and yellow paper
Glue
Paper
Pencil

A. Obtain the materials listed in the margin.

B. Below are the atomic number and mass of four elements.

element	atomic number	atomic mass
helium	2	4
oxygen	8	16
silicon	14	28
silver	47	107

  1. How many protons are in each element?
  2. How many electrons are in each element?
  3. How many neutrons are in each element?

C. The blue circles represent protons, the red electrons, the yellow neutrons.

D. Glue to paper the correct number of circles to represent the nuclei of helium and silicon.

  4. Why are these atoms different if they are made of the same kinds of particles?

E. Oxygen has three natural isotopes. The most common one has a mass of 16 and two others masses of 18 and 17.

  5. Show how to determine the number of neutrons in each isotope.
  6. Construct models of each of the two additional isotopes.
  7. Why is each of these models still oxygen?

Atoms are extremely small. Even so, they are still matter. The atomic number of an atom is the number of protons in the nucleus. It is also the number of electrons circling the nucleus. The atomic mass is the total of protons and neutrons in the nucleus.

1. What information can be obtained from an atom's atomic number?
2. What is meant by the atomic mass of an atom?
3. An atom has an atomic number of 19 and an atomic mass of 39. How many electrons, protons, and neutrons does it have?
4. Describe the properties and location of the neutron.
5. Why was it possible for scientists to predict the existence of unknown elements from the atomic numbers of known elements?
6. How can there be over 100 types of elements if there are only three types of subatomic particles?

3 THE ELECTRON CLOUD

Do you recognize any of these signs? Do you know what they mean? They are all symbols that are used to represent a word, a place, or an idea. Symbols are very useful where information has to be exchanged between people who speak different languages. Road signs use standard symbols for showing where to find service stations, restaurants, and so on. We live in a world of symbols. When you finish this section, you will be able to:

● Explain the use of *chemical symbols* to represent elements.

● Describe the arrangement of electrons in electron shells.

● Describe what happens when an electron changes shells.

▲ Use a simple model to show the complete structure of an atom.

As time went on, more and more information about the nature of matter was gathered. It became necessary to simplify the system of names and symbols being used for the elements. It was so confusing that by Dalton's time there were at least twenty different symbols for mercury. Most of these symbols were drawings. A list of symbols looked like a page of Egyptian writing. Dalton made an attempt to simplify the symbols, but it was a Swedish chemist who developed the system used today.

	1500s	1600s	1700s	1783	1808	1814
Gold	☼	℞	⊙	⊙	Ⓖ	Au
Mercury	♆	♆	☿	☿	⊕	Hg
Lead	♄	♄	♄	♄	Ⓛ	Pb

2-10. *Early symbols for elements were often very confusing.*

TABLE 2-1

First Letter
Hydrogen (H)
Carbon (C)
Oxygen (O)

First Two Letters
Helium (He)
Calcium (Ca)
Aluminum (Al)

Latin Name
Mercury (Hydrargyrum) (Hg)
Copper (Cuprum) (Cu)
Gold (Aurua) (Au)

Chemical symbols: One or two letters used to represent an atom of a particular element.

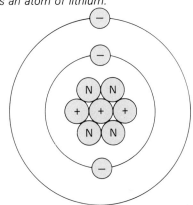

2-11. *The satellite atomic model is inaccurate but still useful. This is an atom of lithium.*

Today's system has three simple rules. 1. The symbol for an element is the first letter of its name. 2. If the names of two elements begin with the same letter, the first two letters can be used. 3. Some elements, in particular those discovered by the alchemists, use the letters of their Latin names. Table 2-1 shows a few examples.

Chemical symbols are a kind of shorthand for chemists. The symbols are used all over the world and are easier to write than the whole name. Symbols are also used when describing how atoms react with one another. This will be shown in a later chapter.

Let's review what we know about atoms so far.
1. An atom is not solid but is composed of several small particles.
2. Protons are particles with positive charges and are found in the atom's center or nucleus.
3. Neutrons have the same mass as protons but no electrical charge. These are also located in the nucleus.
4. Electrons are very light particles that circle the nucleus. Electrons have a negative charge.
5. The atomic number tells how many protons or electrons an atom has.
6. The atomic mass is the combined mass of the neutrons and protons in the nucleus.
7. Atoms of different elements have different numbers of protons in the nucleus.

So far, most of our discussion has centered on the nucleus of the atom. We have not said much about the electrons, other than that they circle the nucleus. One early attempt to describe the electrons' location said that they moved around the nucleus in a set path or "orbit." This would be like the planets orbiting around the sun. This satellite model is shown in Fig. 2-11.

Scientists now know that the satellite model is not really accurate. However, trying to make a completely accurate model is a problem. Electrons move so fast that in one second they make billions of trips around the nucleus. For this reason, scientists have given up trying to pin down the exact location of an electron at any instant. They are content to say where the electron is most likely to be.

The model used today is called the *electron cloud model*. The dots that form the cloud in Figure 2-12 represent possible locations for the electron at any instant in time. As indicated in this model of hydrogen,

Chapter 2-3   The Electron Cloud

2–12. *(left)* The electron cloud of the hydrogen atom.

2–13. *(right)* This atom has two energy levels or shells in its electron cloud.

Electron shells: The area around an atomic nucleus in which electrons move.

its electron is most likely to be closer to the nucleus than farther away, but it could be anywhere in the cloud.

Larger atoms have more electrons. Within the energy cloud, there are definite energy levels where electrons are most likely to be found. Remember, the atom is a three-dimensional object so these levels are more accurately described as **electron shells.** You could imagine the shells to be like a ping pong ball that is inside a tennis ball that is inside a basketball, and so on. The *electron shells* are not solids like the walls of these balls. The electron shells simply represent the "areas" where the electrons are most likely to be. The electrons are more likely to be found on the shells than in the spaces between them. At the very center of these balls or shells would be the nucleus made up of protons and neutrons.

Figure 2-13 shows an atom with two energy shells in its cloud. For simplicity, we can use the satellite model

TABLE 2-2
Atomic Data for the First 12 Elements

Atomic Number	Atomic Mass	Element	Symbol	Shells 1	2	3	4
1	1	Hydrogen	H	1			
2	4	Helium	He	2			
3	7	Lithium	Li	2	1		
4	9	Beryllium	Be	2	2		
5	11	Boron	B	2	3		
6	12	Carbon	C	2	4		
7	14	Nitrogen	N	2	5		
8	16	Oxygen	O	2	6		
9	19	Fluorine	F	2	7		
10	20	Neon	Ne	2	8		
11	23	Sodium	Na	2	8	1	
12	24	Magnesium	Mg	2	8	2	

to represent these shells as long as we realize that the shells are not definite paths for the electrons to follow.

Scientists have found that only a certain number of electrons can "fit" into a shell. The first shell can hold up to two electrons. Therefore, hydrogen with its one electron and helium with two electrons need only the first shell. The element lithium has three electrons. Two of these fit into the first shell. The third electron has to use the second shell. It is known that the second shell can hold eight electrons. How many shells there are in an atom is determined by the number of electrons it has. There is a maximum of seven shells.

Figure 2-14 shows how the electrons of the first 12 elements are arranged. Sodium, with 11 electrons, begins using the third shell. This shell can hold a maximum of 18 electrons. Some of the outer shells can hold up to 32 electrons. The number of electrons in each shell is important because it is related to the way the element behaves chemically.

As mentioned earlier, the electrons are always zipping around the nucleus on one of the shells. They can travel in any direction. Electrons can also move to higher or lower shells. Higher shells are farther away from the nucleus. Lower shells are closer to the nucleus. It takes energy to make an electron move to a higher shell, just as it takes energy to move to a higher step on the stairs. An electron also has to make the complete "jump."

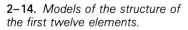

2-14. Models of the structure of the first twelve elements.

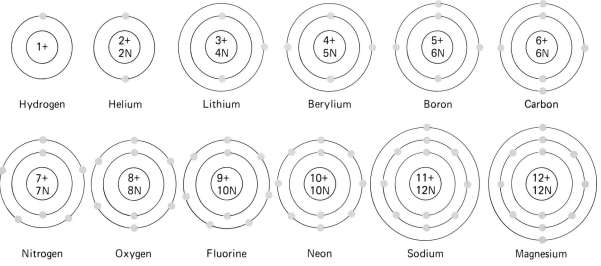

Chapter 2-3 The Electron Cloud

There is no such thing as climbing half a stair or half a shell. If the electron looses energy, it drops to a lower shell. So electrons can either take in or give off energy as they change energy levels. You will learn about the uses of energy in Chapter 3.

Materials
Atomic models from lesson 2
Red circles
Glue
Paper
Pencil

A. Obtain the materials listed in the margin.

B. Refer back to the four elements used in the activity in the last section. See p. 45.
   1. How would the electrons of the following atoms be arranged in shells: helium, oxygen, silicon?

C. Use the red circles as electrons and your nuclear models from section 2. Show how the electrons of each of these atoms might be arranged around the nucleus.

D. Argon's atomic number is 18. Its atomic mass is 40.
   2. How many electrons, protons and neutrons does argon have?
   3. How would argon's electrons be arranged in shells?

E. Use the circles to make a complete model of argon.
   4. What are the symbols for helium, oxygen, silicon and argon?

Chemical symbols are the chemist's shorthand. The electrons circle the nucleus in energy levels within an electron cloud. This cloud contains shells in which the electrons are located. Electrons circle the nucleus billions of times each second.

1. How many electron shells is it possible to have within an electron cloud?
2. How many electrons can each of the first three energy shells hold?
3. An atom has an atomic number of 15. How will its electrons be distributed in the shells?
4. Why are symbols used for the elements?
5. Describe two ways electrons can change shells.

4 NUCLEAR ACTIVITY

In 1977, two spacecraft were launched on a journey to the planets. Both spacecraft passed Jupiter in 1979 and Saturn in 1980 and 1981. One of these is scheduled to pass Uranus in 1986 and Neptune in 1989.

What makes these spacecraft unique is the source of the energy to run their scientific instruments. Too far away from the sun to use solar power, both spacecraft turn the energy of the atom directly into electricity. When you finish this section, you will be able to:

● Explain why some isotopes of elements are *radioactive*.

● Define atomic fission and atomic fusion.

● Describe how the energy of the atom can be controlled.

▲ Demonstrate a chain reaction.

At about the same time some scientists were exploring the atom, others discovered that the elements

2-15. *Marie Curie discovered that radium gives off invisible rays.*

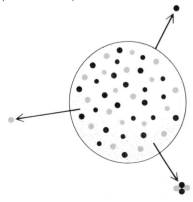

2-16. *Unstable atoms change as particles escape from their nuclei.*

uranium and radium gave off strange energy rays. Some of these rays were able to pass right through wood, glass, and even a person's body. The term *radioactive* was given to elements that actively give off energy rays.

Eventually it was discovered that some of these rays were particles and that they were coming from the nucleus of the radioactive atoms. Several questions now needed to be answered. Why do the atoms of some elements seem to be throwing off pieces of themselves? Why don't the nuclei of all atoms throw off these rays? What holds the protons and neutrons together in the nucleus?

It was already known that particles with the same electrical charge will push away from each other. This pushing away is called *repelling*. If two negatively charged particles are brought close together, they repel each other. The same thing happens if two positively charged particles come together. Why then don't the protons, which all have positive charges, repel each other and leave the nucleus?

To answer these questions, scientists reasoned that there must be some sort of powerful force in the nucleus that held the protons and neutrons together. The more protons there are in the nucleus, the greater the repelling forces between them, therefore, the greater the force needed to keep the nucleus together. This force is now called *binding energy*. It "binds" the protons and neutrons together in the nucleus. Exactly what creates this binding energy is not yet fully understood.

In most atoms, the binding energy is at least equal to the repelling forces between the protons. As a result, the nucleus stays together. These atoms are called *stable*. Some isotopes of some elements do not seem to have enough binding energy. As a result, some protons and neutrons manage to break away from the nucleus and "shoot" out of the atom. These atoms are called *unstable*. The escaping particles make up some of the radiation.

Unstable atoms change. Since some of the particles escaping from the nucleus are protons, the atomic number changes. This means that these atoms are changing into other kinds of atoms. Without as many protons in the nucleus, the repelling forces become smaller. The binding energy can now hold the nucleus together. Thus, the new atom is stable.

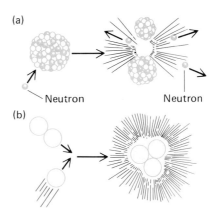

2-17. *a. Atoms are split during fission. b. Atoms combine in a fusion reaction.*

Scientists began to wonder if it might be possible to help this process along. They wondered if they could make one type of atom change into another atom. By directing beams of radiation at atoms, they found that they could indeed change one kind of atom into another. The dream of the alchemists had finally come true!

Sometimes large unstable atoms split into two smaller nuclei. For example, there is an isotope of uranium that has an atomic mass of 235. The symbol for this isotope is $^{235}_{92}U$. This combines the symbol for the element with the number representing the mass of this particular isotope. The $^{235}_{92}U$ nucleus splits into two much smaller nuclei as well as giving off radiation and releasing energy. This kind of change is called *nuclear fission*.

The fission of one atom of $^{235}_{92}U$ does not release very much energy. If billions of $^{235}_{92}U$ atoms were to split in a very short time, tremendous amounts of energy would be released. Scientists reasoned that if this process could be controlled, a new source of energy would be available for us to use.

In order to produce usable amounts of energy, atoms of $^{235}_{92}U$ have to be brought close together. Then the neutrons released when one nucleus splits strike other nuclei, causing those to split. This releases still more neutrons and the splitting process continues to increase very quickly. This is called a *chain reaction*. Uncontrolled chain reactions result in an explosion. An example of this is the atomic bomb.

However, a chain reaction can be controlled. In a *nuclear reactor*, a chain reaction is controlled by rods of a material that trap the neutrons. The speed of the reaction can be controlled by moving these rods into or out of the radioactive material. When the rods are

Chapter 2-4 Nuclear Activity

2-18. *A chain reaction releases tremendous amounts of energy. These are the rods of a nuclear reactor.*

moved in, they trap some neutrons and slow down the reaction. Thus, the tremendous energy of a nuclear chain reaction can be tamed. This energy can be used to make steam, which can then generate electricity.

Such a reaction is the source of power for the *Voyager* spacecraft. Each is equipped with a tiny power plant which changes nuclear energy directly into electricity. However, this reaction cannot continue indefinitely. In time, the radioactive atoms will become stable. There will not be enough energy to run the equipment. Then the *Voyagers* will continue to drift silently out into the universe.

Scientists have also been able to "build" atoms by causing two nuclei to join together. This type of change is called *fusion*. Fusion only occurs when the temperature is several million degrees Celsius. The only place where such temperatures exist naturally is in stars like our sun. At these very high temperatures, atoms move so fast that the normal repelling forces between nuclei cannot keep them apart. The nuclei collide so violently that they *fuse* into a single nucleus. Fusion releases large amounts of energy. The energy we get from the sun is released by the fusion of hydrogen atoms into helium atoms.

Some scientists feel that the tremendous energy released during the fusion reaction can be controlled. If this ever becomes possible, we would have a fantastic new energy source. However, it is not yet possible to maintain the temperatures needed for fusion for any length of time. It will probably be many years, if at all, before energy from fusion becomes practical.

Materials
Ruler
Dominoes (10)

A. Obtain the materials listed in the margin.

B. Arrange 10 dominoes in a line. Leave about 3 cm of space between each domino.
 1. What do you predict will happen if you push the first domino over backward?
 2. Name this reaction.

C. Tip the first domino over to test your prediction.

D. Arrange the 10 dominoes in the pattern shown in Fig. A. Note the spacing involved.
 3. Predict what will happen if the first domino is tipped.
 4. Will this begin a chain reaction?

E. Test your prediction.

F. Arrange the dominoes into the pattern in Fig. B. Again note the spacing.
 5. What do you predict will happen in this case?
 6. Will this begin a chain reaction?

G. Test your prediction.

H. Arrange the entire set in pattern C again. Hold a ruler between rows 3 and 4 so that it blocks off the end two dominoes on the right of row 4.
 7. Predict what will happen in this case. Test your prediction.
 8. How is this chain reaction different from the others?
 9. How is this similar to what happens in a nuclear reactor?

Powerful forces bind the protons and neutrons together in the atom's nucleus. However, some atoms do not have enough of this force and so they throw off particles and energy as forms of radiation. This activity is the source of the nuclear energy scientists have learned to control and make useful.

1. Describe why the nucleus of an atom stays together.
2. Why do some atoms throw off parts of themselves?
3. Define radioactivity, nuclear fission and fusion.
4. Explain how a chain reaction works.
5. How can a chain reaction be controlled?
6. In what ways can the energy of the atom be used?

5 ATOMIC MODELS

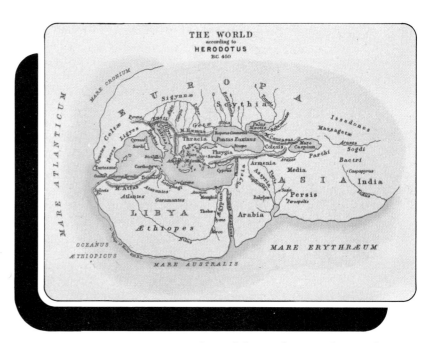

A map is one type of model. It shows what is known about the size, shape, and location of land masses. This early map of the world showed what was known at the time it was constructed. Mapmakers added what they believed to be correct about the areas that were unexplored. Explorers like Columbus and Magellen soon sailed into unknown areas. New maps had to be created based on their discoveries and measurements. When you finish this section, you will be able to:

● Calculate atomic mass or the number of neutrons in an atom when given the necessary information.

● Explain why models are often changed.

▲ Draw satellite models of atoms when given the atomic mass.

Models are based on the information known at the time. As new information is discovered, the model must change to include it. However, sometimes the old model still is useful. This is true of the satellite model of what an atom is like. In this section you will make diagrams of 10 different atoms.

Early ideas about the makeup of matter led to the idea that atoms were like tiny solid balls. This early model was changed when the electron was discovered. A new model was proposed. According to this model, the atom was a positively charged sphere with negative electrons scattered throughout like raisins in a cake. A later model had pairs of positive and negative charges floating in space. As new information about atomic particles and their location in the atom was gathered, the model of the atom changed again and again.

The electron-proton-neutron model shown in Fig. 2-19 is one you will see very often. Atoms do not really look like this. However, this type of atomic model can be helpful. Such models tell you the number of electrons, protons, and neutrons in the atom and their approximate locations.

Niels Bohr first used this type of model to represent paths for electrons. It is usually called the satellite model. Scientists have replaced this model with a modern version based on mathematics. The modern version explains much of the information that exists today concerning the chemical behavior of atoms.

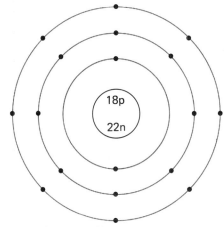

2-19. An argon-40 atom.

Materials
3 sheets of paper
Pencil
Compass

In this activity you will make some satellite models of atomic structure. Even though this model has been replaced by a more modern one, it may still be useful in showing the number of electrons, protons, and neutrons in an atom and their approximate locations.

A. Obtain the materials listed in the margin.

B. Copy the table on page 58 on one sheet of paper.

C. Based on the information given, fill in the blanks in the table.

Chapter 2–5 Atomic Models

Atomic Number	Element	Atomic Mass	Number of Neutrons
1	hydrogen	1	
2	helium		2
3	lithium		4
4	beryllium	9	
5	boron	11	
6	carbon	12	
7	nitrogen		7
8	oxygen		8
9	fluorine	19	
10	neon	20	

D. Draw a diagram of each of the 10 elements listed in the table. Make the diagram similar to the one of argon in Fig. 2-19. Show the number of protons and neutrons in the nucleus. Show the number of electrons in their proper shells around the nucleus.

E. Label each diagram with the name of the element and its atomic mass, for example, hydrogen-1.

F. In the same manner as in step D, draw a diagram of each of the following atoms: magnesium-24, atomic number 12; aluminum-27, atomic number 13; sulfur-32, atomic number 16; chlorine-35, atomic number 17.

SUMMARY & QUESTIONS

Scientists often use models to show things that are not visible to the eye. These models are based on the best information available at the time. As more information is gathered through research, the model has to be changed or discarded completely. In this way, a model is really a hypothesis.

1. A certain atom has an atomic number of 13 and a mass of 27. How many protons, electrons, and neutrons does it have?
2. Diagram the electron arrangement of silicon-14.
3. Potassium has 19 protons, 19 electrons, and 20 neutrons. What is its atomic mass?

VOCABULARY REVIEW

Match the description in Column II with the term it describes in Column I.

Column I
1. scientific model
2. atomic particle
3. atomic mass unit
4. nucleus
5. atomic number
6. electron shell
7. chemical symbol
8. atomic mass
9. isotope

Column II
a. A region around an atomic nucleus in which electrons move.
b. The building blocks of atoms.
c. The central core of an atom containing its protons and neutrons.
d. The number of protons in the nucleus of an atom.
e. Mental picture of something that cannot be seen.
f. One or two letters used to represent an atom of a particular element.
g. Atoms whose nuclei contain the same number of protons but different number of neutrons.
h. Special unit used to express the masses of atomic particles and atoms.
i. The sum expressed in amu of the protons and neutrons in an atom.

REVIEW QUESTIONS

Choose the letter of the ending which best completes each of the following statements.

1. A mental picture of something that cannot be seen describes (a) an optical illusion (b) a scientific model (c) an extrasensory illusion (d) a physical model.
2. The negative, positive, and neutral atomic particles are (a) electron, proton, neutron (b) electron, neutron, proton (c) proton, electron, neutron (d) proton, neutron, electron.
3. What is the same in a proton and a neutron? (a) number of electrons (b) color (c) charge (d) mass.
4. The nucleus of an atom is made up of (a) electrons and protons (b) electrons and neutrons (c) protons and neutrons (d) neutrons.

5. If an atom has a mass of 11 atomic mass units and contains 5 electrons, its atomic number must be (a) 55 (b) 16 (c) 6 (d) 5.
6. In an atom, the electrons are found (a) in shells around the nucleus (b) in shells within the nucleus (c) in clouds within the nucleus (d) everywhere in the atom.
7. The number of electrons found in the first two electron shells in atoms is (a) 1,2 (b) 2,4 (c) 2,8 (d) 2,18.
8. The chemical symbols for hydrogen, helium, and oxygen are (a) Hg, H, and O (b) H, He, and O (c) H, He, and Ox (d) Hy, He, and Ox.
9. The atomic number of lithium is 3. The atomic mass of a lithium atom is 7. The number of protons and neutrons must be (a) 4 and 10 (b) 3 and 10 (c) 3 and 4 (d) 7 and 10.
10. Atoms of the same element can have different atomic masses because they can have different numbers of (a) neutrons (b) protons (c) electrons (d) protons and electrons.

REVIEW EXERCISES

Give complete but brief answers to each of the following exercises.

1. Give an example of a scientific model and explain how it is like the real thing.
2. Name three kinds of atomic particles and give the properties of each.
3. What is meant by the term *atomic nucleus*?
4. Explain what determines the mass of an atom.
5. The atomic number of an element is 5 and the number of neutrons in the nucleus of one of its atoms is 6. How many electrons are in a normal atom? Explain how you found the number.
6. What is an electron shell and how are electrons arranged in the first two shells?
7. Where are electrons in atoms and how do they move?
8. Name five of the first 10 elements and give the chemical symbols for each.
9. Describe what atomic particles account for the mass of an atom.
10. How are atomic mass and atomic number related to the number of protons, neutrons, and electrons in an atom?

CHAPTER 3
ENERGY

1 WHAT IS ENERGY?

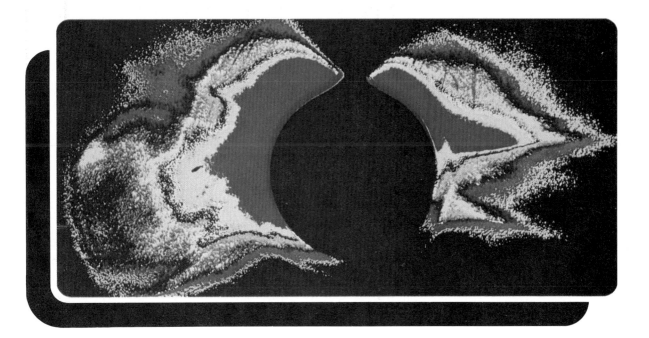

The sun is the powerhouse of our solar system. It is the source of almost all the energy that Earth receives. This super space-furnace changes almost 600 million tons of matter into energy every second. As it gives off life-supporting heat and light, the sun is a constant reminder of the makeup of the universe: matter and energy. When you finish this section, you will be able to:

- Explain why energy is called the ability to do *work*.
- Describe some of the forces of energy at work in nature.

● Discuss the relationship between matter and energy.

▲ Recognize the effects of energy in nature.

3-1. *Where do steam engines get their energy?*

Work: The movement of an object over a distance.

Everything in the universe is either matter or energy. You will remember that matter is defined as anything that takes up space and has mass. But what is energy? In an earlier chapter we defined energy as the ability to do work. This may seem very confusing to you. Energy is not an easy thing to explain.

For example, you might ask "Where do steam engines get their energy?" That seems easy enough to answer. The steam pushing against the pistons turns the wheels. "But where does the steam get its power?" Steam is water that has been changed into hot gas by heat. The heat comes from the burning of coal. "But where does the coal get its energy?" The answer to that question is "from the sun." In later chapters we will study how the sun's energy is trapped by plants and how plants become coal.

But none of this tells us the answer to the big question, "What is energy?" Actually, we've come about as close as we can. Scientists can tell us what energy *does* and what it is not, but not what it *is*. Energy has no mass. It occupies no space. It has no taste or smell. In other words, it doesn't possess any of the properties of matter. But it does move matter. A scientist would say **work** is done if an object is moved through a distance. Energy can move matter. Therefore we define energy as the ability to do *work*.

Perhaps you will understand better what energy is if we examine more of what it does. Later chapters will deal more with what people have learned to do with energy. Here we will take a brief look at energy in nature.

What is the greatest source of energy in our solar system? If you answered the sun, you would be right. The sun changes over 600 million tons of matter into energy every second. This energy is released in the form of heat and light. Heat and light are both types of energy. Earth receives only a small portion of the sun's energy. But the sun is the greatest source of energy we have.

Some of the energy from the sun that reaches Earth warms the earth's surface. However, the energy of the

3-2. (left) Where does wind energy come from?

3-3. (right) Waves carry energy.

sun is stronger in some areas than in others. This causes some areas to be warmer than others. Energy from the sun has become heat energy that warms the atmosphere. Cold air is heavier than warm air. The cold air sinks, pushing the warm air out of the way. This creates winds. Heat energy has become the energy of motion. As you know, strong winds can move objects quite easily. Dust, sand, and leaves are blown by the wind. Sometimes even buildings can be moved by strong winds.

Wind blows across the oceans, pushing the water into waves. If you have ever been at the ocean or on a large lake on a windy day, you may have seen the waves moving sand and rocks along the shore. In violent storms, energy of waves can cause much damage along the shoreline.

These few examples point out some of the characteristics of energy. Energy can be changed from one form to another. Light energy from the sun was changed into heat energy to warm the air. Heat energy in turn was changed into the energy of motion as the wind moved sand and dust. Work is being done in these examples. Objects are being moved.

In these examples also energy is being transferred from one type of matter to others. Energy of the moving wind is transferred to the moving water. The waves in turn transfer some of their energy to the sand and rocks they toss up on a beach during a storm.

Materials
Large sheet of heavy paper
 (10 x 15)
Glue
Scissors
Old magazines and newspapers
Pencil
Paper

A. Obtain the materials listed in the margin.

B. In this lesson we have discussed some of the ways energy works in nature. Use old magazines and newspapers to collect pictures that show energy at work in nature. Do not include any pictures of people using energy.

C. Use these pictures to create a montage on a piece of heavy paper.

 1. Briefly explain how each of your pictures shows energy at work.

D. Create a bulletin-board display from your class's montages.

3-4. *Albert Einstein studied the relationship between matter and energy.*

3-5. *Burning releases energy from matter.*

Albert Einstein stated that matter and energy were very closely related. In fact, he said it should be possible to convert one to the other. Einstein was considering what happened to things as they moved faster and faster. It takes more and more energy to make things move faster and faster. Einstein said that as energy is added to an object, its mass increases, too. Of course, the increase is very small. But it is real. In effect, a golf ball flying through the air has more mass than one not moving.

Einstein stated his ideas in the formula $E = mc^2$. This formula states that the total amount of energy locked in a certain mass is equal to the mass times the square of a constant number. That means that if 1 kg of mass could be completely converted to energy, it would be enough to drive a car around the world almost 100,000 times.

Einstein's ideas remained theories for many years. But in 1942, a group of scientists headed by Dr. Enrico Fermi proved that the energy released by the breakdown of uranium atoms could be controlled. Science had learned how to change matter into energy. This marked the beginning of the Atomic Age. In the years following, scientists also learned how to put that energy to use. Atomic power plants provide electric energy and power ships and spacecraft. Scientists have also learned that energy can be changed back into matter. We will discuss these relationships further in later chapters.

SUMMARY & QUESTIONS

Energy is difficult to define. It is best understood as the ability to do work. The universe is full of examples of how matter is affected by energy.

1. What is work? Why is energy defined as the ability to do work?
2. Give two examples of how energy can be transferred from one type of matter to another.
3. Give two examples of how energy in nature can be changed from one type of energy to another.

2 CONSERVATION OF ENERGY

Why will King Kong have more energy when he reaches the top of the building than he had on the ground? After reading this section, you will be able to answer this question. When you finish this section, you will be able to:

● Use examples to explain the difference between *potential* and *kinetic* energy.

● Demonstrate how one form of energy can be changed into another form.

● Use examples to show that energy is neither created nor destroyed.

▲ Use photographs for explaining the change from potential to kinetic energy.

Chapter 3–2 Conservation of Energy   65

A scientist will often compare an idea being explained with something we already know. This comparison is called an analogy. The analogy here is that a spring can be thought of as a force just as gravity is a force. The two forces do not have to be the same in order for us to feel the "weight" effect. When analogies are used, they only have some things in common with the idea being explained. Analogies are not exactly the same as the idea itself.

Try this experiment: Stand with your feet firmly on the floor. Hold a book in both hands. With your eyes closed, slowly raise and lower the book several times. A force pulls the book down. This force is the weight of the book caused by gravity. Suppose that instead of gravity pulling on the book a spring holds the book to the floor. See Fig. 3-6. If you try hard, you will be able to picture

3-6. *Gravity force pulling down on the book can be thought of as a spring pulling down on the book.*

the spring and see it stretch as you raise the book. Now think about the energy added to the spring as it stretches.

3-7. *Which board would the diver use to hit the water with the greatest energy?*

Potential energy: Energy stored in an object as a result of a change in its position.

How much more potential energy does a book have if it is raised 2 m rather than 1 m?

Suppose you dropped a book on your foot. How many times as much energy would be delivered to your foot if you dropped two books from the same height?

Kinetic energy: Energy that moving things have as a result of their motion.

A spring that is pulled out of its normal shape has energy stored in it. This stored energy cannot be seen. We know it is there because the stretched spring can be used to do work when it pulls an object as it springs back to its normal shape. Energy that is stored in an object as a result of a change of position is called **potential energy** (po-**ten**-shall **en**-err-gee). A spring gains *potential energy* when it is stretched. A book that is lifted against the force of gravity stores energy in much the same way as a stretched spring. The energy spent by your muscles in raising the book is stored in the book as potential energy. More energy is needed to lift the book to a higher position. The amount of potential energy stored in the book increases as the book is raised through a greater distance. Also, it takes more energy to lift a heavy book than a light one. The amount of potential energy caused by gravity stored in an object is a result of its weight and the distance it is lifted. This result is why King Kong has more potential energy on top of the building than on the ground.

When released, a stretched spring returns the potential energy stored in it. A book held above the floor releases its potential energy if it is allowed to fall. The potential energy stored in the book held high is changed into energy of motion as it falls. Energy that an object has as a result of its motion is called **kinetic energy** (ki-**net**-ick **en**-err-gee). An object that is falling or moving in any direction has *kinetic energy*. The kinetic energy of a

3-8. When the book is allowed to fall, the stored potential energy is changed into kinetic energy.

Chapter 3–2   Conservation of Energy

3-9. Which person has the greatest amount of potential energy? kinetic energy?

falling book can be changed into another form if the book stops moving. For example, the kinetic energy of the falling book might result in possible damage to the book, the floor, or your foot!

Materials
pencil
paper

3-10.

Where have you seen energy changed from one form to another? On a roller coaster, potential energy, when the car is high, changes into kinetic energy, when the car swoops down. The kinetic energy can then be changed back into potential energy as the car climbs another hill.

A. Obtain the materials listed in the margin.

B. In Fig. 3-11, you see a roller coaster car going down a steep hill. Use the lettered positions, A, B, C, and D, to answer the following questions.

1. At what position does the car have the greatest potential energy?

2. At what position does the car have the least potential energy?

3. At what position does the car have the greatest kinetic energy?

4. At which point, B or C, does the car have more kinetic energy?

5. Does each increase in kinetic energy occur when there is a decrease in potential energy?

C. Figure 3-12 shows a pendulum swinging. The photo is a time exposure made with a strobe light flashing on at equal

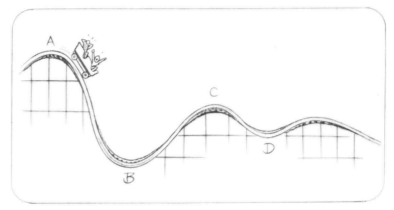

3-11.

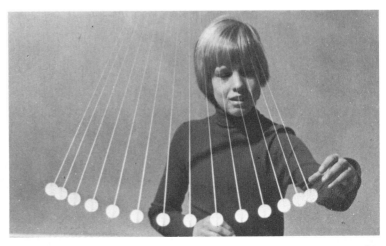

3-12. *Strobe exposure of a swinging pendulum. Since the time between each exposure is equal, greater distance between exposures indicates greater speed.*

Chapter 3—2 Conservation of Energy

time intervals to show the position of the bob.

6. The pendulum bob is moving fastest at the (center, end) of its swing.

7. The pendulum bob is highest above the table at the (center, end) of its swing.

8. Where does the pendulum have the greatest kinetic energy? (center or end of swing)

9. Where does the pendulum have the greatest potential energy? (center or end of swing)

10. Does each increase in kinetic energy occur when there is a decrease in potential energy?

Law of Conservation of Energy: A natural law which says that energy cannot be created nor destroyed but may be changed from one form to another.

The potential energy due to gravity in an object can be changed into kinetic energy. Kinetic energy can be changed back into potential energy or some other form of energy. These energy changes occur throughout the natural world. They are summarized in the **Law of Conservation** (con-ser-**vay**-shun) **of Energy.** This law says that *energy is never created nor destroyed, but is only changed from one form to another.* For example, the potential energy stored in a book held above the floor can be measured by using the weight of the book and the distance it was lifted above the floor. The *Law of Conservation of Energy* says that if the book is dropped to the floor, its kinetic energy when it hits the floor is equal to its original potential energy.

SUMMARY & QUESTIONS

If an object is lifted against the force of gravity, energy is stored in that object in the form of potential energy. That stored energy can be turned into kinetic energy if the object is allowed to fall. When energy changes in form, as from potential to kinetic, the Law of Conservation of Energy tells us that no energy can be gained or lost.

1. Identify each of the following as examples of either potential or kinetic energy.
   a. a book resting on a shelf
   b. a spring pulled out of its normal shape

c. a book being lifted from the table to a high shelf
d. a book falling from a shelf to the floor
e. a pendulum at the highest point of the swing
f. a pendulum at the lowest point of the swing
2. Describe how you could use a pendulum to demonstrate how potential energy changes to kinetic energy which in turn changes back to potential energy.
3. What law of nature helps you to calculate the amount of potential energy converted to kinetic energy as a book falls to the floor?

3 USES OF ENERGY

A waterfall is a natural source of tremendous amounts of energy. When you finish this section, you will be able to:

● Name and describe six forms of energy.

● Describe how energy is used in our daily lives.

● Identify three sources of energy needed by modern civilization.

▲ Make a chart of the forms of energy you use every day.

We live in an ocean of energy. Everywhere there is evidence of the natural ways by which energy is changed from one form to another. As rivers flow to the sea, falling water releases large amounts of kinetic energy. Energy in wind creates waves on the sea or turns windmills. What forms of energy are needed to describe all these changes? Scientists say that at least six forms of energy are needed. These six forms of energy are described below. You will study them more completely in later chapters.

1. *Mechanical energy.* This is energy due to the posi-

tion or motion of an object. An example of mechanical energy is the motion of the pistons and other moving parts of an automobile engine. The energy of a hammer hitting a nail is also mechanical.

2. *Chemical energy.* When gasoline burns in an automobile engine, chemical energy is released. The energy supplied to our bodies by food is also chemical. Chemical energy is associated with the change of one kind of matter into another.

3. *Heat energy.* This form of energy is the cause of changes in temperature in solids, liquids, and gases.

4. *Electric energy.* The flow of electric currents is one form of electric energy. Electric energy is also felt in other ways. Magnetism is related to this kind of energy.

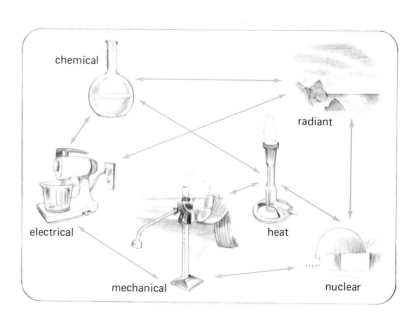

3-13. *Energy can be converted from one form to another.*

5. *Electromagnetic energy.* This form of energy is one of the most common. It spreads out and passes through space. This includes such things as radio waves and Xrays as well as the light we see.

6. *Nuclear energy.* Changes in the nuclei of atoms release this kind of energy. Nuclear energy is the most concentrated of all the forms of energy. The energy of the sun and stars is nuclear.

Materials
paper and pencil

This activity should help make you more aware of the many ways you use energy each day.

A. Obtain the materials listed in the margin.
B. On a sheet of paper, make a table like the one shown below. In the first column, list as many things that you use every day as you can. In the second column, write the kind(s) of energy used by that activity.

Things I use every day	Kind(s) of energy
1. automobile	chemical, mechanical
2. typewriter	
3. lamp	
4.	
5.	

How many forms of energy do you use in everyday life? Even an incomplete answer to this question will quickly show that our lives are built on a foundation of

3-14. *You use energy in many ways each day.*

Chapter 3–3 Uses of Energy

energy. For example, everyone needs the chemical energy supplied by food. We would find it hard to live without heat energy to warm our houses and buildings and for cooking. We use large amounts of energy to run the machines used to produce the necessities of life and for transportation. All over the world, the average person uses about eight times as much energy as could be produced from muscle-power alone. In some highly developed countries such as the United States, each person uses ten times more energy than the average person in most other parts of the world.

Most of these energy needs are met now by burning of fuels such as petroleum and coal. Water power, nuclear energy, wind, and heat from the interior of the earth all together supply only a small amount of the energy needs of the world. The supply of petroleum and coal is limited and will be gone some day. Future civilizations must learn to apply the knowledge of natural energy to help meet growing energy needs.

3-15. *Solar energy can be used to supply electricity as well as heat and hot water.*

Energy exists in at least six different forms. These forms include: mechanical, chemical, heat, electrical, light, and nuclear. Day-to-day living requires a supply of energy that is presently met almost entirely by burning of fuels. Because the supply of fuels is limited, new sources of energy must be developed in the future.

Use the clues given to fill in the following crossword puzzle. DO NOT WRITE IN THIS BOOK.

DOWN

1. ____ can be converted from one form to another.
2. Supplies energy for most of today's living.
3. A hammer hitting a nail.
4. All forms of energy when being stored.
5. Causes a change in temperature.

ACROSS

6. Includes radio waves, X rays, etc.
7. Caused by changes in an atom's core.
8. Energy supplied by food.
9. Most common form of electromagnetic energy.
10. Related to electric energy.

VOCABULARY REVIEW

Match the description in Column II with the term it best describes in Column I.

Column I
1. energy
2. chemical energy
3. heat energy
4. nuclear energy
5. work
6. kinetic energy
7. electromagnetic energy
8. potential energy
9. mechanical energy
10. electric energy

Column II
a. Size of the force multiplied by the distance through which the force acts.
b. Energy that powers the sun and stars.
c. That property of something that makes it able to do work.
d. Any kind of energy that spreads out and passes through space.
e. Connected with the change of one kind of matter into another.
f. Energy that moving things have as a result of their motion.
g. Energy stored in an object as a result of a change in its position.
h. Includes both potential and kinetic energy.
i. The flow of electric current.
j. Causes changes in temperature.

REVIEW QUESTIONS

Choose the letter of the ending which best completes each of the following statements.

1. When a force moves something, the amount of work done depends on (a) only the size of the force (b) only the distance the object moves (c) both the size of the force and the distance the object moves (d) the direction in which the object moves.
2. That which allows a force to do work is called (a) energy (b) power (c) gravity (d) it is not known what allows a force to do work.
3. Magnetic energy is closely related to (a) chemical energy (b) light energy (c) electric energy (d) heat energy.
4. The scientific law that says that energy cannot be created or destroyed but can be changed from one form to another is called the Law of (a) Conservation of Energy (b) Preservation of Energy (c) Supply and Demand (d) Indestructibility.
5. The energy involved when ice melts is (a) light energy (b) nuclear energy (c) chemical energy (d) heat energy.
6. Potential energy is (a) stored energy (b) energy due to motion (c) energy of a speeding bullet (d) both *a* and *b*.

7. The kinetic energy given to a book which is lifted from the floor and placed on the table is changed into (a) force (b) potential energy (c) power (d) gravity.
8. The form of energy which is most concentrated is (a) light (b) electrical (c) nuclear (d) chemical.
9. Day-to-day living requires a supply of energy that is almost entirely supplied by (a) burning of fuels (b) water power (c) nuclear energy (d) heat from the sun.

REVIEW EXERCISES

Give complete but brief answers to the following exercises.

1. List three sources of energy in nature.
2. Explain how heat energy from the sun is changed into the energy of a wave crashing onto the beach.
3. Why can we say that the wind has energy?
4. A 5 kg weight is lifted 5 m above the ground. Another 5 kg weight is lifted 10 m. Which has more potential energy? If both were to fall, which would have the greater kinetic energy? Why?
5. What forms of energy besides light energy make up electromagnetic energy?
6. Given a spring to use, how would you demonstrate to someone else how one form of energy can be changed into another form.
7. State the Law of Conservation of Energy.
8. Using a weight which could be dropped to the floor, point out how the Law of Conservation of Energy affects the amount of energy transformed from one form to another.
9. List the six forms of energy described in Section 3 and give an example of each.
10. Name the present sources of energy used by modern civilization.

CHAPTER 4

CHEMICAL CHANGES

1 TYPES OF ATOMS

In the past 50 years, our whole world has changed because of the new chemical processes and materials science has developed. The study of chemical reactions is a fascinating and ever-changing one. But one question has continued to challenge chemists. "Why do some elements react with each other easily, while others barely react at all?" The answer to this question lies in atomic structure. When you finish this section, you will be able to:

- Distinguish between an atom and an *ion*.

- Explain what is meant by a *stable electron arrangement*.

▲ Predict whether an atom will gain or lose electrons.

Ion: An atom or molecule with an electric charge.

The electrons found in the outer shell of an atom determine how the atom will behave with some atoms. Scientists have experimented to find how the outer electrons are arranged. One experiment involved knocking electrons away from the outer shell of other atoms. This loss of electrons changed the neutral atoms into **ions**. The *ions* have a positive charge since they lost negative electrons. Scientists can determine how much energy is needed to remove an electron from the outer shell of an atom. These energies, ranging from hydrogen (atomic number 1) to calcium (atomic number 20), form a pattern. See Fig. 4-1. The graph shows that the atoms of the elements *helium, neon,* and *argon* require the highest energies to remove an electron.

Helium, neon, and argon atoms have a tighter hold on their electrons than other atoms. Helium (atomic number 2) has a total of two electrons. Neon has 10 electrons and argon has 18 electrons. Is there something special about the number of electrons in an atom? Remember how many electrons fill each shell around the nucleus of an atom. The first electron shell is full with two electrons. The second shell fills with eight electrons. The third shell is also full with eight electrons. Thus, helium has all of its electrons in one completely filled shell. Neon, with 10 electrons, has filled the first and second

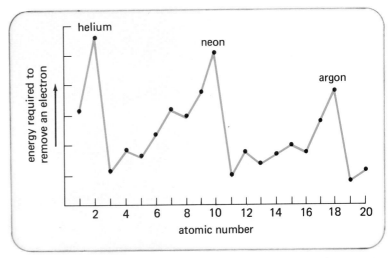

4-1. *This graph shows the energy needed to remove an electron from atoms of the first 20 elements.*

Chapter 4–1 Types of Atoms

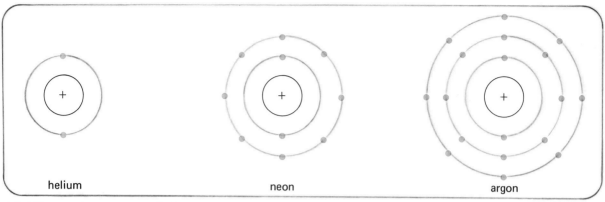

4-2. Why are these atoms called "noble" gases?

shells (2 + 8). Argon, with 18 electrons, has its first, second, and third shells filled (2 + 8 + 8). See Fig. 4-2.

Atoms which have completely filled electron shells do not easily lose electrons. The electron arrangement is **stable**. A *stable* electron arrangement means that an atom does not tend to gain or lose electrons readily since its outermost electron shell is completely filled. The elements with stable atoms are gases. These gases are called the **noble gases**.

Experiments have shown that there are a total of six *noble gases* among all the elements. In addition to helium, neon, and argon, other noble gases are krypton, xenon, and radon. Krypton has 36 electrons, xenon has 54 electrons, and radon has 86 electrons. With these numbers of electrons, the electron shells of krypton, xenon, and radon are completely filled.

The noble gases are not found in great quantities on earth. The most abundant is argon. Argon makes up about 1% of the air. Helium is found mixed with natural gas. Most of the remaining noble gases are found in small quantities in the atmosphere.

Stable electron arrangement: An electron arrangement in which the outermost electron shell is filled.

Noble gases: The six elements whose atoms have completely filled electron shells.

Materials
Pencil
Paper

You know now that having eight electrons in its outer shell makes an atom stable. An atom will either gain or lose electrons to wind up with the stable number of eight. An atom with only one electron in its outer shell will give up that electron. An atom with seven electrons will gain one to fill its

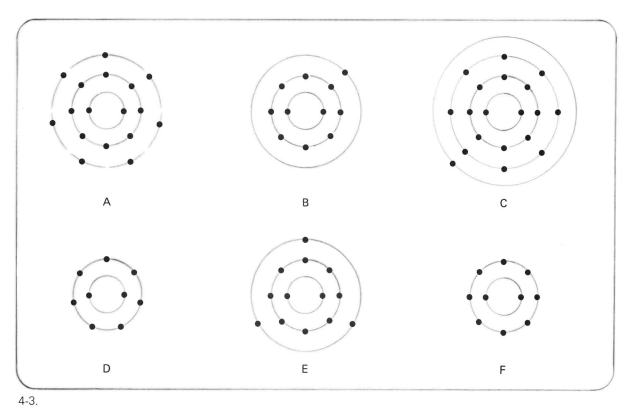

4-3.

outer shell. In general, atoms either lose or gain electrons depending on which requires less work. In this activity, you will predict whether different atoms will gain or lose electrons to become stable.

A. Look at the electron diagrams of six different atoms in Fig. 4-3.

1. Which atoms do you think will lose electrons to become stable? (A, B, C, D, E, F)

2. Which atoms do you think will gain electrons to become stable? (A, B, C, D, E, F)

3. Do any of the atoms shown already have a stable electron arrangement? What would this kind of atom be called?

Can you tell now what makes an atom "noble"? Atoms of the noble gases have completely filled electron shells. The filled shells make these gases stable. Noble gases are not common on earth.

Chapter 4—1 Types of Atoms

1. A certain atom has a stable arrangement of electrons.
   a. Which one of the following number of electrons would it most likely have: 6, 7, 8, 9, 10?
   b. Will the atom gain or lose an electron more easily?
   c. Will it most likely be a gas, a liquid, or a solid?
2. A certain atom has 11 electrons.
   a. Is this a stable electron arrangement?
   b. Will this atom most likely gain an electron?
   c. Will this atom most likely lose an electron?
   d. If the atom lost an electron, what would it become?
3. How does a stable atom become a negative ion? a positive ion?

2 CHEMICAL ACTIVITY

The *Hindenburg* was a famous German passenger airship. In 1937, while attempting to land in New Jersey after crossing the Atlantic, it exploded and crashed in flames. What could have caused this disaster? The *Hindenburg* was filled with hydrogen gas. It was the hydrogen which exploded and burned. Modern airships like the Goodyear blimp are filled with helium. In this section you will find out why helium is a safer gas than hydrogen. When you finish this section, you will be able to:

● Relate the *chemical activity* of an element to the number of electrons in the outer shell of an atom of that element.

● Explain why certain elements can be grouped together.

▲ Classify nine elements into groups on the basis of their characteristics.

To stay in the air, an airship must be filled with a gas that is lighter than air. The airship then floats in the air the same way a piece of wood floats in water.

Hydrogen gas is lighter than air. But, hydrogen is a dangerous gas. It can burn when mixed with oxygen. A much safer gas for airships is helium. Helium is also much lighter than air. Helium does not burn because it does not combine with oxygen. In fact, helium will not readily take part in any chemical changes. Helium atoms usually exist separately. They are seldom part of a molecule. Why is it that helium does not readily take part in chemical changes while hydrogen does?

Hydrogen and helium are the only two elements that have only a single electron shell. This shell can hold only two electrons. Helium already has two electrons. Therefore, helium has a completely filled shell. It does not join readily with other atoms to form molecules. Hydrogen has only one electron. Hydrogen needs one more electron to fill the shell. Therefore, unlike helium, hydrogen reacts with many other atoms to form molecules.

The way an atom reacts with atoms of other elements is called its **chemical activity**. Hydrogen, which reacts readily with other elements, is *chemically active*. Helium, which does not react readily with other elements, is not chemically active. It is **inert**.

Is it possible that the two elements, hydrogen and helium, behave so differently because of the different number of electrons in their outer shells? If this is true, then atoms with the same number of outer electrons should behave in the same way.

hydrogen

helium

4-4. Both hydrogen (H) and helium (He) have only one electron shell.

Chemical activity: The way an atom reacts with other kinds of atoms.

Inert: A description of an atom which does not react readily with other atoms.

Materials
9 element cards
paper and pencil

A. Obtain the materials listed in the margin. Each of the nine cards represents a different element. On the front of the card is the name of the element. Below the element's name is the number of electrons in each shell. For example, see Fig. 4-5 below.

On the back of each card is a short description of the element. The object of this activity is to separate these nine elements into groups on the basis of their chemical and physical characteristics. You will use the chart on p. 85.

B. Begin with card 1. Read the description of argon on the *back* of the card.

C. Look at the chart and find the characteristics that describe argon. Follow the direction of the rows.

The last circle on the chart contains the name of the group to which argon belongs.

D. Write the name of this group on a piece of paper. List argon under this heading.

E. Follow the same procedure for cards 2 through 9.
   1. How many groups did you find?
   2. What are the names of these groups?
   3. Which elements are found in each group?

F. Now look at the front of each card. Beside the name of each element in your lists, write the number of electrons in its outer shell.
   4. Describe any pattern you see in the numbers of electrons.

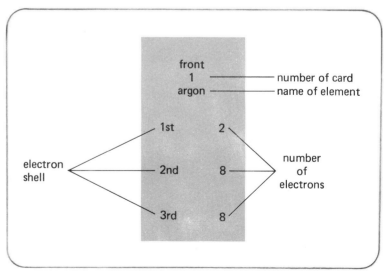

4-5.

Unit 1 What Makes Up Our World?

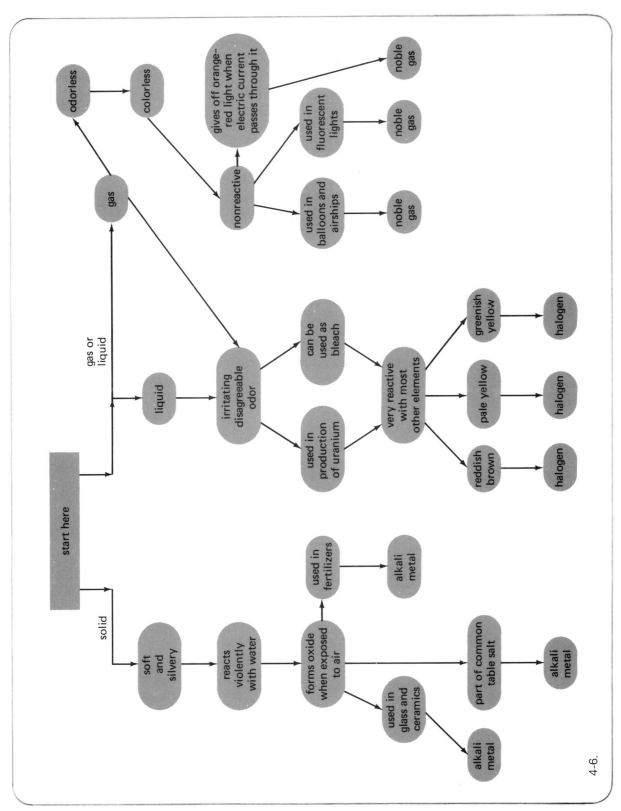

In the previous section, you learned that atoms with a stable number of electrons do not easily join with other atoms. If an atom has 1, 2, or 3 electrons *less* than a stable number of electrons, it will tend to *add* electrons until a stable number is reached. If an atom has 1, 2, or 3 electrons *more* than a stable number of electrons, it will tend to *lose* electrons until a stable number is reached. For example, think about a lithium atom. The atomic number of lithium is 3. A lithium atom has three electrons. There are two electrons in the first shell and one electron in the second shell. See Fig. 4-7. Thus, lithium has one *more* electron than the stable number of two.

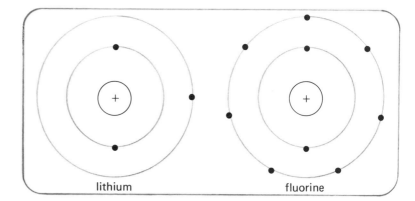

4-7. How does a lithium atom (left) differ from a helium atom?

4-8. How does a fluorine atom (right) differ from a neon atom?

TABLE 4-1
The alkali metals.

Atoms which lose one electron easily	Atomic Number	Stable electron number after losing one electron
Lithium	3	2
Sodium	11	10
Potassium	19	18
Rubidium	37	36
Cesium	55	54
Francium	87	86

Lithium will tend to *lose* that one electron to reach the stable number of two. Table 4-1 lists five other atoms that can also be expected to lose one electron. This group of atoms is called the **alkali** (al-kah-lie) **metals**.

Now think about a fluorine atom. The atomic number of fluorine is 9. Fluorine also has two electrons in its first shell. As shown in Fig. 4-8 fluorine has seven electrons in its outer shell. Fluorine has one electron *less*

Alkali metals: A group of elements whose atoms all have one electron more than the stable number.

TABLE 4-2
The halogens.

Atoms which gain one electron easily	Atomic Number	Stable electron number after gaining one electron
Fluorine	9	10
Chlorine	17	18
Bromine	35	36
Iodine	53	54
Astatine	85	86

than a stable number of eight in its outer shell. Fluorine must *add* one electron to have a stable electron arrangement. Table 4-2 lists four other elements which will gain one electron to become stable. These elements are called the **halogens** (**hal**-oh-jens).

You already know that the elements with a stable electron arrangement are called noble gases. The noble gases do not take part readily in chemical changes. Elements like the *alkali metals* and the *halogens* which lose or gain electrons *do* take part readily in chemical changes.

Halogens: A group of elements whose atoms all have one electron less than the stable number.

SUMMARY & QUESTIONS

Airships like the Goodyear blimp are safe because they are filled with helium gas. The atoms of helium are stable. They do not react with different atoms. Atoms that are not stable tend to gain or lose electrons to reach a stable number. Atoms gain or lose electrons by reacting chemically with atoms of other elements. Atoms with the same number of electrons in their outer shells have similar characteristics.

Look back at the electron arrangements shown on p. 81 and answer the following questions:

1. Which of the atoms would be chemically active: A, B, C, D, E, F?

2. Which of the atoms would not be chemically active: A, B, C, D, E, F?
3. What other atom would belong to the same chemical group as atom A?
4. What other atom would belong to the same chemical group as atom B?

3 CHEMICAL FAMILIES

You probably know people who are all members of the same family and look somewhat alike. They share a family resemblance. They share the same name. Sometimes these family members even behave alike. In this section, you will learn that atoms also belong to families. When you finish this section, you will be able to:

● Describe what is meant by a *chemical family*.

● Explain how you can predict the characteristics of an element by using the *periodic chart*.

▲ Predict the characteristics of a missing member of two different chemical families.

Chemical family: A group of elements which are alike in their chemical behavior.

4-9. Dmitri Mendeleev found a better way to arrange the elements in a periodic table.

4-10. Arrangement of atoms by atomic mass and chemical family.

Scientists have now identified more than 100 different kinds of elements. Additional ones may be discovered in the future. Do you think scientists can keep track of such a large number of elements easily? How would you organize a list of all the different elements?

One way to organize a large number of different things is to put the names in alphabetical order. For example, your teacher probably keeps an alphabetical list of all the students in your class.

In the last activity, the cards representing nine different elements were given to you in alphabetical order. You then found a different system for arranging the cards. The nine elements fell into three groups. Each element was placed in a particular group on the basis of its characteristics. Lithium, sodium, and potassium, for example, had similar properties. They were placed in the same group, called the alkali metals. The general name for a group of elements with similar properties is **chemical family.** The alkali metals, halogens, and noble gases are each examples of a *chemical family.*

During the early 1800's, several scientists discovered that different elements can belong to chemical families. They listed the elements in order of their increasing atomic masses. When they did this, the scientists discov-

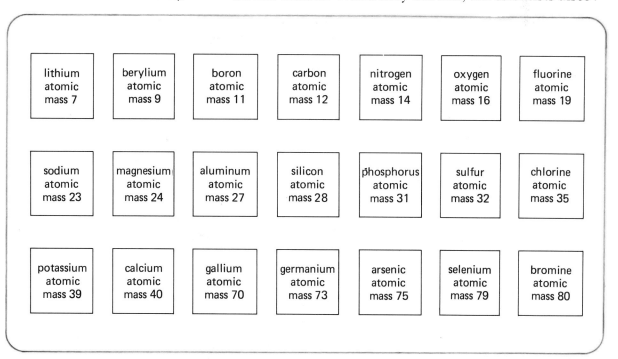

Chapter 4–3 Chemical Families

PERIODIC TABLE

1																	2
H 1.008																	He 4.00
3 Li 6.94	4 Be 9.01											5 B 10.8	6 C 12.01	7 N 14.01	8 O 16.00	9 F 19.0	10 Ne 20.2
11 Na 23.0	12 Mg 24.3											13 Al 27.0	14 Si 28.1	15 P 31.0	16 S 32.1	17 Cl 35.5	18 Ar 39.9
19 K 39.1	20 Ca 40.1	21 Sc 45.0	22 Ti 47.9	23 V 50.9	24 Cr 52.0	25 Mn 54.9	26 Fe 55.8	27 Co 58.9	28 Ni 58.7	29 Cu 63.5	30 Zn 65.4	31 Ga 69.7	32 Ge 72.6	33 As 74.9	34 Se 79.0	35 Br 79.9	36 Kr 83.8
37 Rb 85.5	38 Sr 87.6	39 Y 88.9	40 Zr 91.2	41 Nb 92.9	42 Mo 95.9	43 Tc (97)	44 Ru 101.1	45 Rh 102.9	46 Pd 106.4	47 Ag 107.9	48 Cd 112.4	49 In 114.8	50 Sn 118.7	51 Sb 121.8	52 Te 127.6	53 I 126.9	54 Xe 131.3
55 Cs 132.9	56 Ba 137.3	see below 57-71	72 Hf 178.5	73 Ta 180.9	74 W 183.9	75 Re 186.2	76 Os 190.2	77 Ir 192.2	78 Pt 195.1	79 Au 197.0	80 Hg 200.6	81 Tl 204.4	82 Pb 207.2	83 Bi 209.0	84 Po 209	85 At (210)	86 Rn (222)
87 Fr (223)	88 Ra (226)	see below 89-103	104 Ku (259)	105 Ha (260)	106 (263)												

57 La 138.9	58 Ce 140.1	59 Pr 140.9	60 Nd 144.2	61 Pm (147)	62 Sm 150.4	63 Eu 152.0	64 Gd 157.3	65 Tb 158.9	66 Dy 162.5	67 Ho 164.9	68 Er 167.3	69 Tm 168.9	70 Yb 173.0	71 Lu 175.0
89 Ac (227)	90 Th 232.0	91 Pa (231)	92 U 238.0	93 Np (237)	94 Pu (244)	95 Am (243)	96 Cm (248)	97 Bk (247)	98 Cf (251)	99 Es (254)	100 Fm (257)	101 Md (258)	102 No (255)	103 Lr (257)

4-11. *The periodic table of the elements.*

ered that certain physical and chemical properties were repeated at regular intervals. In other words, elements with similar properties occurred *periodically* in the list. This is like the days of the week (Sunday, Monday, Tuesday, and so on), which occur periodically throughout a calendar month.

In 1872 a Russian chemist, Dmitri Mendeleev (dim-**meet**-tri men-dee-**lay**-eff), tried arranging the elements in another way. He also arranged the elements in the order of their increasing atomic masses. Instead of listing them one after the other, however, he laid them out as you might deal a deck of cards for a game of solitaire. Figure 4-10 shows what he did.

Look at the three elements in the first vertical column: lithium, sodium, potassium. These are the three elements we found belong to the chemical family called the alkali metals. The three elements in the last column are fluorine, chlorine, and bromine. These elements belong to the chemical family called the halogens.

Periodic chart: An arrangement of all the elements which shows chemical families.

Mendeleev discovered that when he arranged all the elements in a table like this, elements in the same chemical family were found in the same vertical column. Elements in each family are found at particular *periods* or places when you put them in order. Mendeleev's chart is called the **periodic chart** of the elements. Modern *periodic charts* are slightly different from Mendeleev's original one. See Fig. 4-11 on page 90. The elements on a modern periodic chart are listed in order of increasing atomic number rather than by atomic mass.

Mendeleev had discovered a valuable tool for scientific research. Each element has its own position on a periodic chart. If you know the chemical properties of two or three elements, you can predict the properties of a neighboring element. You can even predict the chemical behavior of elements not yet discovered. Mendeleev himself predicted the general properties of several elements that had not yet been discovered. In many cases, when the elements were discovered, they were found to behave almost exactly as Mendeleev had predicted.

Materials
cards for: lithium, sodium, potassium, helium, neon, argon, rubidium, and krypton.
pencil
paper

A. Obtain the materials listed in the margin.

You have three cards describing lithium, sodium, and potassium. These three elements belong to the chemical family called the alkali metals.

B. Look at the electron arrangement for each element. Reread the description of the element on the back of each card.

The fourth member of the alkali metal family is called rubidium.

1. How many electrons would you predict rubidium has in its outer shell?

2. On the basis of the known properties of lithium, sodium, and potassium, list three characteristics that you would expect rubidium to have.

You also have the cards for the noble gases helium, neon, and argon.

C. Look at the electron arrangement for each of these elements. Reread the descriptions on the backs of the cards.

The fourth member of the noble gas family is called krypton.

3. Predict how many electrons krypton has in its

Chapter 4–3 Chemical Families

outer shell.

4. On the basis of the known behavior of helium, neon, and argon, predict three characteristics of krypton.

Your teacher will now give you a card for rubidium and one for krypton.

D. Read the descriptions given for rubidium and krypton.

5. Were your predictions in questions 1 and 3 correct?

6. Which properties of rubidium were you able to predict correctly? Of krypton?

You should now be able to understand some of the reasons the periodic table of elements is so useful to scientists.

Just as you belong to a family of people, elements belong to chemical families. You probably resemble the other members of your family in some ways. The members of a chemical family also have a family resemblance. In a periodic chart of all the elements, the members of a chemical family are found in the same vertical column.

An atom of the element beryllium has two electrons in its outer shell. These two electrons are given up rather easily to form an ion having a +2 charge. Beryllium is a solid. It also behaves as do most metals.

1. Choose from the following descriptions the one element which would belong to the same chemical family as beryllium.
    (a) aluminum: solid, metal, +3 ion, 3 electrons in outer shell
    (b) tin: solid, metal, +4 ion, 4 electrons in outer shell
    (c) radium: solid, metal, +2 ion, 2 electrons in outer shell
    (d) polonium: solid, metal, +2 ion, 6 electrons in outer shell
2. Magnesium belongs to the same chemical family as beryllium.
    (a) How many electrons would an atom of magnesium have in its outer shell?
    (b) List three other characteristics of magnesium.

4 CHEMICAL BONDING

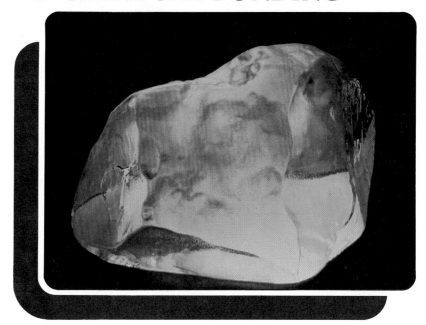

Diamond is the hardest substance known. Diamonds will cut through stone or metal as easily as a knife slices bread. A diamond crystal is made only of carbon atoms. Carbon is also found in other forms. The hardness of a diamond is a result of the strength with which the individual carbon atoms cling to each other. Compare carbon, in the form of a diamond, with helium. As you know, helium is a very light gas. When the temperature drops to −268.8°C, helium changes from a gas into the coldest of liquids. What difference causes carbon atoms to cling so tightly to their neighbors while helium atoms do not? The purpose of this section is to investigate how an atom might join another atom. When you finish this section, you will be able to:

● Explain what is meant by a *chemical bond*.

● Use *electron dot models* to explain why no more than two hydrogen atoms can join to form a hydrogen molecule.

● Explain what is meant by *valence*.

▲ Draw electron dot models for several atoms and simple molecules.

A general rule describes much of the chemical behavior of atoms. The rule says that atoms are of three general types: (1) atoms that tend to lose electrons to become stable; (2) atoms that gain electrons to become stable; and (3) atoms that share electrons to become stable.

Atoms with 1, 2, or 3 electrons *more* than a stable number do not hold these electrons tightly. These atoms are in the first group. Atoms with 1, 2, or 3 electrons *less* than a stable number hold their outer electrons very tightly. These atoms are in the second group. The "strong hold" atoms of the second group can pull electrons away from the "weak hold" electrons of the first group.

For example, sodium atoms have one *more* electron than a stable number. The outer electron is not held very tightly. Chlorine atoms have one electron *less* than a stable number. Its outer electrons are held by very strong forces. As a result, a chlorine atom can pull the outer electron away from a sodium atom. Both atoms will now have a stable electron arrangement.

However, if atoms with the same "hold" react, neither is strong enough to pull electrons away from the other. As a result, they share electrons. These atoms are in

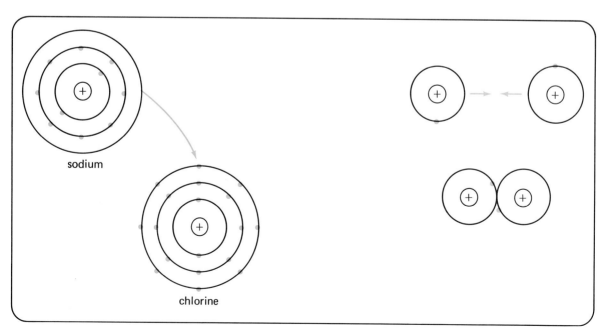

4-12. A chlorine atom fills its outer shell with an electron from a sodium atom; two hydrogen atoms share two electrons.

Unit I   What Makes Up Our World?

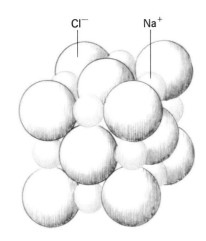

4-13. *A salt crystal is made up of Na^+ ions and Cl^- ions held together in the shape of a cube.*

Chemical bond: A force which joins atoms together.

Ionic bond: A kind of chemical bond formed when atoms transfer electrons from one to another.

Covalent bond: A chemical bond formed when atoms share two or more electrons.

4-14. *Graphite crystal structure.*

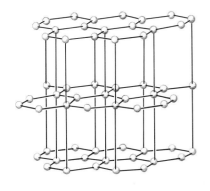

the third group. Some atoms may belong to both group 2 and group 3.

Suppose that two hydrogen atoms come close to each other. Each atom needs one electron. Each atom tends to gain one electron. Neither atom tends to lose one electron to the other. Instead, the two hydrogen atoms *share* the two electrons. In doing so, they form a hydrogen molecule.

When atoms gain, lose, or share electrons, forces develop that hold the atoms together. This joining of atoms is called a **chemical bond**. *Chemical bonds* hold atoms together to form molecules.

When a sodium atom transfers an electron to a chlorine atom, the sodium atom becomes a positive ion, Na^+. The chlorine gains an electron and becomes a negative ion, Cl^-. The two oppositely charged ions are then attracted to each other. Bonds that are formed between atoms as a result of a *transfer* of electrons are called **ionic bonds**. Sodium and chlorine form an *ionic bond* to become sodium chloride, which is common table salt.

Two chlorine atoms form a bond by sharing electrons because neither atom has a tendency to lose electrons. Both atoms tend to gain one electron.

Two hydrogen atoms bond together in the same way as chlorine. This kind of chemical bond is called a **covalent** (co-va-lent) **bond**. When atoms like chlorine and hydrogen *share* electrons to fill their outer shells, they form *covalent bonds*.

A diamond crystal is made up of carbon atoms that are very tightly bonded to four other carbon atoms. All of the bonds in a diamond crystal are equally strong. Carbon can also form bonds in which one bond is weaker than the other three. This type of bonding results in layering. Carbon in this form is called graphite. The "lead" in pencils is graphite. You can write with a pencil because the weak bonds of the graphite break, leaving a trail of flakes on the paper.

In the 1920's, an American scientist, Gilbert Lewis, suggested a way of picturing atoms to help explain how they formed chemical bonds. The symbol for the atom represents the nucleus and the inner electrons. Dots around the symbol represent the outer electrons. This kind of symbol is called the **electron dot model** of the atom. For example, the *electron dot model* of hydrogen is H• You can then write the electron dot model of a hydrogen molecule as H:H.

Chapter 4–4 Chemical Bonding

ACTIVITY

Materials
Pencil
Paper

A. Obtain the materials listed in the margin.

In this activity, you will practice writing the electron dot models for some atoms and several simple molecules. Remember that the electron dot models show the symbol of the element with the outer shell electrons around it. Note that, when there are more than four dots, each additional dot pairs with one already there. For example, nitrogen is ·Ṅ·.

B. Look at the electron models of the atoms shown in Fig. 4-15, below.

C. On your paper, write the electron dot models of the atoms shown in Fig. 4-15.

D. Have your electron dot models checked before going on.

E. Now write the electron dot models which show the molecules you would have if you joined the following atoms: 1 hydrogen atom and 1 fluorine atom; 1 hydrogen atom and 1 chlorine atom; 2 hydrogen atoms and 1 oxygen atom; 3 hydrogen atoms and 1 nitrogen atom.

Remember that when a molecule is formed, the electrons of two different atoms pair up in a way which gives each atom a complete "outer shell."

A particular molecule is always made up of a certain number of atoms. For example, a hydrogen molecule is

4-15.

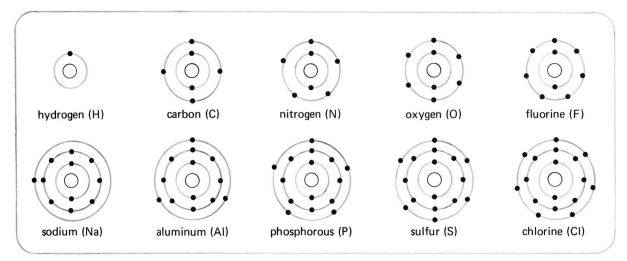

Unit I   What Makes Up Our World?

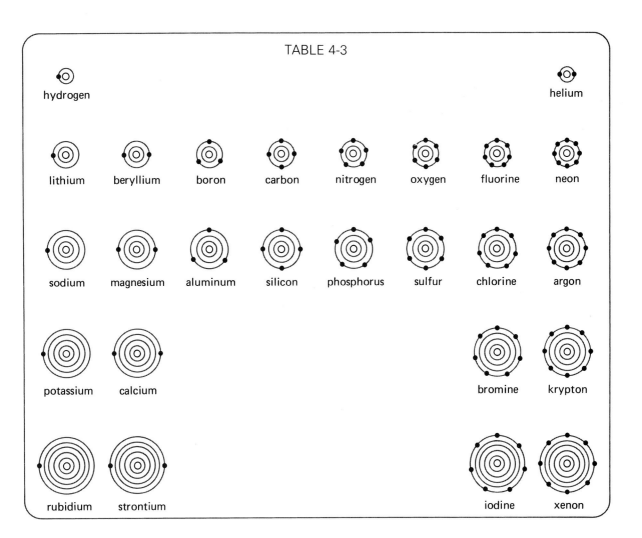

TABLE 4-3

Look at the models of the molecules you prepared in step E of the activity. Can you explain why each molecule contains only the number of atoms shown?

always made up of two, and only two, hydrogen atoms. To see the reason for this, look at the electron dot model for hydrogen. Each hydrogen atom has one outer electron, H·. When a hydrogen molecule is formed, two electrons are shared by two hydrogen atoms, H:H. In a molecule, the outer electron shell of each of the two hydrogen atoms is filled. The two atoms in a hydrogen molecule do not gain any additional electrons from other hydrogen atoms.

Your study has shown that the outer electrons of an atom determine much of the atom's chemical behavior. These outer electrons determine how an atom will join other atoms. Table 4-3, above, shows the outer electron arrangement of a number of atoms. This table is arranged in the same way as the periodic table to show some chemical families. Knowing the outer electron ar-

Valence: The number of electrons gained, lost, or shared by an atom when it forms chemical bonds.

Valence electrons: Electrons in the outer shell of an atom.

TABLE 4-4
Valences of some atom groups.

Name of atom group	Valence
Nitrate (NO_3)	−1
Carbonate (CO_3)	−2
Hydroxide (OH)	−1
Phosphate (PO_4)	−3
Ammonium (NH_4)	+1
Sulfate (SO_4)	−2

rangement of an atom makes it possible to predict how it will form chemical bonds. This can be done by using the **valence** (**vale**-lens) of each atom. *Valence* describes the number of electrons an atom gains, loses, or shares to form chemical bonds. An atom which gives up electrons is said to have a *positive valence*. An atom that receives electrons from a bond has a *negative valence*.

Look at some atoms in the second row of Table 4-3. The first atom, lithium, has one outer electron that will be lost. The valence of lithium is +1. This means that a lithium atom will lose one electron to form a bond. Now go across the second row to oxygen. This atom has six outer electrons and has a valence of −2. An oxygen atom will gain two electrons in a bond.

The number of outer electrons determines the valence of a particular atom. For this reason, the outer electrons in an atom are called **valence electrons**. Lithium has one *valence electron* and oxygen has six. Notice that atoms such as carbon with four valence electrons have a half-filled outer shell. This means that carbon can have valences of both +4 and −4 depending upon the other atoms it combines with.

Experiments show that there are certain groups of atoms that remain together in chemical changes. One such group is made up of one sulfur atom joined to four oxygen atoms. This is called the *sulfate group*. Since this group of atoms acts chemically like a single atom, it is given a single valence of −2. The names and valences of some other common atom groups are given in Table 4-4.

How can valences be used? Perhaps you can already answer this question. Valences can be used to predict how atoms will join together. For example, if you know that the valence of lithium is +1 and oxygen is −2, you can predict how lithium oxide will be formed. Since lithium loses one electron and oxygen gains two, two lithium atoms will join with one oxygen atom to make a stable molecule. The following rule is another way of saying this.

The total number of positive and negative valences of the atoms in a simple compound must add up to 0.

In lithium oxide, the total positive valences of two lithiums is $1 + 1 = +2$. The valence of oxygen is −2 so the sum of the valences is $(+2) + (−2) = 0$. Used in this way, valences can help a person predict how atoms will form chemical compounds.

SUMMARY & QUESTIONS

The two hydrogen atoms that form a hydrogen molecule are joined by a chemical bond. Specific molecules always contain the same number of atoms. That number depends on the number of outer or valence electrons in each atom. The valence electrons of an atom can be shown by an electron dot model for the atom.

1. What is meant by a chemical bond?
2. Draw electron dot models for atoms of hydrogen, carbon, oxygen, sodium, and chlorine.
3. Explain why no more than two hydrogen atoms are needed to form a hydrogen molecule.
4. Using an electron dot model, show how hydrogen (Li•) and oxygen ($\:\ddot{\text{O}}\:\cdot$) would form a molecule of lithium oxide.
5. How many atoms of each element would a molecule of magnesium and chlorine contain?

5 CHEMICAL REACTIONS

How would you describe what is pictured in the margin? You have probably seen this process, or something like it, many times. You might say simply, "Some pieces of charcoal are burning."

Scientists have a different way of describing this same process. When you finish this section, you will be able to:

● Give some examples of a *chemical reaction*.

● Explain how a *chemical equation* describes a chemical reaction.

▲ Use the correct formulas to write and balance a chemical equation.

Chapter 4–5 Chemical Reactions

Chemical reaction: A reaction in which a chemical change takes place.

You have probably seen charcoal burning in a barbecue grill. Did you know that the burning of charcoal is a chemical change? Charcoal is made up of carbon atoms. When charcoal burns, these carbon atoms combine with oxygen. This combination of carbon and oxygen results in a chemical change that is called a **chemical reaction.** When charcoal burns, the following *chemical reaction* takes place:

carbon plus oxygen produces carbon dioxide
carbon + oxygen ⟶ carbon dioxide

Another way of writing this reaction is with the formulas for the substances involved:

$$C + O_2 \longrightarrow CO_2$$

Chemical equation: A description of a chemical reaction using chemical formulas for the substances used and produced.

A description of a chemical reaction using formulas for the substances involved is called a **chemical equation** (e-**quay**-shun). The *chemical equation* for the burning of charcoal says that one atom of carbon reacts with one molecule of oxygen to form one molecule of carbon dioxide.

The following chemical equation shows the formation of water from hydrogen and oxygen:

hydrogen + oxygen ⟶ water

Using formulas, the equation is:

$$H_2 + O_2 \longrightarrow H_2O$$

Note that the small number to the lower right of a chemical symbol shows the number of atoms in that molecule. For example, O_2 means there are two atoms of oxygen in this molecule. Numbers such as these are called *subscripts*.

However, this equation is not complete. Look at the number of oxygen atoms on each side of the equation. There are two oxygen atoms on the left of the arrow but only one oxygen atom on the right. One oxygen atom seems to have disappeared. Scientists have shown that atoms do not disappear during chemical reactions. The *Law of Conservation of Matter* explains what happens during a chemical reaction. This law says that *the same number of atoms exists after a chemical reaction as before the reaction.* To correct the above equation you must *balance* it. Since there are two oxygen atoms on the left of the arrow, there must also be two on the right:

$$H_2 + O_2 \longrightarrow 2H_2O$$

Now the oxygen atoms are balanced but the hydrogen atoms are not. There are four hydrogen atoms on the right. There are only two hydrogen atoms on the left. To correct this, place a "2" in front of H_2:

$$2H_2 + O_2 \longrightarrow 2H_2O$$

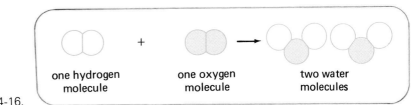

4-16.

This equation is now correctly balanced.

An equation which shows that atoms are not created or destroyed in the reaction is said to be balanced.

This is a balanced equation: $2H_2 + O_2 \longrightarrow 2H_2O$

This is not a balanced equation: $H_2 + O_2 \longrightarrow H_2O$

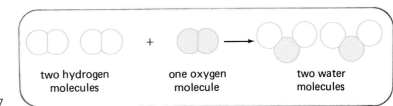

4-17.

In summary, a correctly balanced chemical equation shows the following information about a chemical reaction:
1. The formulas of the molecules which are reacting (reactants) are shown on the left of the arrow (or equal sign).
2. Formulas of molecules which are produced by the reaction (products) are given on the right side of the equation.
3. The total number of each kind of atom used equals the number of those same atoms in the products. No atoms are created or lost during the reaction.

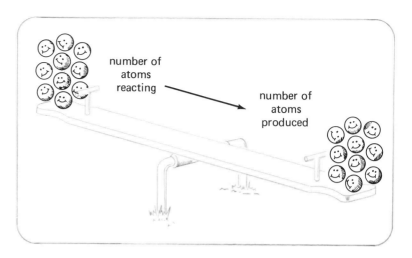

4-18. *The number of atoms reacting equals the number of atoms produced.*

Materials
paper and pencil

A. Look at Fig. 4-19 which shows how seven paper cut-outs were rearranged.

1. Why is this rearrangement incorrect?

The rearrangement does not obey the Law of Conservation of Matter. In chemical reactions, atoms are merely rearranged. This is why they must be balanced in the manner discussed earlier. In this activity, you will practice balancing equations.

Hydrogen gas (H_2) reacts with chlorine gas (Cl_2) to produce a gas called hydrogen chloride (HCl). Applying Rule 1 and Rule 2 previously discussed would give:

$$H_2 + Cl_2 \longrightarrow HCl$$

When you apply Rule 3, numbers are placed *in front* of formulas to balance the number of each kind of atom in the equation.

B. On a piece of paper, copy the equation. Balance the equation so that it shows the same number of hydrogen and chlorine atoms on both sides of the arrow.

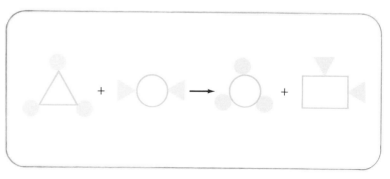

4-19.

Hydrogen gas (H_2) also will react with nitrogen gas (N_2) to form a gas called ammonia (NH_3).

C. On your paper, write the formulas showing the reactants and product as a chemical equation.

D. Balance the equation.
  2. How many hydrogen atoms are used to make the ammonia (NH_3)?
  3. How many nitrogen atoms are shown on each side of the arrow?

E. Copy the following unbalanced equations on your paper. The formulas for the reactants and products are written correctly. Balance each equation.
  4. $Li + Cl_2 \longrightarrow LiCl$
  5. $Ca + Br_2 \longrightarrow CaBr_2$
  6. $CO + O_2 \longrightarrow CO_2$
  7. $H_2O \longrightarrow H_2 + O_2$
  8. $Al + O_2 \longrightarrow Al_2O_3$

SUMMARY & QUESTIONS

A chemical equation is a useful form of shorthand describing a chemical reaction. An equation tells you what substances react and what the products of the reaction are. No atoms are created or destroyed in a chemical reaction.

1. Give at least two examples of chemical reactions.
2. In your own words, describe what the following chemical equations say:
   a. $2H_2 + O_2 \longrightarrow 2H_2O$
   b. $2Na + Cl_2 \longrightarrow 2NaCl$
3. Write the following chemical equations using formulas for the substances. Then balance the equation so that all atoms are accounted for.
   a. hydrogen plus oxygen produces water
   b. carbon plus oxygen produces carbon dioxide
   c. hydrogen plus chlorine produces hydrogen chloride
4. Balance the following unbalanced chemical equations:
   a. $Li + Cl_2 \longrightarrow LiCl$
   b. $Al + O_2 \longrightarrow Al_2O_3$

VOCABULARY REVIEW

Match the description in Column II with the term it describes in Column I.

Column I
1. inert
2. noble gas
3. chemical activity
4. stable arrangement
5. ion
6. chemical family
7. chemical bond
8. valence
9. chemical reaction
10. chemical equation

Column II
a. An atom or molecule which has an electric charge.
b. An electron arrangement of completely filled shells.
c. A gas having a stable electron arrangement.
d. The way an atom reacts with other atoms.
e. Not chemically active.
f. A group of elements with similar properties.
g. The number of electrons gained, lost, or shared by an atom when it forms chemical bonds.
h. A description of a chemical reaction using formulas for the substances used and produced.
i. A force which joins atoms together.
j. A reaction in which a chemical change takes place.

REVIEW QUESTIONS

Choose the letter of the ending which best completes the following statements.

1. When an atom absorbs energy, its (a) electrons move to a greater distance from the nucleus (b) protons move faster (c) electrons move closer to the nucleus (d) protons move to a greater distance from the nucleus.
2. When an atom releases energy, its (a) electrons move to a greater distance from the nucleus (b) protons move faster (c) electrons move closer to the nucleus (d) protons move to a greater distance from the nucleus.
3. If an atom gains an extra electron, the atom becomes (a) a positive ion (b) a neutral atom (c) a negative ion (d) a new element.
4. An atom has a stable electron arrangement if it (a) requires a great deal of energy to release an electron (b) has completed electron shells (c) is a noble gas. (d) All of these are correct.
5. Which one of the following atomic electron arrangements requires one electron to become stable?
   (a) 2, 8 (b) 2, 7 (c) 2, 6 (d) 2, 8, 1.

6. In order to predict how many electrons an atom will gain, lose, or share to form chemical bonds, you must know (a) the atom's atomic mass (b) the number of electrons in the atom's outer shell (c) the atom's atomic number. (d) All of these are correct.
7. Magnesium is atomic number 12, atomic mass 24, and has 12 neutrons in its nucleus. The valence of magnesium is (a) +36 (b) +24 (c) −12 (d) +2.
8. Aluminum has a valence of +3 and oxygen has a valence of −2. Using this information, the most likely compound of aluminum and oxygen would be (a) AlO (b) Al_3O_2 (c) Al_2O_3 (d) Al_2O_2.
9. In order to be complete, the incomplete chemical equation $H_2 + O_2 \longrightarrow H_2O$ needs the number 2 before (a) H_2 and O_2 (b) H_2, O_2, and H_2O (c) O_2 and H_2O (d) H_2 and H_2O.
10. A chemical bond can be defined as (a) a force which joins atoms together (b) a force which causes nuclei to blend together (c) a force due to electrical repulsion. (d) All of these are correct.

REVIEW EXERCISES

Give brief but complete answers to each of the following exercises.

1. Draw electron dot models for hydrogen, chlorine, oxygen, and water.
2. Use an electron dot model to show how two hydrogen atoms join to form a molecule of hydrogen.
3. Using the information in Table 4-3 (p. 97), write the chemical formulas for water, sodium chloride, and carbon dioxide.
4. Explain what is meant by the valence of an atom.
5. Write the following chemical equation out in words:

$$4Al + 3O_2 \longrightarrow 2Al_2O_3$$

6. How does a correctly written chemical equation show that atoms are not lost in a chemical change?
7. Give an example of an electron arrangement which is stable.
8. A certain atom has 2 electrons in its first shell, 8 electrons in its second shell, and 1 electron in its third shell. Will it gain or lose electrons to have a stable arrangement?
9. Describe what is meant by a chemical family.
10. Describe what happens when an atom becomes an ion.

CAREERS IN CHEMISTRY

People who are interested in chemistry as a career can pursue a wide variety of interests. Not only do chemists explore what objects are made of, they also determine what and how substances have been acted upon. Results of their studies are very helpful in the fight to preserve our environment—both natural and synthetic.

(right) **The testing of mined ore has gone far beyond what the assayer did in Gold Rush Days. This chemist is checking the content of copper ore samples. Her analysis will be one of the factors that will determine whether a copper mine will continue to operate.**

(left) **This woman is an environmental protection engineer. She is collecting samples of water from a stream. Some of her tests will determine if the stream is fit for water life.**

(above) **Part of our environment was made by human hands and passed down through the generations. Whether art is ancient and rare, or from modern times, care and skill are required to preserve and protect it.**

Special study must be made of glass, for example, to determine of what the glass was made and how, so that the best methods can be used to restore it. Here the restorers are working on stained-glass windows. An art restorer's craft is one that combines interest in the arts with scientific skill.

(right) **Chemical tests are performed on the water in swimming pools. Some of these tests check for the presence of bacteria that could cause ear and throat infections in swimmers.**

(below) **What chemicals do plants need for healthy growth? In nurseries, special care is taken to give each type of plant the proper chemical mix in its soil and its food.**

UNIT 2 HOW DOES ENERGY AFFECT MATTER?

CHAPTER 5
HEAT AND TEMPERATURE

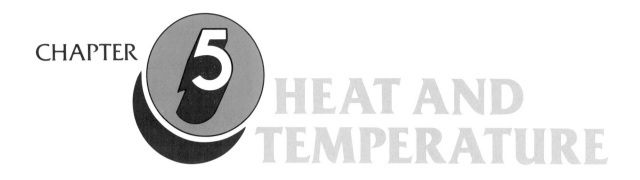

1 HEAT ENERGY

Rub your hands together as fast as you can. Do they get warm? Where does this heat come from? This kind of question puzzled scientists for hundreds of years. Their search for an answer led them to some important discoveries. When you finish this section, you will be able to:

- Explain why heat is considered a form of energy.
- Predict what will happen to particles of matter when heat energy is added.
- ▲ Demonstrate what happens to particles of matter when heat energy is added or removed.

Unit II How Does Energy Affect Matter?

What is heat? About 200 years ago, a scientist would probably have answered this question by saying something like: "Heat is an invisible and weightless fluid. It soaks into an object when the object is heated and drains away upon cooling." According to this theory, water was ice joined with "heat fluid." Steam was water with still more "heat fluid" added. If you burned your hands by sliding too quickly down a rope, the burns were caused by "heat fluid" squeezed out of the rope.

This old theory did explain some observations about heat. For example, heat seems to move away from a hot place to a cold place. Heat leaves a warm house in cold weather and must be replaced constantly. This behavior of heat seems to support the idea that heat "flows" from one place to another. The theory of "heat fluid" did not pass the last and most important test of a scientific theory, however. Experiments showed that heat could not be "heat fluid" or any kind of matter.

One such experiment was done in 1798 by Benjamin Thompson. Thompson was an American who became a government official in Europe where he was called Count Rumford. At one time, he was in charge of a cannon factory. In those days, cannons were made by drilling machines run by horses. Rumford noticed that the cannons became very hot during this process. He set up

5-1. *Count Rumford proved that heat is a form of energy by studying the heat produced in boring a cannon.*

5-2. *As water is heated, its particles move more rapidly.*

the following experiment: A cannon was drilled while surrounded by a wooden box containing water. After several hours of drilling, the water began to boil. It continued to boil as long as the drilling went on. To Rumford, the experiment showed that the supply of heat was without limit and could not be something contained in the metal of the cannon. Rumford decided that heat was actually a form of energy supplied by the work of the horses. Rumford's experiment, and others like it which were done later, showed that heat is a form of energy.

If heat is energy, how is this energy contained in something that is hot? Modern scientists think of all matter as being made up of tiny particles such as atoms and molecules. One of these particles alone is much too small to be seen. Heat energy causes these particles to move faster. For example, the heat energy caused by Rumford's drilling experiment made the particles in the cannon move rapidly. The more heat that is added, the faster the particles in matter will move. For example, a drop of water contains a huge number of individual water molecules. All of these molecules are moving. Heating the water will cause the particles to move faster and bump into each other more often. Thus the heat added to the water is "stored" in the motion of the molecules. When the water loses heat, the molecules slow down.

Materials
2 small glass containers
Blue food coloring (or some other dark color)
Ice cube
Bunsen burner
Water
Pencil
Paper

Water molecules are extremely small. You cannot watch individual molecules to see if heating or cooling changes their speed. It is possible to mix a colored substance with a larger amount of water, however, and watch the color spread throughout the water. If the mixture is not stirred, the spreading of color must be the result of the motion of molecules.

A. Obtain the materials listed in the margin.

B. Fill a small clear glass or plastic container about ⅔ full of water. Use water that is at room temperature.

C. Place one drop of blue food coloring onto the surface of the water. See Fig. 5-3 on page 113.

5-3.

D. Watch what happens to the coloring in 2 to 3 min. Look at the coloring from the top and through the sides of the container.

E. Answer the following:
   1. Describe what happened to the drop of food color just as it hit the water surface.
   2. Did the drop of food color sink to the bottom of the water?
   3. Did the food color tend to spread out as it fell through the water?

F. Now pour out the water and food color mixture. Rinse the container so no color remains. The motion of water molecules causes a mixing of food color with the water. If the water molecules are moving faster in hot water than in cold water you would expect to see a difference in the rate of mixing of the food color. You will carry out an experiment to check this hypothesis.

G. Fill the container about ⅔ full of water and add an ice cube to it.

H. While the water is cooling, fill another container of the same size and shape a little more than ⅔ full of hot water. See Fig. 5-4.
   4. In which container (hot or cold) are the water molecules moving faster?

I. Place the two containers side by side.

J. Swirl the ice cube to cool the water a little more. Remove any of the ice which remains.

K. Wait a minute for the water currents to stop. Now add one drop of blue food color to each container, one right after the other.

L. Watch the food color spread out in both containers.

M. Compare the mixing rate in the two containers.
   5. In which container (hot or cold) did the mixing appear faster?
   6. Explain your results on the basis of the motion of the water molecules.

5-4.

Chapter 5-1 Heat Energy

5-5. *One of the first steam engines (1642).*

Heat engine: A machine which changes heat energy into mechanical energy.

Because heat is a form of energy, it can do work. In a steam engine, for example, the motion of water molecules in hot steam runs the engine. In an automobile engine, the burning fuel produces hot gases. The hot gases run the engine. A steam engine and an automobile engine are examples of **heat engines.** A *heat engine* is a machine which changes heat energy into mechanical energy. See Fig. 5-5.

However, there is a problem in using all heat engines. Much of the heat energy in a heat engine is wasted. There is no way to use all of the energy contained in the moving particles of the heated gases which run heat engines. An automobile engine, for example, has a radiator to take care of this wasted heat energy. The wasted heat escapes from the engine through the radiator. No one knows how to make a heat engine that will change all of the heat energy supplied to it into mechanical energy.

SUMMARY & QUESTIONS

What happens when you rub your hands together to warm them? Experiments show that heat is a form of energy. When a substance is heated, energy is added to its particles. This causes the particles of matter to move faster. The heat energy is the energy of these moving particles. When you rub your hands, you cause the particles in the outer parts of your skin to move faster. These particles bump others until the entire thickness of the skin in your hands is heated. The motion of your hands is changed into heat energy in the skin. Unfortunately, when we try to change heat energy into mechanical energy with heat engines, some of the heat energy is wasted.

1. Describe the experiment done by Count Rumford which proved to him that heat was not a fluid contained within a substance.
2. Explain what happens in matter as:
   a. heat energy is added to it.
   b. heat energy is removed from it.
3. Describe how you could demonstrate that heat energy added to water causes water particles to move faster.
4. How does the steam engine provide evidence that heat is a form of energy?

2 HEAT TRANSFER

A fire in a fireplace is a cheerful source of heat energy. An iron poker put into the fire takes on some of the fire's heat. People sitting near the fire feel the heat. A bird perched on top of the chimney receives some of the heat. How does heat move from one place to another? When you finish this section, you will be able to:

● Use examples to show how heat energy can move from one material to another.

● Explain *conduction* and *convection* in terms of movement of particles in matter and radiation in terms of infrared rays.

▲ Demonstrate the transfer of heat energy by conduction, convection, and radiation.

Conduction: Transfer of heat by direct contact.

A metal poker put directly into a fire is heated by **conduction** (con-**duck**-shun). Transfer of heat by direct contact is called *conduction*. It is the simplest method of heat transfer. See Fig. 5-6. The metal in the poker, like all other substances, is made up of particles. The rapidly moving particles of the burning fuel in the fireplace

5-6. *The heat from the fire causes the particles in the poker to move rapidly. The handle, which is not touching the fire, is heated by conduction.*

bump the particles in the poker and make them vibrate faster. The particles in the end of the poker, in turn, hit particles in the cooler part. This continues on up the poker until the particles in the handle are also heated. If you touch the handle of a hot poker, you will burn your fingers. Heat is transferred to your skin by conduction.

The metal in a poker conducts heat very well. Not all materials conduct heat as well as metals. Wood, for example, is a poor *conductor* of heat. This is why the handles of pokers, pots and pans, and other metal objects that must be heated can be made of wood. Nonmetallic solids, liquids, and gases are all poor conductors of heat. They are called **insulators**. *Insulators* are often used to prevent the loss of heat. Houses and other buildings must be insulated against heat loss. See Fig. 5-7 on page 117. Without insulation, it becomes very expensive to provide the heat needed to keep the house warm during cold weather. Clothes effectively shield our bodies against loss of heat. Trapped air under clothing acts as an insulator. Many layers of cloth containing air spaces would be a better insulator than one heavy layer.

Insulator: A material that does not allow energy to flow through it easily.

5-7. *Insulation can prevent loss of heat in a home.*

Convection: Transfer of heat by movement of the heated part of a gas or liquid.

Radiation: Transfer of heat through space by infrared rays.

How was the bird on the chimney heated by the fire? Below, air directly over the fire was heated. This hot air expanded because its particles started moving faster and took up more space. Expansion of the heated air made it lighter than the surrounding air. The warm air moved up the chimney and warmed the bird. See Fig. 5-8. Transfer of heat in this way is called **convection** (con-**veck**-shun). When a gas such as air or a liquid like water is heated unevenly, the heated part rises. This movement of a heated gas or liquid is *convection*. Heat transfer by convection can take place only in gases or liquids.

A fireplace is a poor heater for a room because convection carries most of the heat up the chimney. On the other hand, other kinds of room heaters work fairly effectively by convection. For example, steam or hot water is moved through a device which heats some of the air in the room. This warm air rises and moves across to the cooler parts of the room. Cool air sinks and moves toward the heater where it is heated. See Fig. 5-8.

Convection transfers heat when a part of the heated material, a gas or a liquid, moves. A third method of heat transfer is very different. It is called **radiation** (ray-dee-**ay**-shun). *Radiation* does not cause particles of matter to move. *Infrared radiation* is a form of electromagnetic radiation.

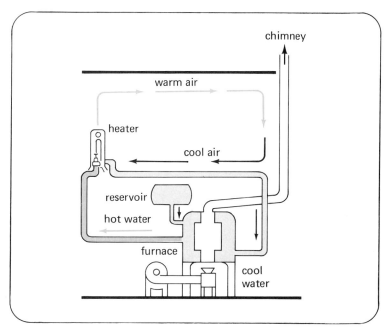

5-8. *This system heats a home by conduction, convection, and, to some extent, radiation.*

Chapter 5–2 Heat Transfer

Sources of heat, such as fire in a fireplace, send out these invisible infrared rays. When these rays reach your skin or any other substance, the rays are changed back into heat. Infrared rays are responsible for some of the warmth felt by the people around a fireplace. Radiation is a very important method of heat transfer. The earth receives energy from the sun by radiation. The sun is 150 million kilometers from the earth. This distance is mostly empty space. Heat cannot be transferred across empty space by conduction or convection. However, infrared radiation, as well as light and ultraviolet radiation, can travel through empty space from the sun to the earth. You feel warm while sitting in the sun because you are receiving heat from the sun by radiation.

5-9. *A hot air balloon makes use of the fact that hot air rises.*

5-10. *The earth receives heat from the sun by radiation. Even your body radiates heat, as indicated in this infrared photograph. Red indicates the warmest areas, blue the coolest.*

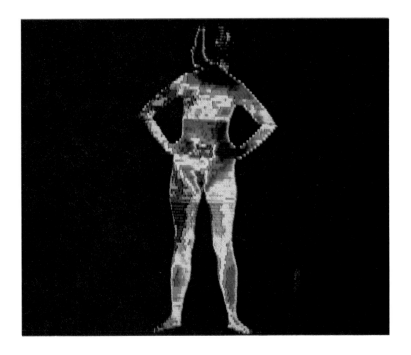

Materials
candle
matches
nail (16d, finishing)
index card
metric ruler
tin can
pencil
paper

CAUTION: IN THIS ACTIVITY, BE CERTAIN YOU BLOW OUT MATCHES BEFORE DISPOSING OF THEM. Do not leave hot objects where someone else may pick them up.

A. Obtain the materials listed in the margin.

B. Light the candle and drip some wax on the card. Blow out the candle. While the wax is still soft, stand the candle upright in it. See Fig. 5-11.

C. Hold your fingers about 30

5-11.

Chapter 5–2 Heat Transfer

5-12. **CAUTION:** *Do not bring your finger too near the flame. You should just be able to feel the heat of the flame.*

5-13.

cm above the flame. See Fig. 5-12. Move your fingers closer until you can feel heat from the flame.

1. How close to the flame must you be to feel heat from the flame?

D. Hold your fingers to the side of the candle the same distance as in your answer to question 1.

E. Move your fingers closer to the candle until you can feel heat from the flame.

2. How close to the side of the flame must you be to feel heat from the flame?

3. What word describes the method by which you received heat to the side of the flame?

4. What word describes the method by which you received heat above the flame?

5. Would any one method cause you to receive heat from both the top and the side of the flame?

F. Hold the nail near its head. Hold the point of the nail in the flame for about 15 sec. See Fig. 5-13.

G. Move the nail to more than 30 cm from the candle.

6. How long a time is it before you feel any heat at the head end of the nail?

7. What word describes the method by which you received heat at the head of the nail?

SUMMARY & QUESTIONS

Heat can be transferred in three ways. (1) Direct contact between a source of heat and a cooler substance will speed up the motion of the particles in the cooler material. (2) Gases or liquids become heated unevenly. The heated part moves, carrying with it a supply of heat. (3) Every source of heat gives off invisible rays which heat any material they fall upon. Most common sources of heat, such as the fire in a fireplace, transfer heat in all three ways.

1. Each of the following are examples of the transfer of heat energy.

I. Which are examples of conduction?
II. Which are examples of convection?
III. Which are examples of radiation?
   a. You feel the steam rising from a cup of hot chocolate.
   b. You feel the heat from the hot cup handle just before you touch it.
   c. You burn your tongue drinking hot chocolate.
   d. A spoon in the hot chocolate becomes warm.
   e. You are warmed standing in front of a fireplace.
   f. You are warmed as you stand over a furnace register.
2. Conduction and convection transfer heat energy through the movement of particles. How do they differ?
3. Given a candle, match, and nail, describe how you could demonstrate transfer to energy by a. conduction, b. convection, c. radiation.

3 TEMPERATURE AND HEAT

Which would you rather have accidentally spilled on you—a bucketful of boiling water or a cupful of boiling water? You know by experience that boiling water contains energy in the form of heat. If the hot water is spilled on your skin, the heat energy may cause serious burns. Why would a bucketful of boiling water be more dangerous than a cupful? Both are at the same temperature. There must be more heat energy, however, in the larger amount of water. Temperature alone does not indicate the amount of heat in water. You can continue studying heat energy by finding out how to measure amounts of this form of energy. When you finish this section, you will be able to:

● Distinguish between *temperature* and heat.

● Explain the difference between the *Fahrenheit* and *Celsius* temperature scales.

● Define a *calorie*.

▲ Show that the same heat transfer does not always produce the same temperature change.

Temperature: A measurement of the movement of particles in matter.

What is the difference between heat and **temperature** (**tem**-pur-ah-ture)? You now know that heat energy is the result of the movement of particles in matter. When you measure *temperature*, you are measuring the amount of movement of the particles.

Suppose you touched a glass of warm water and then touched an ice cube. You could probably tell that the temperature of the water is higher than the temperature of the ice cube. The water particles in the glass are moving faster than the particles in the ice cube.

Do you think other people would always agree with your observations of temperature made after touching an object? More accurate measurements of temperature are made with an instrument. Then observations are the same for everyone. A *thermometer* is an instrument used for measuring temperature. Most materials expand when heated and shrink when cooled. The most commonly used thermometer is based on this principle. A liquid is sealed in a glass tube. Alcohol is often used and so is the liquid metal mercury. When a thermometer is heated, the liquid expands and rises in the tube. Cooling causes the liquid to shrink and fall. See Fig. 5-14. The thermometer tube must have a scale.

One kind of temperature scale was invented by a scientist named Fahrenheit in the 1700's. On the Fahrenheit temperature scale, water freezes at 32° and boils at 212°. The Fahrenheit (F) temperature scale was widely used in the past. The temperature scale now used most often all over the world is the **Celsius** (**sell**-sea-us) **scale**. In scientific work, the *Celsius temperature scale* is always used. The Celsius temperature scale is sometimes also called *centigrade*. On the Celsius scale, water freezes at 0° and boils at 100°. A comparison of the Celsius and Fahrenheit scales is shown in Fig. 5-15 on page 123.

Not all thermometers are made of a liquid in a glass tube. In some kinds of thermometers, a metal strip bends with changes in temperature. See Fig. 5-16 on page 124. The amount of bending then moves a needle on a dial to indicate temperature. Another kind of thermometer measures temperature by its effect on the flow of an electric current. See Fig. 5-17 on page 124. These thermometers are often used to measure body tempera-

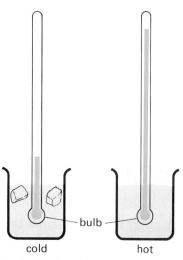

5-14. *The liquid in a thermometer is affected by temperature changes. The length of the column of liquid is a measure of the temperature at the bulb.*

Celsius (C): The name of the most commonly used temperature scale. The Celsius scale is always used in science.

5-15. *A comparison of the Celsius and Fahrenheit temperature scales.*

To change a temperature reading in degrees Fahrenheit to degrees Celsius, use the formula:
°C = ⁵⁄₉(___°F − 32).

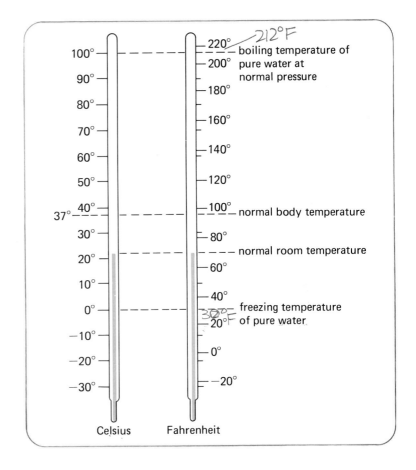

ture since there is no time spent waiting to read the temperature.

Suppose you had a bucketful of water at a temperature of 25°C. This is about the temperature of water coming from a faucet on a warm day. It would take a large amount of heat to raise the temperature of this water to 100°C. How much heat would be required to change the temperature of a cupful of water from 25°C to 100°C? The cupful of water would need much less heat than a bucketful. The amount of heat in the water depends not only on its temperature but also on the amount of water. For example, a teapot full of boiling water and a cupful of boiling water both have the same temperature, 100°C. The movement of the water particles in both the pot and the cup is the same. However, the pot contains more water than the cup. The amount of heat is greater in the pot than in the cup.

In order to measure heat, we must include the amount of material heated as well as its temperature change.

5-16. *(below) The two metal strips in a bimetallic thermometer respond differently to temperature changes.*

5-17. *(right) An electronic thermometer is often used in hospitals.*

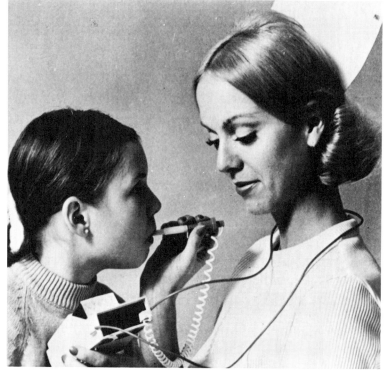

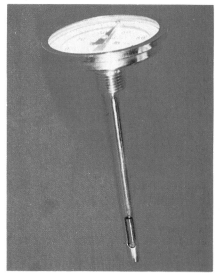

Materials
2 Styrofoam cups
2 ice cubes
thermometer (°C)
stirring rod
water
pencil
paper

How do the two terms temperature and heat differ in meaning? In this activity, you can find out.

A. Obtain the materials listed in the margin.

B. Fill one cup about ¼ full of tap water. Call this cup A.

C. Fill the other cup about ¾ full of tap water. Call this cup B.

D. Measure the temperature of the water in each cup. Record.
1. What is the temperature of the water in cup A? Cup B?

E. Take 2 ice cubes of equal size.
It will take the same amount of heat to melt each of these pieces of ice. If they are put in the cups of water, can you predict which cup will change its temperature the most?
2. What is your prediction?

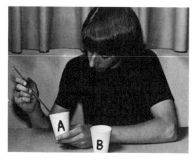

5-18.

Unit II How Does Energy Affect Matter?

F. Place one piece of ice into each cup of water.

G. Use a stirring rod to stir the water in cup A until the ice melts.

H. Read the temperature of the water.

  3. What is the temperature? Record.

I. Use the stirring rod to stir the water in cup B until the ice melts.

J. Read the temperature of the water.

  4. What is the temperature? Record.

  5. How did the temperature change in cup A compare with the temperature change in cup B? Remember, the amount of ice in both cups was the same.

  6. Compare the amount of heat removed from the water in cup A to melt the ice to the amount of heat removed from the water in cup B to melt the ice.

  7. Do you think that the thermometer correctly measured the amount of heat removed from the water in each cup? Explain your answer. How could you test whether the thermometer is correct?

Calorie: An amount of heat equal to that needed to raise the temperature of 1 gram of water by 1°C.

A unit called a **calorie** (cal-lor-ee) takes into account the amount of material and the temperature change. The amount of heat required to raise the temperature of 1

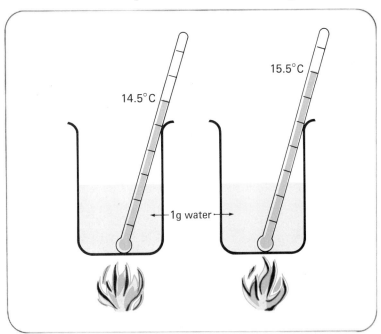

5-19. *A calorie is the amount of heat needed to raise the temperature of 1 g of water 1°C, from 14.5°C to 15.5°C.*

Chapter 5–3 Temperature and Heat

gram of water by 1°C is 1 *calorie*. See Fig. 5-19 on page 125. One of the most familiar uses of the calorie is in expressing the energy content of foods. The "calorie" used to measure food energy is actually a *kilocalorie*. A kilocalorie is a thousand times larger than a calorie. Sometimes a kilocalorie is written "Calorie" to distinguish it from the ordinary calorie.

SUMMARY & QUESTIONS

A bucketful of boiling water can be dangerous because of the heat energy it contains as well as its high temperature. Thermometers measure temperature in degrees Celsius or degrees Fahrenheit. They are not used to measure heat energy. The amount of heat energy in a material depends upon both temperature and the amount of the material. Both the amount of material and the change in temperature are used to measure heat changes in calories.

1. Use the clues given to complete the following word puzzle. DO NOT WRITE IN THIS BOOK.

    a. F _ _ _ _ _ | _ | _ _
    b. _ | _ S _ _ S
    c. _ | L _ _ _ _
    d. _ E _ _ E _ _ | _ | _ _ E

    Clues

    a. A scale on which ice measures 32°.
    b. A scale on which ice measures 0°.
    c. Heat required to raise 1 g of water by 1°C.
    d. Measurement on a thermometer.
2. Write a definition for the key word found within the box of the word puzzle in question 1.
3. One hundred calories of heat are transferred to a spoonful of cold water and 100 calories to a bucketful of cold water. Which water would be hotter? Why?

4 BEHAVIOR OF GASES

Matter is either a solid, a liquid, or a gas. All matter is made of particles. To understand how those individual particles behave, a comparison might be helpful. Think of yourself and everyone in your class as individual particles. At times the entire class is seated in some kind of orderly arrangement. This situation is somewhat like the behavior of particles in a solid. When the members of the class are out of their seats and moving around in the room, they resemble the motion of particles in a liquid. Finally, the class ends and everyone leaves the room, moving apart to fill the space outside in the hall. This movement is similar to the way particles spread out to fill space made up of gas. The properties of matter depend upon the behavior of its particles. When you finish section 4, you will be able to:

- Describe how particles move in gases, liquids, and solids.
- Use the *kinetic theory* to explain how gases behave.
- Explain why nothing can be colder than *absolute zero*.
- ▲ Show how temperature affects the behavior of a gas by making an air thermometer.

Kinetic Theory of Matter: The scientific principle which says that all matter is made of particles whose motion determines whether the matter is solid, liquid, or gas.

The scientific belief that all matter is made of moving particles is called the **Kinetic Theory of Matter**. The *Kinetic Theory* is one of the most important theories of modern science. By using the kinetic theory, scientists have been able to explain and predict the properties of matter. For example, the kinetic theory can be applied to observations about gases in the following ways:

1. Particles in a gas are moving very fast with large average distances between them. See Fig. 5-20.

A gas is mostly empty space. Gas particles move like ping-pong balls in a box you are shaking. Each particle collides with others many times each second. The particles are not affected by these collisions. Gas pressure is the result of the rapidly moving particles colliding against their container. Gases have no natural shape but will expand to fill any space available.

2. Gases can be compressed.

Gas particles can be crowded together. When a gas is squeezed into a smaller space, the pressure of the gas goes up if the temperature remains constant. See Fig. 5-21. This rise in pressure is a result of the rapidly moving gas particles hitting the walls of the container more often. See Fig. 5-22 on page 129. In the same way, a gas will

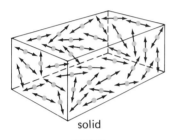

5-20. *The movement of particles in a solid, liquid, and gas.*

5-21. *This bicycle pump forces air under pressure into a bicycle tire.*

Unit II How Does Energy Affect Matter?

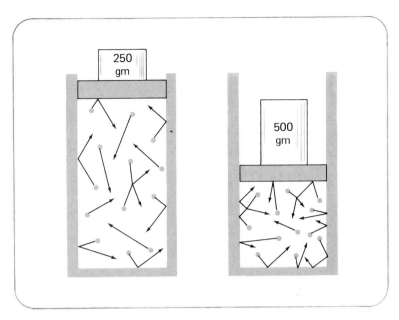

5-22. *Cutting the volume of a gas in half will cause the pressure to double if the temperature remains the same. The pressure goes up as a result of the crowding together of the gas particles. Twice as many particles will now collide with the walls of the container.*

have less pressure if it is allowed to expand to fill a larger space at the same temperature.

3. The temperature of a gas determines how fast its particles move.

If heat is added to a gas, the heat energy causes the gas particles to move faster. See Fig. 5-23. Faster moving particles hit the container holding the gas more often. The more frequent collisions cause higher pressure. When the temperature of a gas increases, its pressure also increases, if the volume is kept constant.

5-23. *Heating a gas in a closed container causes the particles to move faster and collide more often against the walls.*

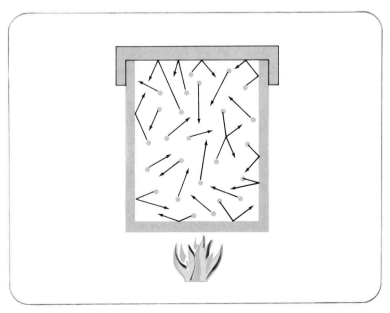

Chapter 5–4 Behavior of Gases

Materials
small container
soda straw
paperclip
ice cube
water
pencil
paper

Galileo, one of the great scientists of the 16th century, invented the air thermometer. The air thermometer is simply a quantity of air trapped by water in a tube. You will use modern materials to make an air thermometer. You will then be able to study the behavior of the trapped air when it is cooled or warmed.

A. Obtain the materials listed in the margin.

B. Fill a small container about ½ full with water.

C. Place one end of a soda straw in the water, resting it on the bottom.

You may note that the water rises up inside the straw to the level of the water in the container.

D. Now bend the top end of the straw over and fasten it with a paperclip. See Fig. 5-24. Closing the top traps some water at the bottom of the straw and some air at the top of the straw. You now have an air thermometer.

E. Now raise the straw straight up out of the water, keeping it over the container. Do not squeeze the straw. Hold it near the top under the paperclip.

1. Does the water drip out of the bottom of the straw?

If you have carried out the procedure carefully, the top end of the straw is sealed air tight.

2. Why do you suppose the water tends to slowly drip out the bottom?

F. Make this test. Watch a drop

5-24. *(left)*

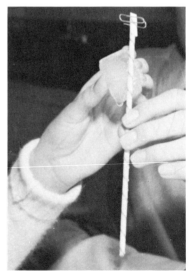

5-25. *(right)*

Unit II How Does Energy Affect Matter?

of water form at the bottom of the straw. See Fig. 5-25 on page 130. Just before the water drops off, touch an ice cube to the side of the straw near the top where you are holding it.

3. What happened?
The results you are looking for depend on the top of the air thermometer being air tight so check for leaks.

G. Remove the ice cube. Hold the straw by cupping your hand around it. Take care not to squeeze it!

4. What will warming this do to the trapped air?
5. Does a drop of water start to form again?
6. What result does warming have on the trapped air?

If the trapped air is expanding, it must be doing so by taking on the heat from your hand. This is why the water drips even though the tip is air tight. The expanding air forces the water out of the straw. You can check this hypothesis in the following way.

H. Hold the straw between two fingers of your left hand. Warm the trapped air in the straw

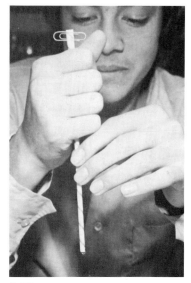

5-26.

with your right hand to form a drop of water. See Fig. 5-26.

I. Just before the water drips, remove your right hand. Be careful not to shake the drop off the straw.

J. Watch the drop carefully as the straw slowly cools.

7. What happens to the drop as the straw cools?
8. What effect does cooling have on the trapped air?
9. Explain why an ice cube was used to stop the dripping at the beginning of this activity.

A gas such as air will expand when its temperature is raised if the pressure is kept constant. In the same way, cooling a gas will cause it to shrink as its particles move more slowly.

Can you predict what will happen if a gas is made very cold? If a gas is cooled, its particles move slower and slower. At what temperature would the particles stop moving? Experiments have shown that this temperature is $-273°C$. At a temperature of 273° below zero Celsius, particles in matter stop moving. The particles do not

5-27. *A comparison of the Celsius and Kelvin temperature scales.*

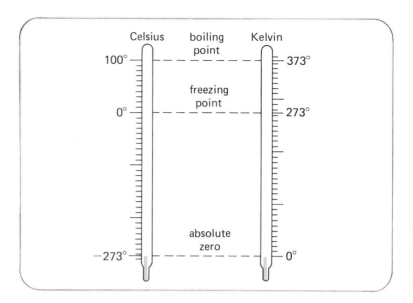

Absolute zero: The temperature (−273°C) at which particles in matter stop moving.

Kelvin temperature scale: A scale of temperature on which 0° is equal to absolute zero.

have the heat energy needed for motion. The temperature at which all motion of particles stops (−273°C) is called **absolute zero** (**ab**-sole-lute **zear**-row). Since temperature actually measures the amount of particle motion in a substance, there can be no lower temperature than *absolute zero*. Absolute zero is the lowest possible temperature. A temperature scale based on absolute zero is often used in scientific observations. This scale is called the **Kelvin (K) temperature scale.** On the *Kelvin scale*, 0° is equal to absolute zero. A comparison of the Kelvin and Celsius scales is shown in Fig. 5-27.

SUMMARY & QUESTIONS

You cannot see separate particles in matter. The kinetic theory describes how particles move in solids, liquids, and gases. The properties of gases can be explained by the rapid motion of their separate particles. Temperature is a measure of how fast particles move. At the lowest possible temperature, there is no motion of particles in matter.

1. In which form of matter are the individual particles likely to be moving the fastest?

2. Describe a gas in terms of the Kinetic Theory of Matter.
3. Explain why you would not be able to cool anything below absolute zero.
4. Describe how you would make an air thermometer.

5 LIQUIDS AND SOLIDS

By watching a diamond cutter at work, you could learn something about gems and about solids. You might be surprised to learn that the diamond "cutter" actually *breaks* a large diamond into smaller pieces. Diamonds are too hard to be cut. A large diamond to be "cut" is carefully examined. The diamond is then marked with lines. The diamond cutter gently taps on a line. If everything has been done properly, the diamond splits evenly. A mistake may cause a valuable large diamond to shatter into many pieces too small and too irregular to be used for jewelry. How does a diamond cutter know exactly where to tap the diamond? The trained eye of a diamond cutter can see in a diamond a characteristic that is found in most solids. They have natural lines along which they will split. This property of solids is one result of the way particles are arranged. When you finish section 5, you will be able to:

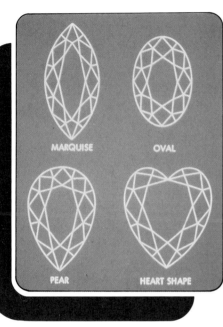

- Describe how particles are arranged in most solid materials.

- Explain what happens when a solid melts.

- Explain what happens when a liquid boils.

▲ Examine crystals of several solids and describe their appearance.

A piece of solid matter cannot change its shape by itself. A diamond, for example, holds its shape unless it is split by a blow. The particles of a solid are not free to

5-28. *(left) An example of an ice crystal—a snowflake.*

5-29. *(right) Rock candy has large crystals which are clearly visible.*

Crystal: A solid whose orderly arrangement of particles gives it a regular shape.

move about. They stay in position. The arrangement of particles forming a solid is usually very orderly. For example, when particles of water freeze to form solid ice, they form a **crystal** of ice. See Fig. 5-28. A *crystal* is a piece of solid matter with a regular shape. A large piece of ice is made up of many small ice crystals fitted together like the pieces of a jig-saw puzzle. Almost all solids are formed in this way.

When a diamond is split, it is separated along the surfaces which join the separate crystals that make it up. In many solids, such as rock candy, the individual crystals are large enough to be seen. See Fig. 5-29. However, the crystals are usually too small to be seen with the eye alone. The shape of an individual crystal is determined by the way its particles are arranged. For example, in common table salt each crystal is in the form of a cube. See Fig. 5-30. If you could see the particles in a crystal, each particle would be found in a definite position.

5-30. *This is how crystals of ordinary salt look under a microscope.*

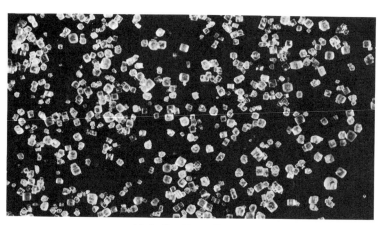

Unit II How Does Energy Affect Matter?

ACTIVITY

Materials
table salt
plastic sandwich bag
rock salt
small magnifying glass
piece of plastic (spoon or fork)
pencil
paper

Crystals are easily recognized by the beauty of their structure. Their flat surfaces meet, forming definite angles which give the crystals their particular shape. The angles and shape are determined by the kind of particles which make up the solid. For example, you already know the shape of the crystals of common table salt.

A. Obtain the materials listed in the margin.

B. Shake a few crystals of salt into a plastic sandwich bag. See Fig. 5-31 below.

C. Look at the salt crystals with a magnifier.
   1. Are most of the crystals cube shaped?
   2. Describe any crystals you see which are not cube shaped.

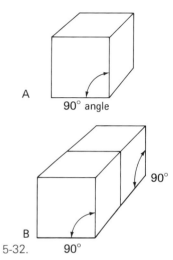

5-32.

5-32. *(above right)* **A** *The surfaces of a cube meet at 90° angles.* **B** *Two cubes joined together form a rectangular solid. The surfaces still meet at 90° angles.*

   3. Do all the crystals have flat surfaces?

A cube is formed when six square surfaces meet forming 90° angles. See Fig. 5-32. If you placed two cubes together, you would have a rectangular solid. The surfaces of the rectangular solid would still meet to form 90° angles.

D. Look at the table salt again.
   4. Did you find any rectangular crystals?

Another feature of crystals is their ability to break apart forming pieces with flat surfaces. The broken pieces form the same angles as the original crystal. This can be seen with rock salt. Rock salt is made of the same substance as table salt.

E. Pour the table salt from the sandwich bag back into the container provided.

5-31.

Chapter 5–5 Liquids and Solids

5-33.

F. Place a few crystals of rock salt in the plastic bag and look at it with the magnifier. See Fig. 5-33.

   5. Are there any cube-shaped crystals?

   6. Do the crystals have flat surfaces?

   7. Do the surfaces meet at 90° angles?

G. Now move several rock salt crystals which are *not* cubes to one side of the plastic bag away from the others.

H. Roll your pencil over these crystals to break them up. You may need to press very hard to do this.

I. Look at these pieces of crystals.

   8. Are there any pieces shaped like a cube?

   9. Do the pieces have flat surfaces?

   10. Do the surfaces meet at 90° angles?

Not all substances form crystals. There are a number of substances whose particles are not arranged to give natural flat surfaces. These substances can be made to have flat surfaces. However, the particles under the surface and throughout the solid are not lined up as they are in a crystal. All plastics are of this type.

   11. Would you predict that a piece of plastic would break in the same manner as a crystal?

J. Look at the piece of plastic provided. This piece should be large enough to see without using the magnifier.

5-34.

12. Does it have flat surfaces?

K. Pick up the plastic and bend it until it breaks. See Fig. 5-34 on page 136.

Breaking the plastic will probably require more effort than crushing the salt.

L. Look at the pieces of plastic. Use a magnifier for a closer look.

13. Are the broken edges flat surfaces?

14. If you see any flat surfaces, do they meet at the same angle as before you broke the plastic into pieces?

Heat of fusion: The amount of heat required to change 1 gram of solid to a liquid at the same temperature.

Melting point: The temperature at which a solid becomes a liquid.

The particles in a solid are always vibrating. If heat is added to a solid, the particles in that solid vibrate faster and faster as the temperature goes up. At some definite temperature, the motion of the particles becomes so great that the particles can no longer hold their orderly arrangement. When this happens, the solid melts and becomes a liquid. See Fig. 5-35. The heat required to melt 1 g of a solid is called **heat of fusion**. The temperature at which a solid changes into a liquid is called its **melting point**. Each kind of solid made of crystals has a particular temperature at which it melts. *Melting points of some common substances are given in Table 5-1.* Solids like glass or plastic that are not made of crystals do

5-35. *When a solid is heated, the particles of the solid begin to lose their orderly arrangement. The solid melts and becomes a liquid.*

not have a sharp melting point but soften gradually. The particles in solids like glass are not held in an orderly pattern. A few materials, such as dry ice and moth balls, do not melt under ordinary conditions. They change directly into a gas.

TABLE 5-1
Melting Points of Some Common Substances.

Substance	Melting Point (°C)
Iron	1535
Salt	801
Lead	327
Sugar	186
Water	0
Mercury	−39

5-36. *In a geyser, hot water and steam gush into the air.*

As you have seen, the particles of a liquid move more rapidly than the same particles in a solid. In most liquids at ordinary temperatures, a few particles have enough energy to escape and become a gas. For example, a pan of water left uncovered in a room will slowly evaporate because a few of its particles are constantly leaving the water in the pan and entering the air as water vapor. See Fig. 5-2 on page 112. If more heat is added to a liquid, the speed of evaporation will increase. Continuing to add heat will finally give all the particles of the liquid enough energy to become a gas. The heat required to change 1 g of liquid to a gas is **heat of vaporization** (vay-pore-i-**zay**-shun). If you measure the temperature when the liquid is changing to a gas, you will have found the **boiling point** of the liquid. The exact *boiling point* depends upon two factors: (1) the amount of heat energy needed to make the particles of the liquid separate to become a gas; (2) the pressure of the air. For example, water boils at 100°C when the air pressure is normal. On a mountaintop where the air pressure is lower, water boils at a temperature below 100°C.

Heat of vaporization: The amount of heat required to change 1 gram of a liquid to a gas at the same temperature.

Boiling point: The temperature (at ordinary air pressure) when the particles of a liquid have enough energy to become a gas.

SUMMARY & QUESTIONS

Why can something as hard as a diamond be broken with only a gentle tap? The answer is that the particles in most solid materials like diamonds are held together in an orderly pattern. This pattern allows crystals to be separated and causes most solids to have a particular melting point. Increasing the temperature of a liquid will cause it to boil and become a gas.

1. Most solids are formed as ____. Their particles are arranged in an ____ way giving them a regular shape.
2. Describe what happens to the particles of a solid as it is heated until it melts.
3. What two factors determine the exact boiling point of a liquid?
4. Describe what must happen to the particles that make up water to cause them to become a gas.
5. What are some of the properties you would look for if you had to decide whether a solid was a crystal or not?

Chapter 5–5 Liquids and Solids

CAREERS IN HEATING

Fire has been a source of heat to people through the ages. Open fires wasted fuel and caused smoke and fumes to spread throughout the living quarters. In time, methods of conserving fuel and eliminating pollution were devised. Stoves that conserved fuel were used in China by the 6th century B.C. Chimneys, which drew off the smoke and fumes from the fires were in use in Europe by the 13th century A.D.

We are still faced with the same challenges. One is to make the most efficient use of existing fuel sources and the other is to prevent pollution of our environment.

(below) **One method of conserving fuel is by insulation. The home-insulating industry is growing rapidly. The development of new and efficient materials and methods of insulation provides a growing field for career opportunities.**

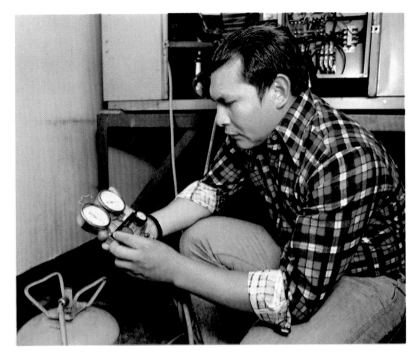

(right) **Air conditioning involves control of temperature, relative humidity, circulation of air, and dust particles.** As more people live in urban areas, situations that increase pollution, including smoke, exhaust gases, and chemical fumes will increase. As a result, more air conditioning equipment will be needed to provide livable environments.

(below) **This environmental scientist is attaching an instrument to an airplane.** The instrument scans the ground beneath the plane as it flies. The device is capable of monitoring heated water discharges from steam electric power plants. By its use, heat losses from building rooftops and forest fires are detected.

VOCABULARY REVIEW

Match the description in Column II with the term it describes in Column I.

Column I
1. absolute zero
2. boiling point
3. calorie
4. conduction
5. convection
6. crystal
7. heat engine
8. heat of fusion
9. Kinetic Theory
10. radiation

Column II
a. Changes heat energy to mechanical energy
b. Transfer of heat by infrared rays
c. Transfer of heat by direct contact
d. Transfer of heat by movement of a gas or liquid
e. Heat needed to raise temperature of 1 g water 1°C
f. Matter is made of particles whose motion determines whether the matter is solid, liquid, or gas
g. The temperature at which particles in matter stop moving
h. A solid whose orderly arrangement of particles gives it a regular shape
i. The temperature at which the particles of a liquid have enough energy to become a gas
j. The amount of energy which must be removed to change 1 g of water into ice at the same temperature.

REVIEW QUESTIONS

Choose the letter of the phrase which best completes each of the following statements.

1. Compared to the Fahrenheit scale, the same temperature change in the Celsius scale would contain (a) more degrees (b) the same number of degrees (c) fewer degrees (d) these scales are not comparable.
2. Temperature tells how hot an object is while heat tells (a) how cold the object is (b) something of how much energy the object has (c) something of how much matter is in the object (d) how much the object has changed temperature.
3. The number of calories needed to raise the temperature of 3 grams of water by 2°C is (a) 1 calorie (b) 1.5 calories (c) 5 calories (d) 6 calories.

4. The three ways of transferring heat from one object to another are (a) convection, conduction, and radiation (b) convection, contraction, and radiation (c) contraction, conduction, and radiation (d) convection, contraction, and conduction.
5. Heat is held in matter as (a) heat fluid added (b) heat fluid squeezed out (c) moving particles (d) crystals.
6. From the results of his cannon-boring experiment, Count Rumford reasoned that (a) the water boiled as a result of the work done by the horses (b) heat is a form of energy (c) the supply of heat was without limit (d) all of these.
7. For a substance, the particles are moving slowest when it is in which form? (a) gas (b) liquid (c) solid (d) There is no relation between movement and the form of matter.
8. The temperature at which all motion of particles stops is (a) 0° centigrade (b) 0° Kelvin (c) 0° Fahrenheit (d) 0° Celsius.
9. The melting point of a crystal is reached when (a) the substance has absorbed all the energy it can (b) the crystals of the solid become visible (c) the substance changes to a gas (d) the particles can no longer hold an orderly arrangement.
10. Which of the following affects the boiling point of a liquid? (a) The rate energy is added. (b) The temperature of the surrounding air. (c) The air pressure. (d) The melting point of the substance.

REVIEW EXERCISES

Give complete but brief answers to each of the following.

1. Explain what is meant by a calorie.
2. In your own words describe the difference between temperature and heat.
3. What are the features of the Fahrenheit and Celsius temperature scales that make them different?
4. State what happens to the particles in an object when it is heated.
5. Explain briefly the three ways heat is transferred from one material to another.
6. Describe what is meant by the Kinetic Theory of Matter.
7. Discuss the features of the Kelvin temperature scale.
8. State briefly how the motion of particles differs in gases, liquids, and solids.
9. Other than turning to a liquid, what happens when something melts?
10. Explain what happens when a liquid boils.

CHAPTER 6

FORCES AND MOTION

1 SPEED AND ACCELERATED MOTION

While you are reading these words, you are moving faster than a jet plane. Do you feel as if you are moving that fast? Very likely you feel as if you are sitting in one place. The motion is the result of being carried around with the earth as it makes one complete turn each day.

It is hard for you to make observations about the speed of the earth. We prove something is moving only by comparing it to something that is not moving. When you finish section 1, you will be able to:

● Using examples, explain what you mean when you say something is moving.

- Give examples to show that the speed of a moving object may change from one period of time to another.

- Explain how to find the average speed of a moving object.

- Demonstrate some ways the speed of a moving object can be measured.

Motion: A change in position of an object when compared to a reference point.

Motion is always observed by comparing the moving thing to something that appears to stay in place. The object that appears to stay in place is a *reference point*. Look at the photo on page 144. If you were watching a real game, what reference points would tell you that the ball was moving? Every moving thing covers a certain distance in a certain period of time. In one second the ball would travel a certain number of meters from the pitcher's mound toward the catcher. Speed is expressed as a measurement of distance moved during a period of time. When the speed of a moving object is measured, distance and time are given in convenient units. The speed of a running person might be measured in meters covered each second or meters per second (m/sec). The speed of a car is usually measured in miles per hour. In the metric system, a similar unit is kilometers per hour (km/h). A km is equal to 0.621 mile. A speed of 50 miles per hour is equal to 80 km/h.

Can you do this?
Distance/Time/Speed

26 m	8 sec	?
96.3 m	3 sec	?

Materials
metric ruler
marble
pencil
paper

A. Obtain the materials listed in the margin.

B. Find a table top or floor space that is smooth and level for a distance of about 1.5 m. You must be able to see the second hand on a watch or clock from the place you choose.

  1. Using a marble, how can you tell if the table is level?

C. Place a ruler so that one end

Chapter 6–1 Speed and Accelerated Motion

6-1.

is raised about 1.5 cm and the other end is on the table top.

D. Roll a marble down the groove and onto the table. Start the marble from the upper end of the ruler.

E. Determine how far the marble rolls from the bottom end of the ruler in 2 sec. Record.

F. Repeat this procedure at least three times. Record the distance each time. Finding the distances are your observations.

To find the average distance rolled in 2 sec, add together all the distances measured. Divide this total by the number of measurements.

2. What was the average distance the marble rolled in 2 sec?

To find the average speed of the marble, divide the distance traveled by the time to travel that distance. For instance, if the marble rolled 30 cm in 2 sec its average speed would be $\frac{30 \text{ cm}}{2 \text{ sec}}$ or 15 cm per second (15 $\frac{\text{cm}}{\text{sec}}$).

3. What was the average speed of the marble?

G. Determine how far the marble rolls from the bottom end of the ruler in 3 sec. Make several more measurements. Record your observations.

4. What was the average distance in 3 sec?

5. What was the average speed of the marble?

6. How does this average speed compare with the average speed during the 2-sec runs? (greater, same, less?)

Friction: A force that opposes or slows down motion.

When you ride a bicycle, your speed does not stay constant. Usually, you pump for a while then coast. Your speed first increases then, while coasting, drops because of **friction** (**frick**-shun). All moving objects, in this case

Acceleration: The change in speed during a given interval of time; calculated by dividing the change in speed by the time it took for the change in speed to happen.

the tires, axles, and wheel bearings, have *friction* when they touch another surface. Friction acts to slow moving objects. If you peddle uphill on the bicycle, your speed will also be slower. The speed, on the other hand, will increase when you go downhill. Because of these changes in speed, your motion on a bicycle trip would be called **accelerated** (ack-**cell**-er-ate-ted). If the speed of a moving object is found to change from one time period to another, its motion is called *accelerated*. Another very common example of accelerated motion is the speed of something which is falling. Any change of motion, either speeding up, slowing down, or changing direction, is called acceleration.

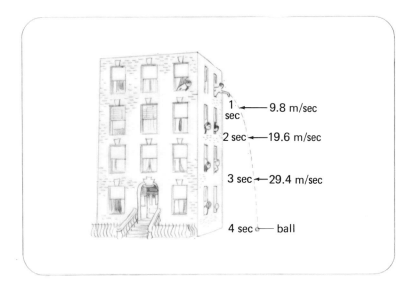

6-2. *What is the speed of the ball at 4 sec?*

SUMMARY & QUESTIONS

Motion is observed by comparing the moving object to some reference point that is not moving. Speed is a calculated measure of motion. Average speed is calculated by measuring the distance an object moves and dividing the distance by the measured time it took to go that distance. Speed, distance, and time are measured in convenient units. Objects in motion do not always move at the same speed. Many factors such as friction and moving up and down hills act to change the speed of an object. An object that changes from one speed to another is accelerating.

Chapter 6–1 Speed and Accelerated Motion

Use the following list of terms to fill in the numbered blanks for completing the paragraph. A term may be used more than one time. You will not use all of the terms.

a. accelerating
b. friction
c. motion
d. reference point
e. speed
f. time
g. 2m/sec
h. 5m/sec

In order to observe (1) of an object, its position would need to be compared from time to time with a (2) which is not moving. The distance moved by an object divided by the (3) it took to move that distance is called (4). For example, a skateboard rider who moved 10 m in 5 sec would have a speed of (5). When an object's speed changes it is (6). A marble rolling downhill would roll faster and faster and thus is (7). Most moving objects slow down due to (8). This slowing down of an object also means it is (9).

2 FORCES

The time is many years in the future. A spaceship is taking off to travel to a star beyond the sun. It is a journey of many years. The distance is too great even for our imagination. Once the spaceship is clear of the earth, the rocket motors are turned off. The ship then continues to move at the same speed until it reaches its goal.

How can this happen? Does a car continue to coast along for years after the engine is shut off? Certain natural rules control the motion of spaceships and automobiles. To understand these rules, you must remember the observations you have already made of moving objects. When you finish section 2, you will be able to:

● Briefly explain the three Laws of Motion.

● Predict the behavior of a moving object when a *force* acts upon it.

▲ Demonstrate that a force is needed to cause a moving object to change its speed or direction.

Unit II How Does Energy Affect Matter?

6-3. *Sir Isaac Newton (1642–1727) was one of the greatest scientific thinkers of all time.*

Force: Any push or pull which causes something to move or change its speed or direction of motion.

A book is resting on a level desk top. You would not expect the book to begin moving unless someone pushes or pulls on it. If that happens, the book will not move very far. Friction will make it stop. If friction was not present, the book would continue to move at a certain speed in a straight path until it ran into something.

On earth, it is very hard to escape friction. Most of our common observations of moving objects show that it is natural for them to slow and then come to a stop. A spaceship would feel almost no friction. Once started, it would continue to move in a straight path. To change its direction, a push would be necessary. Any push or pull is called a **force**. A *force* can cause something to change its speed or direction. Because our natural environment is filled with friction, it took many centuries for scientists to discover this principle. It was first put in the form of a scientific law by one of the greatest scientists, Sir Isaac Newton. Newton stated this principle as his First Law of Motion: *Every object remains at rest or moves with a constant speed in a straight line unless acted upon by some outside force.*

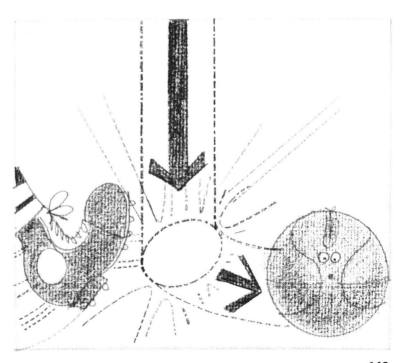

6-4. *An object remains at rest or moves with constant speed in a straight line unless acted on by an outside force.*

Chapter 6–2 Forces

PART I

Materials
Marble
Metric ruler
Pencil
Paper

A. Obtain the materials listed in the margin.

B. Raise the end of a ruler about 1.5 cm.

C. Starting at the top, roll a marble down the groove.

Measure the distance from the bottom of the ruler that the marble rolls in 2 sec. Repeat at least three times. Record.

  1. What was the average distance the marble rolled in 2 sec?
  2. What was the average speed of the marble?

D. Repeat step C but raise the end of the ruler about 1 cm higher. Keep everything else the same. Roll the marble several times and record your measurements.

  3. What was the average distance the marble moved in 2 sec with this set-up?
  4. What was the average speed of the marble?
  5. How does the speed this time compare with the speed from the 1.5-cm height? (smaller, same, greater)

The difference in speed between these two set-ups shows that a force was acting on the marble.

  6. What force do you think was acting on the marble?
  7. What change caused the force to be greater in the second set-up?

6-5. *The mass of an object determines the size of the force needed to move it.*

Some of the most common forces around us are gravity forces. Gravity acts upon all objects and pulls them toward the earth's center. The force of gravity acts on falling objects to cause them to pick up speed or accelerate. Gravity force on a particular object is called weight.

Common observations will show that the mass of an object is important when it is moving. For example, think about the force required to cause a bicycle and an automobile to accelerate to 8 km/h. See Fig. 6-5. A much larger force would be needed for the heavy car than for the lighter bicycle. The mass of an object determines the size of the force needed to produce a certain change in speed. Everything falls with the same acceleration if the effects of air resistance can be removed. For instance, a light and a heavy rock are both dropped at the same time. Both will accelerate at the same rate and reach the ground at the same time. The rock with the larger mass requires more force to accelerate it. That larger force is supplied by its greater weight.

Materials
Book
Rubber band
Metric ruler
String loop, 60 cm around
2 rubber erasers or stoppers
Heavy string, 1m
Pencil
Paper

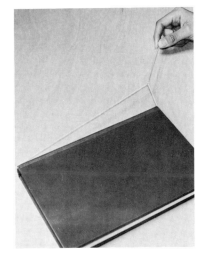

6-6.

6-7. *The mass of an object determines the amount of force needed to pull it out of a curved path.*

PART II

A. Obtain the materials listed in the margin.

B. Measure the length of the folded rubber band loop. Stretch it until it measures 5 cm more than its relaxed length. Notice the feel of the force (pull) you must apply. Now stretch it until it measures 10 cm more than its relaxed length.

8. Which stretch (5 cm or 10 cm) required the greater force?

C. Attach the rubber band to the string loop and fasten the string to your book by looping it over the cover as shown in Fig. 6-6.

D. While measuring its stretched length, pull on the rubber band until the book slides across the level desk. After the book is moving slowly, pull only hard enough to keep the book moving at a constant, slow speed.

9. What is the length of the rubber band while the book moves slowly?

E. Apply a force to the book by stretching the rubber band 3 cm more than your answer to question 9. Notice about how long it takes for the book to move 50 cm. Don't try to measure the time.

F. Apply a force to the book by stretching the rubber band 6 cm more than your answer to question 9. Notice about how long it takes the book to move 50 cm.

10. Which force (pull) (3 cm or 6 cm) takes the least time to pull the book 50 cm?

11. Since the book is speeding up while it is being pulled, which force (pull) (3 cm or 6 cm) produces the greatest change in speed?

12. Which force produces the greatest acceleration?

The mass of an object is also related to its motion in a curving path. A force is needed to make an object follow a curving path. The greater the mass of the object, the greater the amount of the force needed to pull the object out of a straight path. This relationship between masses and the forces acting on them was summed up by Newton in his Second Law of Motion: *The acceleration of an object of a certain mass is determined by the size of the force acting and the direction in which it acts.*

To show how Newton's Second Law of Motion works, think about a tug of war in your gym class. Imagine first of all, that all the students are of equal strength. The tug of war begins with four students on each side. As long as this number is maintained, and everybody pulls equally, the rope will not move. Now imagine that two students join the team on the right side. The rope will begin to move to the right because the greater force is pulling in that direction. If two students join the left hand team, they will be able to stop the movement to the right because the forces will be balanced again. However, they will not be able to reverse the movement. If still another student joins the left team, the rope will move to the left. The more students that join the left team, the faster the rope will move to the left because the force is greater in that direction.

The fact that forces never exist alone was also discovered by Newton. It is the principle of his Third Law of Motion: *For every action, there is an equal and opposite reaction.* A book rests on the table. The weight of the book pushes down on the table with a certain force. This force is balanced by an equal but opposite force. The table pushes up on the book. A rocket accelerates because the hot gases pushed out behind produce an un-

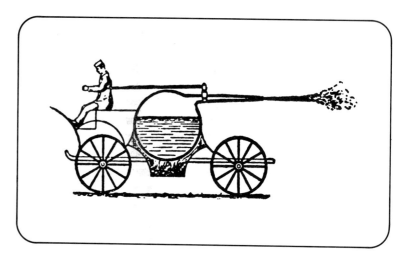

6-8. *For every action there is an equal and opposite reaction.*

balanced force on the rocket. The force makes the rocket change its speed and possibly its direction. Every unbalanced force is accompanied by an acceleration.

Newton's Third Law of Motion can be used to explain how gravity force works. For example, a force is required to make the planets follow a curving path around the sun. This force is supplied by the sun's gravity pulling the planets toward it. Newton also studied gravity forces. He determined that if the sun attracts a planet, the planet must also attract the sun. Two objects such as the sun and a planet like the earth must pull on

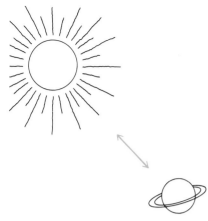

each other with a gravity force. The size of the gravity force depends on the mass of the planet, the mass of the sun, and the distance between them. Gravity must pull two objects toward each other with a force whose size is related to the masses of the two objects. The sun and the earth pull each other with equal but opposite forces. The force of gravity also becomes smaller as the distance between objects becomes greater. The force of gravity increases as the objects come closer together. The planets that are far away from the sun feel less attraction toward the sun. These planets follow less curving paths than if they were closer to the sun.

6-9. *The sun and a planet pull on each other with equal, but opposite, forces.*

Sir Isaac Newton's Laws of Motion help us account for the effects of forces on objects in motion or at rest. Newton also showed that friction is a force that affects motion. An understanding of these laws helps us explain such things as the flight of a balloon or a spaceship.

1. Why does a spaceship move in a straight line without a change of speed after its engines are shut down?
2. The same force is applied for the same length of time to each of three objects, one at a time. Object A does not move. Object B accelerates to a speed of 2 m/sec. Object C accelerates to a speed of 5 m/sec.
   a. Which object has the greatest mass?
   b. Which object has the least mass?
3. Which of the following is an example of the Third Law of Motion? (a) a model racer powered by a CO_2 cartridge (b) an ice skater throws a basketball and slides backward (c) a rocket ship lands on the moon (d) all of these.

3 WORK AND POWER

For thousands of years, humans used their own muscle power to do work. They then found that horses and oxen were more powerful. These animals pulled wagons and plows, carried people and goods, turned machines and lifted heavy loads. Then in 1770, James Watt developed an efficient steam engine and opened the door to the industrial world of today. When you finish this section, you will be able to:

6–10. *Steam power was more efficient than horsepower.*

● Define *work* and *power*.

● Calculate the amount of *work* done in metric units.

● Identify the units used to measure *power*.

▲ Make measurements and calculate work done in several examples.

James Watt had a problem to solve. During the eighteenth century, most of the heavy lifting and pulling at the mines was done by horses. To lift coal out of a deep mine, ropes would be sent up out of the mine and tied to a horse. The horse would then do the work of pulling the coal up to the surface. Mine owners were always

154 Unit II How Does Energy Affect Matter?

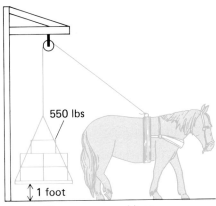

6–11. *The power used by a horse to lift 550 lbs. one foot in one second equals one horsepower.*

Work: The force applied to an object multiplied by the distance the object moves. W = F × d

Newton: The unit used to measure force in the metric system.

6–12. *If effort is equal to resistance, no work is done.*

asking Watt to compare the power of his new engine to the power of their horses.

To make this comparison, Watt invented a new measurement unit. At that time, the English system of measurement was used. So, Watt determined that a good horse could lift 550 pounds to a height of one foot in one second. This amount of work he called one *horsepower*. We still use this unit to measure the power of machines. For example, a lawn mower might have a 5 horsepower engine. This means that the motor can lift 5 × 550 pounds to a height of one foot in one second.

To the scientist, **work** means using a force to move an object through a distance. In the last section, we looked at a tug of war as an example of Newton's Second Law of Motion. Let's look at it again in terms of *work* and power. We will call your team, Team A, and your opponents Team B. The purpose of the contest, of course, is to move an object (Team B) through a distance. In other words, your team is trying to do work. Your team has to apply a force in order to move Team B. This involves Newton's First Law: An object at rest tends to remain at rest unless acted on by an outside force. We call this outside force the effort force or *effort*. Team B is resisting your force so we call their force the resistance force or *resistance*.

As long as the effort force is equal to the resistance force, nothing moves. According to the scientist, no work has been done. That might seem strange to you and your teammates. You have been pulling and straining for five minutes or so, yet we say you haven't done any work. It certainly feels like work. You have been using energy and are probably tired. Yet, strictly speaking, since nothing has moved, no work has been done.

Finally, your effort begins to pay off. Your team moves Team B closer to the line. The effort force is now greater than the resistance force. This causes the resistance to move. Therefore, work is being done.

How much work? The formula: *Work = Force × Distance* is used to determine this. To solve this problem, we have to know how much force Team B was using to resist your team. Remember, they were trying to win too.

In the metric system, forces are measured in units called **newtons.** If you were holding a 1 kg bag of sugar in your hand, you would be using a force of about 9.8 *newtons (N)* to support this mass.

Let's assume that the members of Team B are resisting

Chapter 6–3 Work and Power

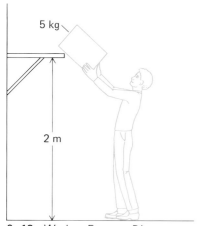

6-13. *Work = Force × Distance. In this case, 10 newton-meters.*

Power: How fast work is done. Power = Work ÷ Time

Watt: The metric unit used to measure power. 1 watt = 1 newton-meter per second.

6-14. *Human muscle power isn't very effective over long periods of time.*

your team with a force equal to 10,000 newtons. Your team applies enough force to move the resistance 3 meters. The work done can be calculated as:

$$W = F \times d$$
$$W = 10{,}000 \text{ N} \times 3 \text{ m}$$
$$W = 30{,}000 \text{ N-m (newton-meters)}$$

Let's calculate the work done in the following problem. A student uses a 5 newton force to lift a mass from the floor to a shelf that is two meters high. $W = F \times d$, so:

$$W = 5 \text{ N} \times 2 \text{ m}$$
$$W = 10 \text{ N-m}$$

Notice that we have not said anything about how long it took to do this work. When we talk about how long it takes to do a certain amount of work, we are talking about **power**. Both the horses and Watt's engine could do the same amount of work, but the engine could do it faster. Therefore, the engine was more *powerful*. It was also cheaper and didn't need to rest.

Power is the *rate* at which work is done. The English system measures power in *horsepower* units. In the metric system, power is rated in **watts.** One horsepower equals 746 watts. If 1 newton-meter of work is done in 1 second, 1 *watt* of power is generated. Power depends on time. The faster the work is done, the greater the power generated. Power = Work ÷ Time (in sec). This is written as $P = \frac{W}{T}$. So, if 15,000 N-m of work are done in 5 seconds, 3,000 watts of power are developed.

Imagine the amount of work that was done in building the ancient Egyptian pyramids. Each of the great stones in the pyramids had a mass of thousands of kilograms. Yet because it took many years to build a pyramid, the power generated was very small. Look back at the tug of war example. If it took Team A one minute to move Team B a distance of 3 meters, Team A would have developed about 500 watts of power. This small amount is only enough power to light one 100 watt light bulb for 5 seconds. It is obvious that human power isn't very effective in doing large amounts of work quickly. This is why machines were invented. The next section looks at the simple machines from which all machines were eventually developed.

SKILL BUILDING ACTIVITY

Materials
Pencil
Paper

PROBLEM SOLVING

A. Obtain the materials listed in the margin.

B. Use the information given in this lesson to solve the following problems.

C. It takes 9.8 newtons of force to lift 1 kg. A carton that has a total mass of 15 kg is lifted 2 m off the floor.
 1. How many newtons of force must be used? (Assume no friction.)
 2. How much work is done?

D. A crane lifts a slab of concrete with a mass of 1,000 kg to a height of 30 m up the side of a new building. The work takes 10 minutes.
 3. How many newtons of force must be used? (Assume no friction.)
 4. How much work is done?
 5. How many watts of power does the crane generate?

E. A boy who has a mass of 75 kg carries a 5 kg mass up a flight of stairs 10 m high, in 30 seconds.
 6. What is the *total* amount of work done?
 7. How many watts of power were generated to do this work?

SUMMARY & QUESTIONS

To a scientist, work is done if a force moves an object through a distance. The amount of work done in a certain amount of time is called power. In the metric system, work is measured in newton-meters and power in units called watts.

1. How is work calculated?
2. If a 20 N force lifts an object 3 meters, how much work is done?
3. An object has a mass of 25 kg. How many newtons of force will be needed to lift it?
4. If the object in question 3 is lifted 2 meters, how much work is done?
5. How is power different from work?
6. What does the term *4 watts* mean?
7. If a 5 watt engine ran for 2 minutes, how much work could it do?

Chapter 6–3 Work and Power

4 USING SIMPLE MACHINES

The huge power shovel in the picture is a wonder of mechanical engineering. Able to lift large amounts of coal in each scoopful, it dwarfs the worker inside its scoop. Yet for all its size, it is but a combination of the same simple machines humans have used for thousands of years. When you finish this section, you will be able to:

- State the Law of Machines.
- Describe six *simple machines*.
- Calculate the advantage *simple machines* give us.
- ▲ Demonstrate the use of *pulleys*.

Simple machine: A device that changes the size, direction, or speed of a force.

Mechanical advantage: The number of times a simple machine multiplies an effort force.

6–15. *Did you know the pop-top can is an example of a simple machine?*

Many parts of this power shovel are similar to the **simple machines** we use every day. These *simple machines* are so common we often do not realize we are using a machine. Every time you open a pop top can, turn a doorknob, or climb the stairs, you are using a simple machine. Since there are several different types of machines, it is necessary to define what machines are and why we use them.

We use machines because they make it easier to do work. Work involves using a force to move an object through a distance. A machine is a device that: 1. increases the force being applied, 2. speeds up the force, or 3. changes the direction of the force. How much help a machine gives us is called its **mechanical advantage (M.A.)**. This advantage is calculated in different ways depending on the machine.

Machines do not increase the work being done. In fact, the *Law of Machines* states that the work you get out of a machine is always less than the work you put into it. This is because some of the energy (force) put into the machine is used to overcome friction. This energy changes to heat energy and is lost.

If you use a hammer to pull a nail out of a piece of wood, feel the nail as soon as it is out. It should feel warm. This heat is from the extra force you used to overcome the friction between the wood and the nail. Have you ever tried to turn a screw into a piece of wood? If the threads of the screw are rubbed with soap, the friction will be reduced. It will take less effort force to turn the screw. Oil reduces friction between moving parts of machines. Thus, more force goes toward doing work.

Machines make work easier. Try this simple activity. Slide the end of a wooden ruler under the right side of this book until only 10 cm of the ruler sticks out. Place your pencil under the ruler, 7 cm from the end that sticks out. Press down on this end hard enough to lift the book. Now pull the ruler out so that the pencil is 15 cm from the end. Keep the book 3 cm from the pencil. Press down on the end with enough force to lift the book. Now pull the ruler out until the pencil is 22 cm from the end. Again press down on the end. Which time was it easier to lift the book?

You have just used one of the oldest simple machines, the **lever**. A *lever* consists of a solid bar that can move about a fixed point. The fixed point is called the *fulcrum*. The effort and resistance are applied at other points on the bar. The lever you made is called a *first-class lever* because the fulcrum is between the effort and the resistance. Your pencil formed the fulcrum. The book was the resistance. You provided the effort. See Fig. 6-16. A first-class lever multiplies the effort force and decreases speed.

A second-class lever has the resistance between the effort and the fulcrum. A wheelbarrow and a bottle opener are examples of second-class levers. A second-class lever also multiplies the effort force and decreases speed.

Your forearm and a shovel are two examples of third-class levers. The effort is applied between the resistance and the fulcrum. An object supported in your hand provides the resistance. Your elbow is the fulcrum. Your muscles apply the effort in the middle. A third-class lever decreases the effort force but increases speed. Examples of first-, second-, and third-class levers are shown on page 162.

Levers provide a mechanical advantage (M. A.). This is a measure of how much the effort force is multiplied. The

Lever: A rigid bar that moves on a fixed point and can change the direction and size of the force applied.

6-16. *This simple experiment can teach you some facts about levers.*

Chapter 6-4 Using Simple Machines

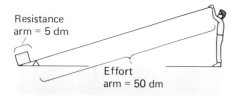

6-17. Why does the mechanical advantage of this lever equal 10?

mechanical advantage of a lever is found by dividing the length of the effort arm by the length of the resistance arm.

$$\text{M. A.} = \frac{\text{effort arm}}{\text{resistance arm}}$$

So, if the effort arm of a lever is 50 dm long and the resistance arm is 5 dm long, the M. A. is 10. See Fig. 6-17.

The mechanical advantage of a machine can be used to determine how much force is needed to move a given resistance. To do this, simply divide the resistance by the M. A. Using a lever having a mechanical advantage of 10, an object having a resistance force of 50 N could be moved with an effort force of only 5 N. However, the effort force must move a greater distance than the resistance moves. For example, a 5 N effort force must move through a distance of 10 m in order to move a 50 N resistance a distance of 1 m. See Fig. 6-18. This is because for all levers:

effort force × effort distance =
resistance force × resistance distance
50 N × 10 m = 50 N × 1 m

This is the same as saying the work you put into a machine is equal to the work you get out of the machine. Actually, slightly more effort would have to be put in to overcome the force of friction. For a third-class lever, the M. A. is less than one, so the effort force is decreased. However, the resistance moves a greater distance than the effort.

A door knob and a steering wheel are examples of a **wheel and axle.** This is a machine similar to a lever. The fulcrum is the center of the axle. The resistance distance is the radius of the axle. The effort distance is the radius of the wheel. A small effort applied to the wheel results in a large force turning the axle. Did you ever try to turn the shaft, or axle, of a doorknob when the handle, or wheel, was missing? If so, you know how the size of the handle multiplies the force. See page 163.

The **pulley** is similar to the wheel and axle except that the axle does not turn. A rope runs over the *pulley* with the resistance hanging down on one side and the effort applied to the other side. Since the effort and resistance distances are the same, there is no gain in force. A single fixed pulley only changes the direction of the force. It is easier to pull downward than to lift upward.

Wheel and axle: A large wheel fixed to a small axle so that they turn together.

Pulley: A grooved, freely turning wheel over which runs a rope or chain.

6-18. A lever allows you to use less force, but you must apply it through a greater distance.

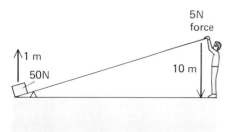

6-19. *A fixed pulley (left) changes the direction of the effort force. A movable pulley (right) multiplies the effort force.*

6-20. *The mechanical advantage of this system of pulleys equals 5. Why?*

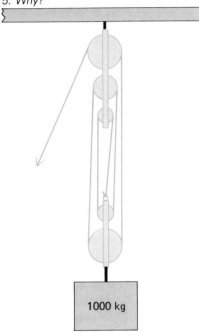

Inclined plane: A surface that slopes from one level to another.

Wedge: Two inclined planes attached base to base.

Screw: An inclined plane wrapped around a cylinder.

Compound machine: A machine containing more than one simple machine.

Combinations of pulleys can reduce the effort needed. Fig. 6-19b shows a single movable pulley. To lift the resistance with this arrangement, it only takes half as much effort as with a single fixed pulley. Why? In Fig. 6-19a, all the resistance is hung on one rope. In Fig. 6-19b, two ropes support the mass. It only takes half as much effort to pull up on one rope.

In a pulley system, the M. A. is equal to the number of ropes that support the resistance. In Fig. 6-19b, the M. A. is 2. The M. A. in Fig. 6-20 is 5. Therefore, it would only take 200 N of effort to lift 1,000 N of resistance. However, as with the lever, the effort has to travel farther than the resistance. The person in Fig. 6-19b has to pull up two meters of rope to lift the resistance one meter.

Another very ancient simple machine is the **inclined plane.** This is more commonly called a ramp. A flight of stairs or a stepladder are also *inclined planes*. How does the inclined plane help us to do work? It is easier to roll a barrel up a plank than to lift it from the ground. The M. A. of the inclined plane is found by dividing the length of the ramp by its height. The longer the inclined plane, the higher its M. A.

A **wedge** is simply two inclined planes back to back. Ax blades, knife blades, and nails are all *wedges*. **Screws** are really inclined planes wrapped around a cylinder. Bolts, nuts, and even jar lids are examples of *screws*.

These simple machines can be combined in many ways to make the **compound machines** we use in our homes and at work. A pair of scissors is two levers held together by a screw. The cutting edge of each blade is a wedge. A telephone dial is a wheel and axle. How many other compound machines can you name? See page 163.

Chapter 6–4 Using Simple Machines

Using Simple Machines

The Lever

$$\text{M.A.} = \frac{\text{effort arm}}{\text{resistance arm}}$$

The class of a lever depends on where the force is applied in relation to the fulcrum. For each type, the effort force times the effort distance equals the resistance force times the resistance distance (without friction).

First class lever

Second class lever

Third class lever

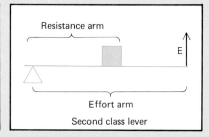
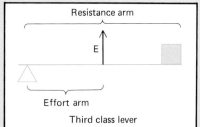

The Pulley

The M.A. of a fixed pulley is one. It makes work easier by changing the direction of the force. Movable pulleys or combinations of pulleys have an M.A. equal to the number of ropes supporting the mass.

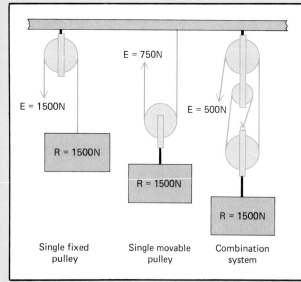

Single fixed pulley — E = 1500N, R = 1500N
Single movable pulley — E = 750N, R = 1500N
Combination system — E = 500N, R = 1500N

The Wheel and Axle

The wheel and axle is a type of lever. The radius of the axle is the resistance distance. The radius of the wheel is the effort distance. The M.A. is found by dividing effort distance by resistance distance. If the effort is applied to the axle, the wheel can be made to turn faster.

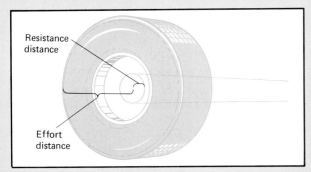

Inclined Plane, Wedge, and Screw

An inclined plan is a ramp connecting one level with a higher level. Its M.A. is found by dividing the length of the ramp by its height. A wedge is two inclined planes placed back to back. A screw is an inclined plane wrapped around a cylinder.

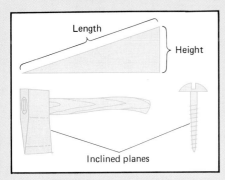

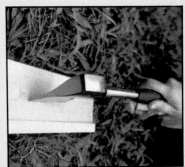

Compound Machines

A compound machine such as this 10-speed bicycle is a combination of several simple machines. The handlebars are part of a wheel and axle. So are the gear wheels and chain. The brake handles and the pedals as well as the shift are levers. The various nuts and bolts that hold everything together are types of screws. The brakes on the bicycle and the tires are examples of places where friction is a useful force.

Chapter 6-4 Using Simple Machines

Materials
Pulley
String
Support stand
500 g mass
Paper, pencil
Ruler

A. Obtain the materials listed in the margin.
  1. How much force (in newtons) is needed to lift a 500 g mass?

B. Set up a single fixed pulley. Attach the 500 g mass.
  2. What is the M.A. of a single fixed pulley?
  3. What force do you predict will be needed to lift the mass with this pulley?

C. Pull slowly on the spring balance until the mass is 10 cm off the table.
  4. Does the reading on the balance agree or disagree with your prediction? Explain.
  5. What is the advantage of a single fixed pulley?
  6. How much work was done?

D. Set up a single movable pulley. Attach the 500 g mass.
  7. What is the M.A. of this pulley?
  8. What force do you predict will be needed to raise the mass 10 cm?
  9. Test your prediction. Explain your answer.
  10. How much work was done?
  11. How does the distance the mass moved compare with the distance the effort moved?
  12. If you used a system of pulleys that had a mechanical advantage of 4, how much force would you need to lift 4,000 g?

We all use many different machines every day. Each of these is either one type of simple machine or a combination of simple machines. Machines make work easier or faster to do. However, machines cannot increase the amount of work being done.

1. State and explain the Law of Machines.
2. Name six simple machines. Give one example of each.
3. How can you find the M.A. of a pulley system?
4. How can you find the M.A. of a lever?
5. Compare and contrast the three classes of levers.
6. Give three reasons for using machines.
7. A 20 kg mass is 2 m from the fulcrum of a first-class lever. How far from the fulcrum should a 98 N force be applied to move the mass if there is no friction?

VOCABULARY REVIEW

Match the description in Column II with the term it describes in Column I.

Column I
1. force
2. acceleration
3. mechanical advantage
4. friction
5. motion
6. energy
7. simple machine
8. lever
9. power
10. work

Column II
a. A change in position of an object when compared to a reference point.
b. The number of times a simple machine multiplies an effort force.
c. Any push or pull.
d. A force that slows the motion between two objects in contact with each other.
e. The change in speed during a given interval of time.
f. The size of a force multiplied by the distance through which the force moves.
g. How fast work is done.
h. That property of something that makes it able to do work.
i. A device that can change the direction and size of forces.
j. A rigid bar moving about some fixed point. It usually changes the direction or size of the force applied to it.

REVIEW QUESTIONS

Choose the letter of the ending that best completes each of the following statements.

1. When a force moves something, the amount of work done depends on (a) how fast the object moves (b) the size of the force (c) the distance the object moves (d) both the size of the force and the distance through which the force is applied.
2. Machines *do not* (a) increase the force being applied (b) increase the work being done (c) make work easier (d) change the direction of a force.
3. Objects accelerate as they fall to the earth since (a) the earth's mass is pulling them (b) there is no force being applied (c) they are lighter than air (d) there is no frictional force present.
4. Since the speed of a falling object changes as it falls, the motion is called (a) acceleration (b) gravity (c) free fall (d) constant motion.

5. The wedge and the screw are really types of (a) pulleys (b) wheel and axles (c) inclined planes (d) levers.
6. The ability of something to do work is called (a) energy (b) power (c) gravity (d) a machine.
7. Power is a measure of (a) how great a force is being used (b) how fast work is being done (c) force and distance (d) how much work is done.
8. In the metric system, work done is measured in (a) ft-lbs (b) kilowatts (c) horsepower (d) newton-meters.
9. All of the following are simple machines *except* a (a) lever (b) pulley (c) screw (d) bicycle.
10. The M. A. of a pulley system is equal to (a) 2 (b) the number of pulleys (c) the number of ropes supporting the resistance (d) the number of ropes supporting the effort.

REVIEW EXERCISES

Give complete but brief answers to each of the following.

1. Compare the amount of work done when (a) a book is moved 20 cm and the same book is moved 40 cm; (b) one book weighing 500 g is moved 20 cm and two books weighing 250 kg each are moved 20 cm.
2. Explain why a machine cannot produce more work than is put into it.
3. Explain the difference between work and power.
4. Give two examples of accelerated motion.
5. Why does a heavy rock fall with the same acceleration as a light one?
6. Describe and give examples of each of the three Laws of Motion.
7. Give three examples in which friction is a useful force.
8. What would happen to a moving object if there were no friction?
9. Diagram and label each of the three classes of levers.
10. How do you find the mechanical advantage of (a) a lever (b) a pulley system (c) an inclined plane?

CHAPTER 7

WAVES AND SOUND

1 ENERGY AND WAVES

Each year hurricanes and other tropical storms threaten the coastal areas of the world. Much of the damage they do is caused by the strong waves these storms create. The waves have tremendous energy that can destroy many objects as they pound the shore. When you finish this section, you will be able to:

● Describe two types of *waves*.

● Identify four properties of all *waves*.

● Explain the units used to measure these properties.

▲ Demonstrate examples of two types of *waves*.

Wave: A disturbance caused by energy moving from one place to another in a substance.

You may have watched the last drops of a heavy rainstorm fall from a tree or roof and splash into a puddle below. The impact of the raindrop starts a series of circular **waves** that move out from the drop. These *waves* are transferring energy from one point to another.

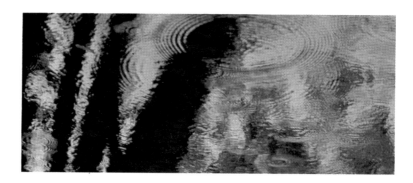

7-1. *The energy of a falling drop is transferred to the waves.*

To better understand this idea, think of the waves started by the wind of a large storm. The moving energy of the wind was transferred to the water. Now the waves are transferring that energy to the boats, beaches, buildings, and piers as they slam ashore. When it struck the water, the raindrop had kinetic energy from its fall. This energy was transferred to the water, transferred away from the impact point by the waves.

Waves are involved in the transfer of many forms of energy from one place to another. Both sound and light energy can travel in wave form. The broadcast beams of radio and television stations are in wave form. Even some of the energy of an earthquake travels in waves.

Not all waves are the same. Some, like the ripples on a puddle or the waves on an ocean, involve what seems to be an up and down motion. Think of a leaf floating in the puddle or a bob on a fishing line. As the ripples move across the water, the bob moves up and down at right angles to the wave motion. It does not ride along with the wave like a surfer. This is also true of the water par-

7-2. In a transverse wave, the particles move up and down but the wave moves along the rope.

ticles. It is the wave pattern that moves outward. The individual water particles do not.

This motion can be seen by tying a rope to a doorknob. If you hold the other end of the rope and move it up and down in a snapping motion, a wave will move along the rope. If you tie several knots of colored string to the rope, you will see that the knots move up and down when the waves reach them, but the knots do not move along with the wave. This is called a **transverse wave**.

The water particles in the puddle also seem to move up and down, but studies have shown that they actually travel in a small circle. Fig. 7-3 shows the motion of an individual water molecule as the wave passes it. As each particle starts its circular motion, it causes the next particle to move, and so on. In this way, energy is passed on.

A second type of wave is called a **longitudinal wave**. In this type of wave, the particles are pushed or pulled back and forth in the direction the wave is traveling. They are squeezed together, then stretched apart. This is somewhat like a long line of students waiting for lunch. If someone at the end of the line starts to shove, the en-

Transverse wave: A wave in which the molecules vibrate at right angles to the direction of the wave.

Longitudinal wave: A wave in which the particles vibrate back and forth in the direction of the wave.

7-3. Diagrams 1 through 5 show the forward motion of a single wave (AB). The dotted lines show the circular movement of an individual particle.

Chapter 7-1 Energy and Waves

169

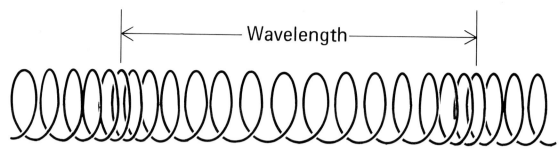

7-4. *In a longitudinal wave, particles are squeezed together, then pulled apart.*

ergy is passed up the line. The student who was shoved bumps into the next student, who bumps into the next student, and so on. After being shoved, each student returns to their original position. The students are squeezed together then stretched apart.

This can be demonstrated with a screen door spring. If the spring is first stretched slightly and then a few coils are squeezed together, then released, the wave is transferred down the length of the spring. Each coil squeezes against the next coil, then returns to its original position, stretching the coils out again. See Fig. 7-4.

All types of waves have the same properties and they can be described in terms of these properties. The first property of waves is called **wavelength.** In the *longitudinal* type of wave, the *wavelength* is the distance between two neighboring areas of squeezing or stretching. Long wavelengths are usually measured in meters and short wavelengths in centimeters.

Wavelength: The distance between two successive identical points on a wave.

Think again about the bob on the fishing line. As the waves pass it, it rises to the high point on the wave. Then it drops to the low point. Finally, it rises to the high point on the next wave. The high point on each wave is called the *crest*. The low point between crests is called the *trough*. The complete wave consists of a crest and a trough. In this type of wave, the wavelength is the distance between two crests or two troughs. See Fig. 7-5.

Imagine a pond without waves or ripples. The water surface is completely flat. This is the normal level of the water. If waves form, the crests rise above the normal level. The troughs drop below it. The distance the crest rises above the normal level is called the **amplitude** of the wave. The trough falls the same distance below the normal level. *Amplitude* is the second property of waves. Twice the amplitude is called the *height* of the wave. This is the total distance from the bottom of the trough to the top of the crest.

Amplitude: The strength of a wave. One half the distance from the bottom of the trough to the top of the crest.

Unit II How Does Energy Affect Matter?

7-5. *A transverse wave showing the properties of wavelength, amplitude, and height.*

Amplitude is a measure of the strength or energy a wave has. Think again about the example of the rope tied to the doorknob. The harder you shake the rope, the larger the resulting wave will be. The greater the energy applied to the rope, the greater the amplitude of the wave.

Frequency: The number of complete waves that pass a given point in one second.

A third property of waves is called **frequency.** The *frequency* of a wave is a measure of how many wave crests pass a given point in one second. Frequency is measured in units called **hertz.** This unit is named after a scientist who led the study of certain types of waves. One *hertz* equals a frequency of one complete wave passing a given point in one second.

Hertz: A unit used to measure frequency of a wave. One hertz means that one complete wave passes a given point each second.

A fourth property of waves is how fast they travel. *Speed* is equal to the frequency of the wave times its wavelength. Speed is usually measured in meters per second (m/sec). Sound waves travel at 330 m/sec in air. Light travels at over 300 million m/sec.

Materials
Slinky or similar spring
Pencil
Paper
Yarn

A. Obtain the materials listed in the margin.

B. Work with another student on this activity. Place the spring on a table or the floor. Stretch it fairly tightly to a two meter length. This will be the normal position.

C. One student snaps the end of the spring from side to side quickly. A complete snap is from the normal to a point 0.5 m to the left, back to a point 0.5 m to the right, then back to the normal position. Observe the wave as it travels down the spring. Repeat this until you are sure you know the form of the wave.

D. Snap the spring *twice* without stopping. Make a sketch of these waves on your paper.

Chapter 7–1 Energy and Waves

1. Label the crests and troughs of the waves.

2. Indicate one complete wavelength on your sketch.

3. Show where the amplitude and height of the wave would be measured.

E. Tie a piece of yarn in the middle of the spring. Snap a single wave on the spring and watch the yarn carefully.

4. On your diagram, show how the yarn moves as the wave passes it.

5. What is this type of wave called? Explain.

6. What happens when the wave reaches the other end of the spring?

7. Why does the wave eventually "fade out"?

F. Lift the spring off the floor. Stretch it out just enough so that it does not sag.

G. Have your partner pinch together the first 10 coils of the spring, then release them quickly. Observe this wave carefully.

8. How is this wave different from the first type?

9. What happens to the yarn in this case?

10. What is this type of wave called? Define this type of wave.

SUMMARY & QUESTIONS

Waves transfer energy from one point to another. All waves can be described by the same basic properties. These properties include wavelength, amplitude, speed, and frequency. The frequency of waves is measured in units called hertz.

1. How does the movement of particles in a transverse wave differ from movement in a longitudinal wave?
2. What is meant by the wavelength of a wave?
3. What is meant by the frequency of a wave?
4. Explain the term *amplitude*.
5. What property of waves is measured in hertz? What does two hertz mean?
6. A wave has a wavelength of 5 meters and a frequency of 4 hertz. What is its speed?

2 WAVE MOTION

A ball bounces off a wall. You see yourself in a mirror. A shout echoes in a gym. A laser beam from Earth bounces off a reflector left on the moon. These examples show one way energy waves can behave when they strike a barrier. Do you know what this behavior is called? When you finish this section, you will be able to:

● Explain and give examples of the *reflection* of waves.

● Describe and give examples of the *refraction* of waves.

● Discuss the *Doppler effect*.

▲ Identify examples of *reflection* and *refraction*.

Reflection: The bouncing of an energy wave off a barrier that does not absorb all the energy of the wave.

Have you ever watched or played a handball or racquetball game? A good player knows how to use the walls to score. Such a player knows that if there is no spin, the ball will bounce off the wall at the same angle at which it hit the wall. Scoring depends on using these angles to put the ball where the opponent cannot reach it.

Waves bounce off barriers in the same manner as a ball does. This is called **reflection.** Sound waves bounce off a wall, producing an echo. Light waves bounce off mirrors producing *reflections*. Water waves bounce off walls and rocks. The wave traveling down a rope tied to a post will bounce back from the post. In each case, some of the energy is absorbed by the barrier. However, most of the energy is reflected.

If the wave strikes a barrier straight on, the reflected wave will bounce straight back the way it came. See Fig. 7-6 on page 174. If the wave strikes the barrier at an angle, it bounces off at the same angle in the opposite direction. See Fig. 7-7 on page 174.

Think of a wave, striking a smooth barrier and reflecting from it. The result is shown in Fig. 7-8. Notice how the direction of the wave is changed in a predictable way.

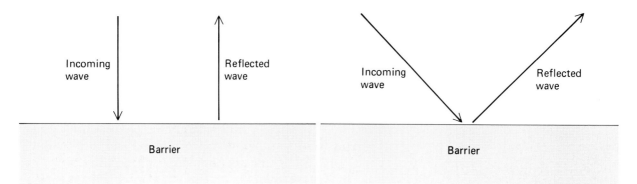

7-6. If a wave strikes a barrier straight on, it reflects back the way it came.

7-7. (right) A wave bounces off a barrier at exactly the same angle as it strikes the barrier.

Refraction: The bending of a wave due to a change in its speed.

7-8. A smooth barrier gives a uniform but reversed reflection of waves.

7-9. (right) A rough barrier scatters wave reflections in many directions.

However, if the wave strikes a rough barrier, the wave is bounced away in many directions. See Fig. 7-9. The funny mirrors in amusement parks are based on this idea. They have curved surfaces that reflect the light waves in different directions. This causes your reflection to be stretched out of shape.

Waves can also be **refracted.** *Refraction* is the bending of waves when they speed up or slow down. For example, look at the pencil in the glass of water in Fig. 7-11. The pencil is not really bent. Why does it look bent? Because the light waves that allow you to see the pencil bend when they enter the water.

Refraction of some waves depends on the density of the materials the wave is passing through. Sometimes a wave passes at an angle from a material of low density into a material of high density. This causes the wave to slow down and to bend. If a wave passes from a high density to a low density, it speeds up. It will then bend in the opposite direction. See Fig. 7-10. The light waves reflecting off the pencil in the water behave this way. As they pass from the more dense water into the less dense air they are refracted. This makes the pencil appear bent.

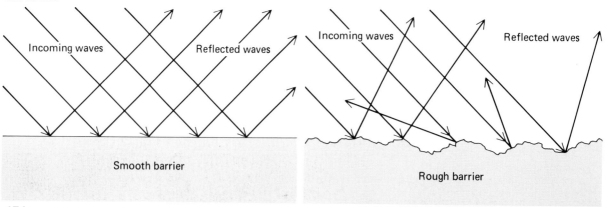

Unit II   How Does Energy Affect Matter?

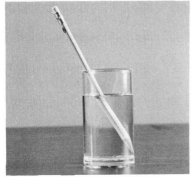

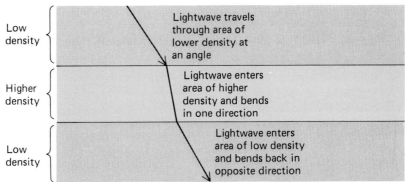

7-10. *Refraction of light waves makes this pencil appear bent (left). This is because waves refract as their speed changes. See diagram (right).*

Pitch: The level of a sound, either high or low. Pitch is related to frequency.

Doppler effect: An apparent change in the frequency of waves that comes from the fact that the observer or the source of the wave is moving.

7-11. *The Doppler effect. As you get closer to the source of the sound, the pitch seems to increase. As you get farther away, the pitch seems to decrease.*

As waves in the ocean approach the shore, they enter shallow water. The wave particles drag on the bottom and slow down. This causes the wave to "break." If the wave approaches the shore at an angle, the end closest to the shore slows down first. As a result, the wave bends toward the beach. This is a type of refraction.

Sometimes the frequency of waves can appear to change. This is best shown by sound waves. Imagine a train is approaching you with its whistle sounding. Since the train is moving toward you, the crest of each sound wave has less distance to travel to reach you. Thus, each sound wave reaches your ear faster. As a result, the sound waves crowd together. Even though the whistle is giving off sound waves at a steady frequency, the frequency at which they are reaching your ear is increasing. This causes you to hear the sound at a higher level or **pitch** as the train approaches.

As the train moves away, the *pitch* becomes lower. Each wave crest has farther to travel to reach you. The waves are stretched farther apart. The frequency seems to decrease. This is known as the **Doppler** (dop-lur) **effect.** How might you notice a Doppler effect if you were rowing a boat against the current in choppy water?

Chapter 7–2 Wave Motion

ACTIVITY

Materials
Pencil
Paper
Ruler

A. Obtain the materials listed in the margin.

B. Copy each of the diagrams shown in Fig. 7-12 onto your paper. DO NOT WRITE IN THIS BOOK.

C. Hidden beneath each of the squares shown in Fig. 7-12 is a material that either reflects or refracts the light beams entering from the left. Each beam is numbered. The arrows with the same number show where the entering beam comes out.

D. In each case, show how the material hidden under the square must be arranged to produce the results shown.

E. Label each example as either reflection or refraction. In the case of reflection, state whether the hidden surface is smooth or rough.

7-12.

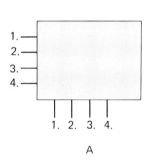
A

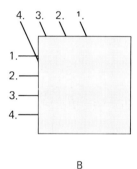
B

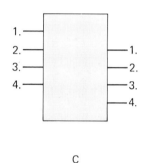
C

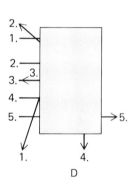
D

SUMMARY & QUESTIONS

All types of waves can be reflected or refracted. Waves bounce off barriers at exactly the same angle with which they strike the barrier. Waves are refracted or bent when they slow down or speed up. If the source of the wave or the observer is moving, the frequency of the waves seems to change.

1. Define reflection. Diagram a reflecting wave.
2. Define refraction. Diagram a refracting wave.
3. Explain why waves often bend as they approach a beach at an angle.
4. Why does the level of a sound increase as you approach it and decrease as you move away? What is this called?

3 SOUND WAVES

You have probably seen movies or television programs dealing with humans in space in the future. You have seen models of spacecraft large enough to hold hundreds of people traveling between planets or solar systems. Many of these programs are very accurate in detailing what scientists feel space travel will be like in the future. But one detail of these programs is not accurate. They portray space in which there is sound. When you finish this section, you will be able to:

● Describe how *sounds* are produced.

● Explain the properties of sound waves.

● Distinguish between *music* and *noise*.

▲ Demonstrate several properties of sound waves.

Sound: A wave consisting of vibrations of the particles of the material through which the wave passes.

Sound is a type of energy that is transferred by wave motion. *Sound waves* are started by back and forth motion of an object. This motion is called vibration. If you pluck the string of a guitar, it starts moving back and forth rapidly, making sound. We say it is vibrating. The

7-13. *Why does blowing across the mouth of a bottle produce a sound?*

end of a ruler sticking over the edge of your desk will vibrate if you snap it. This will also make a sound. If you blow across the mouth of a soda bottle, the air inside will vibrate, making a sound. In short, all vibrating objects can start sound waves. However, the vibrating object must be in contact with some material that will carry or transmit the sound waves.

For example, as a rocket is launched, the motors vibrate. These vibrations cause molecules in the air to vibrate. Energy is thus passed through the air in the form of longitudinal waves called sound waves. Eventually, these waves reach the ears of people watching the launch and are translated into sound.

To create sound waves, molecules must vibrate. Vibrating molecules give off energy that then causes neighboring molecules to vibrate. This type of chain reaction causes sound waves to be transmitted. On Earth, the molecules of the atmosphere are close enough together for this chain reaction to take place. Therefore, sound can travel through the air. However, in outer space the molecules in the atmosphere are very far apart. The energy of vibration cannot be transmitted to neighboring molecules. The atmosphere is not dense enough to transmit sound waves. So in space, the sound of a spacecraft engine would not be heard by people in another spacecraft. These people would hear the sounds made within their own ship because there is air inside to transmit the sound waves. However, no sounds can travel through the emptiness of space.

What are sound waves like? Each time a vibrating guitar string moves upward or downward, it squeezes the air molecules on that side closer together. These molecules crowd into the molecules next to them causing them to be squeezed, and so on. When the string moves in the

7-14. *Crests and troughs of transverse waves are like the squeezed and stretched areas of longitudinal waves.*

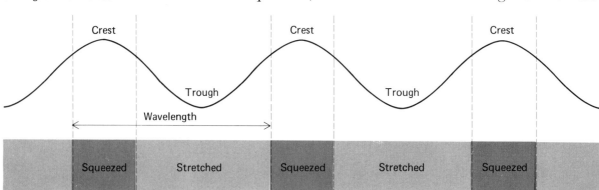

opposite direction, it leaves an area into which the crowded molecules can spread out. As the string continues to vibrate, a series of waves are created by the crowding together and spreading apart of the air molecules. These are sound waves. Sound waves, then, are a longitudinal type of wave. Remember, the individual molecules only vibrate back and forth in a very small space. It is the energy of the wave that is passed on and spread out in all directions. Each time air molecules are crowded together and then spread apart, one complete wave is created. See Fig. 7-14.

All sounds are caused by something that vibrates. Your vocal cords vibrate as you speak. Strings or other surfaces on violins, pianos, and drumheads vibrate as they are struck or plucked. Some wind instruments have a thin reed in the mouthpiece that vibrates as air is blown past it. The lips of a trumpet player vibrate as air passes between them.

Sound waves have all the properties of other waves. The distance between identical points on two neighboring waves is the *wavelength*. The *frequency* is determined by the number of vibrations the object makes each second. For example, if the lips of a trumpet player are tightened they vibrate faster as the air passes between them. The frequency of vibration increases and so does the frequency of the sound waves created. This causes the *pitch* of the sound to get higher. Most humans can hear sound with a pitch between 60 and 10,000 hertz. Some can hear sounds as low as 20 or as high as 20,000 hertz.

Loudness: The amplitude or amount of energy contained in a sound wave.

If a piano key is struck gently, the sound produced is soft. If the same key is struck harder, the amplitude of the vibration is greater. The pitch does not change but the **loudness** does. We found earlier that amplitude depends on the amount of energy in a wave. If you strike the key with greater energy, the piano string vibrates with greater energy. The amplitude of the wave is greater. Thus, the sound is *louder*.

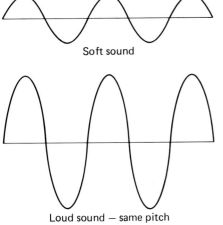

7-15. *Loudness is determined by the amplitude of the waves.*

Soft sound

Loud sound — same pitch

The sounding board of a piano or the box of a guitar or violin serve to increase the amplitude of the sound waves and make them louder. The vibrations of the strings are transferred to the walls of the box. The walls vibrate at the same pitch but with a greater amplitude. Thus, the sounds are louder.

Sound waves will travel through any material that will

Chapter 7-3 Sound Waves

How We Hear Sound

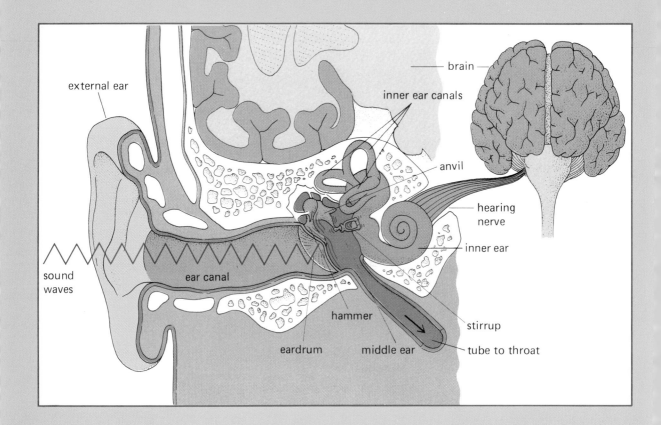

Sound waves enter the ear canal and travel toward the eardrum. When these waves reach the eardrum, they cause it to vibrate. Low tones produce a slow vibration while high tones produce a rapid vibration. These vibrations are then transmitted to the bones of the middle ear: the hammer, the anvil, and the stirrup. These bones act as a lever; the joint between the anvil and the stirrup serves as the fulcrum. Therefore, the vibrations sent from the stirrup to the inner ear are magnified just as effort force is magnified by a lever. This causes a buildup of pressure in the inner ear. Pressure waves are then sent toward the hearing nerve. At the nerve, the pressure waves are changed into electrical impulses, or signals, which travel to the brain. The brain of a normal adult is able to distinguish about 400,000 signals and translate these into different sounds.

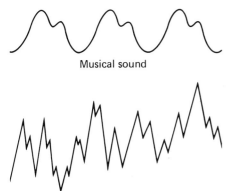

7-16. *Musical sounds form a regular pattern of sound waves. Noise forms an irregular pattern.*

Music: Any tone produced by a regular pattern of vibrations.

Noise: Sounds produced by irregular vibrations.

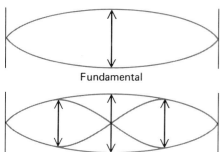

7-17. *A vibrating string.*

vibrate. However, some materials transmit sound better than others. Sound waves travel through steel at 5,200 m/sec. Water transmits sound at about 1,420 m/sec. Air is less dense so sound waves travel through it at about 340 m/sec. Some materials do not vibrate well, therefore, they can trap or slow down sound waves. These materials, such as cloth and other fibers, can be used to make soundproofing tiles.

Have your parents ever said that the **music** you were listening to was not music but **noise**? They probably said this because they did not like the way it sounded. However, there is a scientific difference between noise and music. To the scientist, music is a regular pattern of vibrations, noise is an irregular pattern. Music also has *pitch* and *quality* that are not found in noise.

If a tuning fork is struck, it vibrates as a whole, creating a sound wave of a certain frequency. If a guitar string is plucked, it vibrates as a whole, but it also vibrates in parts. This produces waves of different frequencies. The frequency produced by the vibration of the string as a whole is called the *fundamental*. The frequencies produced by the vibration of the parts are called *overtones*. The blending of the fundamental with the overtones gives music the property of quality not found in noise. Different instruments produce different overtones when playing the same note. This is why a violin sounds different from a guitar and a trumpet sounds different from a tuba even when they are playing the same tune.

Materials
Wood, 22 cm × 3 cm × 1 cm
2 nails
Elastic band
Pencil
Paper
Hammer
Goggles

A. Obtain the materials listed in the margin.

B. Draw a 20 cm line on the wood. Mark it at 0, 5, 10, 15, and 20 cm. CAUTION: WEAR GOGGLES.

C. Drive a nail into the wood at the 0 mark and another at the 20 cm mark.

D. Loop the elastic around one of the nails. Twist several times before hooking it around the other nail. The band should be about 3 cm above the wood.

E. Hold the wood in your hand. Grasp the elastic at the midpoint and pull it about 2 cm to one side. Release the elastic.

1. Record your observations.

F. Place the wood on your desk and hold it firmly against the surface. Repeat step E.

2. What, if any, difference did you notice.

3. Compare this result with the purpose of a guitar box.

G. Keep the wood pressed against your desk. Pull the elastic 2 cm to one side and release it. Pull the elastic about 5 cm to the side before releasing it.

4. Compare and contrast the two sounds.

5. What did you notice about the size of the vibration of the elastic in each case.

6. What is this property of a wave called?

H. Pluck the elastic in the middle and listen to the sound. Hold the elastic at the 5 cm mark. Pluck the larger part of the elastic.

7. What do you notice about the vibration of the elastic?

8. What did you notice about the level of the sound?

I. Hold the elastic at the 10 cm and then the 15 cm marks. Pluck the smaller portion of the elastic in each case.

9. What did you notice about the vibrations?

10. What did you notice about the sound?

11. What two properties of sound waves were changing in these examples? Define each.

12. Make a general statement that shows the relationship between frequency and pitch.

SUMMARY & QUESTIONS

Sound energy is carried by waves through some substance that can vibrate. Sound waves have the same properties as all other kinds of waves. Music is a regular pattern of vibrations while noise is an irregular pattern. Musical instruments add overtones to the sound they make, which make it easier to identify the instrument's sound.

1. Describe two conditions necessary for the production of sound.
2. Name and explain four properties of sound waves.
3. Explain how sound waves travel through air.
4. Name and describe the three properties of music that are not found in noise.

VOCABULARY REVIEW

Match the description in Column II with the term it describes in Column I.

Column I
1. wave
2. transverse
3. wavelength
4. frequency
5. refraction
6. reflection
7. Doppler effect
8. longitudinal
9. pitch
10. music

Column II
a. An apparent change in frequency of waves because observer or source is moving.
b. Waves bouncing back after striking a barrier.
c. A regular pattern of vibrations.
d. A wave in which particles travel back and forth in the direction of the waves.
e. The distance between identical points on two neighboring waves.
f. A wave in which the particles vibrate at right angles to the direction of the wave.
g. A disturbance caused by energy moving from one place to another in a substance without disturbing the substance permanently.
h. A characteristic of a sound wave that is closely related to frequency.
i. The number of complete waves that pass a given point each second.
j. A process that causes a wave to change its direction because its speed changes.

REVIEW QUESTIONS

Choose the letter of the ending that best completes each of the following statements.

1. Energy moving across water, rope, or spring is best seen (a) by the work it does (b) as waves (c) as the circular motion of the particles of water, rope, or spring (d) by the back-and-forth motion of the particles of water, rope, or spring.
2. The property of a wave that is a measure of the strength of the wave is the (a) crest (b) wavelength (c) amplitude (d) frequency.
3. An irregular pattern of vibrations makes a sound called (a) music (b) pitch (c) the Doppler effect (d) noise.
4. A water wave is an example of a (a) transverse wave (b) Doppler wave (c) reflection wave (d) longitudinal wave.
5. When a wave changes direction because of a barrier in its path, the change is called (a) Doppler effect (b) reflection (c) refraction (d) sound-proofing.

6. A sound wave is an example of a (a) transverse wave (b) water wave (c) longitudinal wave (d) refraction wave.
7. The particles of a substance that is transmitting sound waves (a) move up and down as the waves travel (b) carry a great amount of energy (c) do not move (d) move back and forth as the waves travel.
8. It is not difficult to recognize that the sound of a guitar is different from that of a piano since (a) their pitches differ (b) one is louder than the other (c) the mixture of fundamentals and overtones each produces is different (d) a piano is larger than a guitar.
9. The pitch of a sound is increased if (a) the amplitude is increased (b) the frequency is increased (c) the frequency is decreased (d) the amplitude is decreased.
10. An increase in the loudness of a sound is due to an increase in (a) pitch (b) amplitude (c) frequency (d) wavelength.

REVIEW EXERCISES

Give complete but brief answers to each of the following.

1. Describe two types of particle motion possible during wave motion.
2. Draw a transverse wave. Label the wavelength, crest, trough, amplitude, and height. What is meant by the frequency of this type of wave?
3. Describe how a water wave would change when:
    a. it strikes a barrier in its path.
    b. it enters shallow water.
4. Explain how a guitar string can cause sound and how that sound reaches your ear.
5. Compare the properties of sound waves to the properties of other types of waves.
6. Explain how the motion of a cork floating on water is similar to the motion of the individual water particles.
7. Explain how pitch is related to the frequency of sound.
8. Explain three ways in which music and noise differ.
9. Why does the pitch of a siren seem to increase as you get closer to it and decrease as you move away from it?
10. Why do guitar players put their fingers on the strings at different places?

CHAPTER 8

LIGHT

1 BEHAVIOR OF LIGHT

Often, a doctor needs to look inside a person's body to find out what is wrong with the person. The doctor can do this by means of an X-ray machine. X rays are a form of energy similar to light. In this section, you will learn about light and other similar forms of energy. When you finish section 1, you will be able to:

- Use examples to show that light can behave as if it is made of waves.

- Explain what is meant by the *electromagnetic spectrum*.

▲ Name and give characteristics of five types of waves similar to light.

Some of you might try this experiment. Make a small pinhole in an index card. Look through the pinhole at a light some distance away. If the hole is very small, you will see the light spread out. That is, the light will appear to be larger than the pinhole through which it is coming. This strange effect can be explained if light is thought of as a series of waves.

The light will appear spread out because the light waves are **diffracted** when they pass through the pinhole. This is shown in Fig. 8-1. Light seems to be a result of energy moving in the form of waves.

Light has many of the properties of waves, such as *diffraction*, reflection, and refraction. See Fig. 8-2. Light waves are different from other waves, however, in some important ways. For example, unlike sound, light can move through empty space. See Fig. 8-3. Light waves also move faster than sound waves. The speed of light in a vacuum is 3×10^8 m/sec. Light could travel a distance equal to more than seven complete trips around the earth in 1 sec.

Waves which can move through empty space at the

Diffraction: The ability of waves to bend around an obstacle in their path.

8-1. *The light from a distant source will spread out around the pinhole.*

10^8 is a way of writing a number which is equal to 1 followed by 8 zeros. $10^8 =$ 100,000,000.

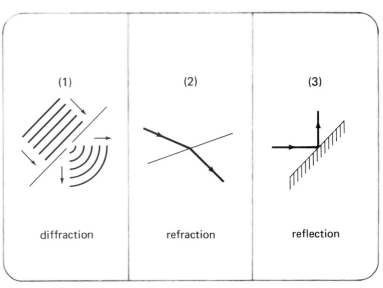

8-2. *Waves can be diffracted, refracted, or reflected.*

Electromagnetic waves: A form of energy able to move through empty space at very high speed.

Recall the definition of frequency: The number of complete waves passing a point in 1 sec.

Electromagnetic spectrum: The series of waves with properties similar to light.

speed of light are called **electromagnetic** (e-**leck**-tro-mag-net-ick) **waves.** Light waves are not the only form of *electromagnetic waves.* Radio waves also move through empty space and travel at great speed. Experiments show that light and radio waves travel through space at the same speed. So, radio waves are also electromagnetic waves. There are many other electromagnetic waves. For example, microwaves used to cook food are electromagnetic waves.

Electromagnetic waves are usually described by their frequency. All of them have rapid frequencies, ranging from 10^4 to 10^{21} hertz (Hz). Light has a frequency of about 10^{14} Hz. The frequency of microwaves is about 10^9 Hz.

All types of electromagnetic waves known to exist are called the **electromagnetic spectrum.** A chart of the *electromagnetic spectrum* is shown in Table 8-1. The major parts of the electromagnetic spectrum are:

1. *Radio waves* cover a low frequency range from about 10^4 to 10^{12} Hz. Ordinary radio broadcasts use frequencies of 5.5×10^5 to 15×10^5 Hz. Television and FM radio operate at higher frequencies up to about 3×10^{10} Hz. Radar and microwaves use frequencies between 10^6 and 10^{12} Hz.

2. *Infrared waves* have frequencies between 10^{12} and about 10^{14} Hz. Infrared waves striking an object cause it to heat up.

3. *Light* falls into a narrow band of frequencies between 4×10^{14} and 8×10^{14} Hz. We see all electro-

8-3. *There is no air on the moon. On the moon, you would still be able to see by reflected sunlight. However, you would not be able to hear a sound.*

TABLE 8-1

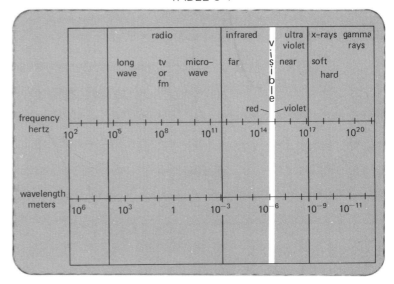

magnetic waves of this frequency as visible light.

4. *Ultraviolet waves* have frequencies between 8×10^{14} and 10^{16} Hz. Ultraviolet waves in sunlight give you a suntan.

5. *X rays* have frequencies from about 10^{16} to 10^{19} Hz. X rays are used by dentists to take pictures of teeth or doctors to take pictures of the inside of your body.

6. *Gamma rays* have frequencies from about 10^{11} to 10^{21} Hz. Gamma rays are very much like X rays and are very dangerous.

Materials
copy of word game
pencil

The words in this activity are all terms used in describing the electromagnetic spectrum. Follow the directions carefully. Read the clues carefully.

A. Copy the puzzle on a separate sheet of paper.

B. Spell out the words described by the following numbered clues. Write the letters of each word in order in the blank spaces after the number of the clue. When you have correctly identified the clues, you will spell out the key word in the column shown. DO NOT WRITE IN THIS BOOK.

CLUES

1. The series of waves to which light belongs is called the electromagnetic ____.
2. Waves that cause objects to become heated when they strike them.
3. Waves that cause suntan.
4. The term whose unit is hertz (Hz).
5. Waves that can be used for cooking.
6. The waves we use to see.
7. Waves which are used by dentists to examine teeth.
8. The waves we use to see. (Same as number 6.)
9. Waves used for AM and FM broadcasts.
10. The name for waves including all of the waves above.
11. What is the key word?
12. How is the key word related to the electromagnetic spectrum?

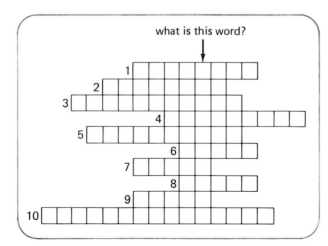

SUMMARY & QUESTIONS

Light has some characteristics which show that it is made of waves. Light waves can be diffracted, reflected, and refracted. Light can also move through empty space and travel at high speeds. Waves like x rays with properties similar to light waves make up the electromagnetic spectrum.

1. Describe one experiment which you could do to show that light behaves as waves.
2. Which letter of Fig. 8-4 matches each of the following kinds of electromagnetic waves?
   a. visible light   d. ultraviolet rays
   b. X rays   e. gamma rays
   c. infrared rays   f. radio waves

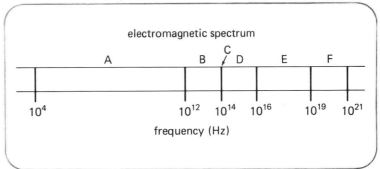

8-4.

2 MOVEMENT OF LIGHT WAVES

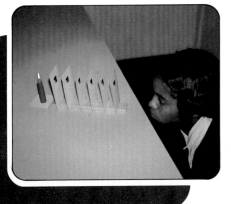

The picture shows several index cards with holes in them in front of a candle. If all the holes are lined up, you can look through all of them at once and see the candle. If one card is moved to the side, however, you will not be able to see the candle. You might try this experiment yourself. What do you think this experiment proves about light waves? When you finish section 2, you will be able to:

● Show that light waves ordinarily move in straight paths.

- Show that light may change its direction when it passes from one material to another by giving two examples.

- Explain how the bending of light waves can produce *mirages*.

- Demonstrate, using a glass of water, the fact that light may change its direction when it passes from one material into another.

8-5. *Sunlight would not appear like this unless light traveled in straight lines.*

Light rays: Straight lines showing the path followed by light.

Suppose you are standing on one side of an open ball field. You can see one of your friends standing on the other side of the field. You want to attract your friend's attention. It is too far to shout and be heard. Can you think of an easy way to solve this problem? You might use a mirror. You could reflect the sun's light from a mirror into your friend's eyes. This should get your friend's attention! Why can you easily direct the reflected light at your friend?

Light seems to move in a straight path as it travels through air. See Fig. 8-5. Because light travels in a straight path, you can easily direct reflected light. The motion of light can be shown by straight lines called **rays**. The *rays* show the path that light follows as it moves through a substance such as air. See Fig. 8-6. What do you think happens to light rays as they move from water into air?

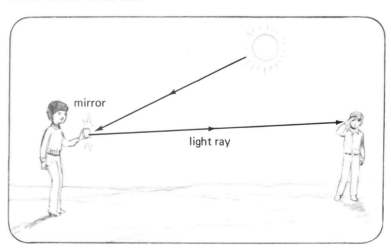

8-6. *Because light travels in straight lines, it can easily be reflected by a mirror.*

Unit II How Does Energy Affect Matter?

Materials
2 glasses
Rubber band
Coin
Water
Pencil
Paper

A. Obtain the materials listed in the margin.

B. Fill one of the glasses about ¾ full of water.

C. Put a rubber band around the other glass about 4 cm from the bottom.

D. Place a coin in the bottom of an empty glass, making sure it is in the center. Hold the coin in place with a pencil. See Fig. 8-7.

8-7.

8-9.

E. Move your head until you see the edge of the coin lined up with the rubber band. Figure 8-8 shows the path followed by the light from the coin to your eye.

F. Watch the coin while you pour the water from the first glass into the glass with the coin. See Fig. 8-9. Don't move your head while pouring. Always look at the coin through the top surface of the water.

1. Does the coin appear to move?

Part of the path of the light which allows you to see the coin is shown in Fig. 8-10. Only the path from the water surface to your eye is shown.

G. Copy Fig. 8-10 on a sheet of paper. Now draw in a line to show the path the light must follow from the coin to the water surface.

2. Does putting water in the glass change the path of the

8-8.

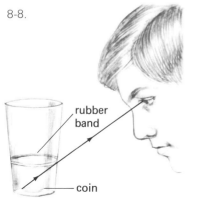

8-10.

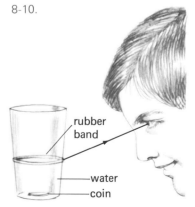

Chapter 8–2 Movement of Light Waves

light from the coin to your eye? If so, how is the path changed?

Try another experiment to see if the direction of light changes in going from water to air.

H. Put a pencil at an angle into the glass of water.

I. Look along the length of the pencil.

3. Does the pencil look straight?

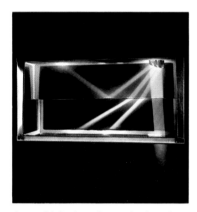

8-11. *Light is refracted when it moves from one material, like air, into another, like water.* (Courtesy Time/Life Books)

Mirage: An illusion caused by the refraction of light in which distant objects are seen upside down or floating in the air.

When light waves move from air into water, their speed changes. Light waves travel a little slower in water than in air. If the rays of light enter the water at an angle, they are *refracted* or *bent* into a new path. See Fig. 8-11. Refraction of light makes objects under water appear to be closer to the surface than they really are. See Fig. 8-12. Refraction of light also makes water seem not so deep as it really is.

Refraction of light explains the appearance of **mirages** (mear-**ahjes**). *Mirages* are illusions or false impressions. These illusions cause distant objects to seem to be upside down or floating in the air. Mirages occur when

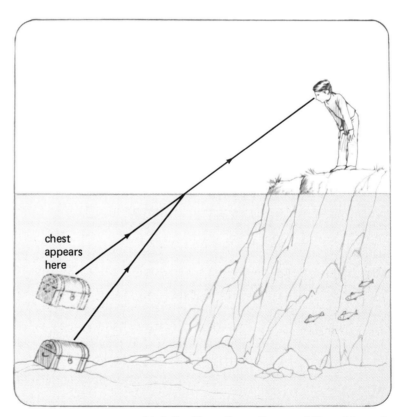

8-12. *Anything seen under water is really lower than it seems because the light rays reflected from the object are refracted as they move from the water into air*

8-13. *An inferior mirage.*

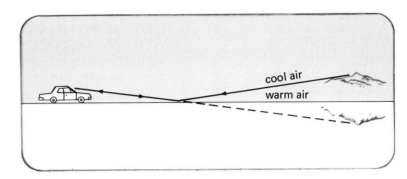

8-14. *A superior mirage.*

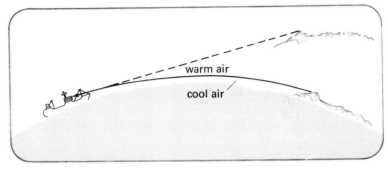

light is refracted in passing through air layers of different temperatures. On a warm day, the air close to the ground is warmer than air at higher levels. Light is refracted by the warm air. This refraction makes distant objects appear upside down. See Fig. 8-13. Sometimes the air near the ground is cooler than the air at higher levels. When this happens, a mirage is formed as shown in Fig. 8-14.

SUMMARY & QUESTIONS

The experiment with the index cards and the candle shows that light moves in a straight path. You can reflect light from a mirror easily because light travels in a straight path. Light moves in straight paths in any particular substance. A ray is a straight line showing the path followed by light rays. When light passes from one substance to another, its speed changes. This change in speed causes the light to be refracted. Refraction of light explains the appearance of objects under water and mirages.

Chapter 8–2 Movement of Light Waves

1. Give several examples which show that light waves travel in a straight line.
2. Explain why the depth of water seems to be less than it really is.
3. Why do mirages appear?

3 COLOR

The phenomenon shown is a rainbow. You may have seen a rainbow right after a rain shower when the sun comes out. The study of colors, like those in the rainbow, is an important part of the study of light. When you finish section 3, you will be able to:

● Interpret diagrams of the *spectrum* to show that white light is made of colors.

● Relate the different colors of light to differences between light waves.

● Explain why objects can have many different colors.

▲ Show that white light is made of colors by using a diffraction grating.

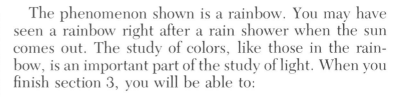

8-15. *Can you imagine how this scene would look if the sky were pink instead of blue?*

Why is the sky blue? Where do the different colors come from at sunrise and sunset? The answers to these questions involve the properties of light. All color comes from light. The bright blue of the sky and the reds and yellows of sunrise and sunset are the result of white light from the sun being separated into different colors.

8-16. *Why does the light from the sun appear to have a different color at sunrise and sunset?*

A simple experiment can be performed to show that white light is made of different colors. This experiment was first performed by the great scientist Sir Isaac Newton. See Fig. 8-17 on page 196.

When a ray of light passes from air into glass, its speed is slowed. This change in speed causes the ray to be bent, or *refracted*. The amount of bending of the ray depends on its frequency. See Fig. 8-18. Light rays with high frequency are bent the most. A specially shaped clear material called a **prism** will spread out the rays according to how much they are bent. White light passed through a *prism* is separated into a band of colors. This band of colors is called the **visible spectrum**. A *visible spectrum* produced by a prism is shown in Fig. 8-19 on page 197.

Prism: A specially shaped piece of clear material. A prism divides white light into its separate colors.

Visible spectrum: The band of colors produced when white light is divided into its separate colors.

Chapter 8–3 Color   **195**

8-17. *Sir Isaac Newton performed many important experiments with light.*

All light that we see contains color. As Newton's experiment with the prism showed, white light is all colors added together. Sunlight, for example, ordinarily appears white. Sometimes you can see the colors in sunlight spread out in a rainbow. A rainbow is made when droplets of water from rain act as natural prisms. Each raindrop spreads out the white sunlight into a tiny spectrum. As we look up from the ground at a rainbow, we can see only one color from each little spectrum. Looking highest, we can see only the reds. As a result, the top of the rainbow is seen as a band of red color. Below the red band we see bands of orange, yellow, green, blue, and violet. All of these colored bands have a curved shape because of the round shape of the raindrops.

It would be easy for you to make a real rainbow. You could use a shower of droplets from a garden hose to take the place of the rain. The experiment should be done in the late afternoon when the sun is about halfway down from directly overhead. With your back to the sun, turn on a fine spray from the hose. Point the spray high in the air. The rainbow can be seen best if you look at it toward

8-18. *If two rays of light, A and B, have different frequencies, they will be refracted by different amounts when passing through glass.*

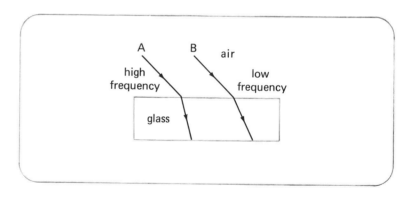

a patch of dark shade. Your small rainbow will be made in the same way as the giant ones you have seen in the sky.

We know from the *electromagnetic spectrum* that violet light has a higher frequency than red light. See Table 8-2. Rays with high frequency are bent more than rays with low frequency when they are refracted by a prism.

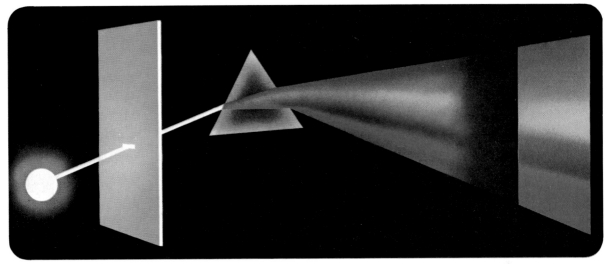

8-19. *A visible spectrum produced by a prism. Does this spectrum remind you of the rainbow on page 194?*

Violet light should be bent more than red light when both are passed through a prism. This is what actually happens. See Fig. 8-20. Passing light through a prism shows that the different colors of light are made up of waves of different frequencies.

TABLE 8-2

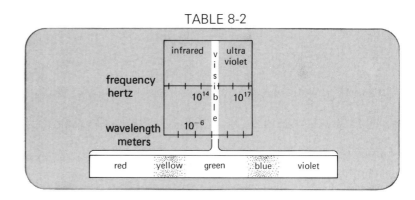

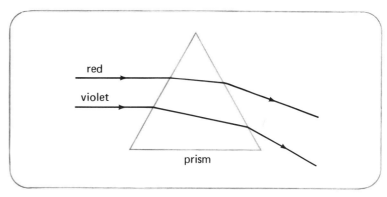

8-20. *Violet light has a higher frequency than red light. Violet light is bent more than red light when both are passed through a prism.*

Chapter 8–3 Color

197

Materials
diffraction grating
pencil
paper

A. Obtain the materials listed in the margin.

B. Look through the plastic diffraction grating at a light bulb.
 The colorful display you see is called the spectrum of visible light.

C. Turn the diffraction grating so that you see all the colors to the left and right of the bulb.
  1. Do the colors repeat themselves to the right of the light?
  2. What is different about the colors to the left of the light?

D. Look at the colors to the right of the light.
  3. Name the colors you see from left to right.

We see light waves of different frequencies as different colors. At midday the sun is high in the sky. The white light from the sun passes through a relatively narrow band of atmosphere. Dust particles in the atmos-

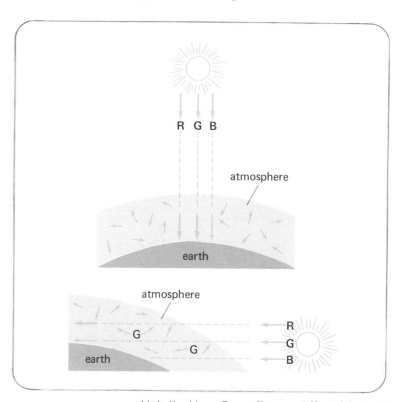

8-21. *The color of the sky is the result of the scattering of different wavelengths of light. At noon, the sky overhead appears blue. At sunset, the sky near the horizon appears reddish.*

phere reflect and scatter mostly blue light waves. The other frequencies of light are not affected as much. This scattering of blue light waves makes the sky appear blue.

At sunrise and sunset, the sun is lower in the sky. White light from the sun comes from an angle and must pass through more of the atmosphere. The additional particles scatter light waves of other frequencies like red and yellow. At sunrise and sunset, the sky near the horizon appears reddish or orange. See Fig. 8-21.

Most light which reaches our eyes has been reflected from the surface of some object. For example, if white light falls upon a red object, all frequencies except red are absorbed. Red light is reflected. See Fig. 8-22. Only the frequency of light we call red is reflected to our eyes. We say the object is red.

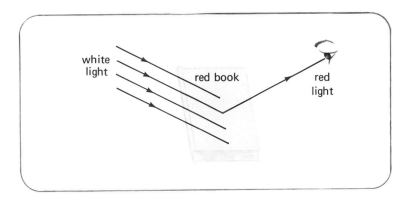

8-22. *A red book reflects only red light.*

Most objects reflect more than one frequency. The color we see is the combination of those colors which are reflected. If light of a particular color is to be reflected, it must be present in the light falling on the object. White light contains all colors. If only red light shines on a red

8-23. *If there is no red light present in the light falling on the red book, no light is reflected. The book will appear black.*

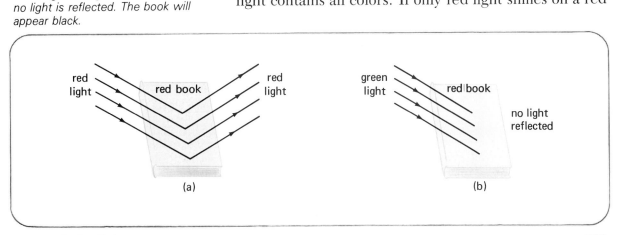

Chapter 8–3 Color

object, the object appears red. If only green light shines on a red object, the object appears black. See Fig. 8-23. No red light is present to be reflected. For example, suppose you wear a red coat in a room in which there is only green light. There is no red light for the coat to reflect. The green light is absorbed. The coat does not reflect any light. The coat will appear black.

SUMMARY & QUESTIONS

White light can be shown to be made of different colors with a prism. Sometimes you can see many of the different colors of sunlight in a rainbow. Each color of light is made of light waves with a particular frequency. The colors we see depend on how the different frequencies of light are refracted or reflected.

Use the following diagram, which represents light passing through a prism, for answering questions 1–3:

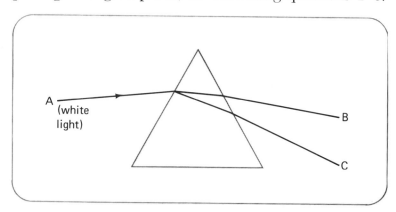

1. If A is a beam of white light, what color will be found at B?
2. If A is a beam of white light, what color will be found at C?
3. How would the results differ if A were a beam of red light?
4. Explain how grass appears green.
5. Why does the sky have a blue color?

4 BENDING LIGHT RAYS

Cameras similar to the one shown first came into use early in the 19th century. Today, modern cameras produce full color photographs in seconds. In this section, you will see how cameras use light to make pictures. When you finish section 4, you will be able to:

● Interpret *lens* diagrams to show that light is refracted in passing through a lens.

● Explain how a lens produces an *image*.

● Name two kinds of lenses and describe their properties.

▲ Use a lens to demonstrate the refraction of light passing through it.

Image: The picture formed by a lens.

Each of your eyes is like a camera. In a camera, light reflected from or produced by an object falls on film which records the picture. In an eye, reflected light is also received. See Fig. 8-25. This light falls on a special nerve which causes the brain to see an image. Both a

8-24. *Developing film in a dark room.*

Chapter 8—4 Bending Light Rays 201

Lens: A piece of transparent material with curved surfaces which refracts light passing through it.

8-26. *All of these instruments contain lenses. We use eyeglasses, telescopes, and microscopes to improve our own vision.*

camera and the eye have an important part that makes the formation of an *image* possible. This important part is a **lens**. A *lens* bends light rays. Light rays are *refracted* by the lens. See Fig. 8-25.

Lenses refract light in different ways, depending on the purpose of the device in which they are used. When the lenses in our eyes do not refract light properly, we need eyeglasses. Lenses in eyeglasses help the lenses in our eyes to produce clear images of what we see. Telescopes use lenses to help us see distant objects clearly. The lenses in microscopes allow us to see small things that are not visible to the eye alone. See Fig. 8-26. We are able to do all this with lenses because lenses bend light rays.

To see how lenses work, look at a simple magnifying glass. Like all lenses, a magnifying glass has at least one

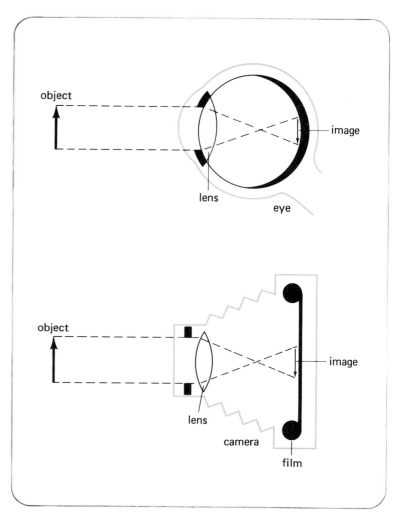

8-25. *(right)*

8-27. *A magnifying glass has at least one curved surface.*

8-28. *A lens bends light rays in the same way as prisms.*

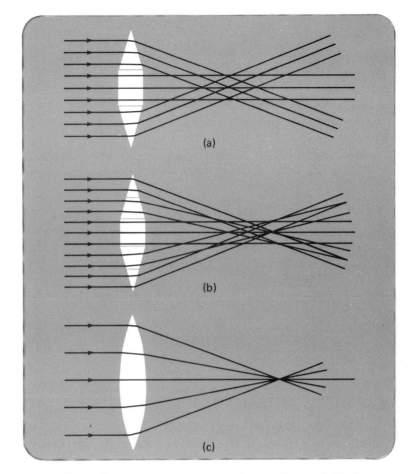

Convex: A lens shape in which the edges are thinner than the center.

8-29. *The refraction of parallel light rays by a convex lens.*

curved surface. See Fig. 8-27. The surface of the lens is curved so that the edges are thinner than the center. This shape is called a **convex** lens. The refraction of light through a *convex* lens can be compared to the way light is refracted by a prism. The convex lens has a shape similar to two prisms combined as shown in Fig. 8-28. The light rays which pass through both prisms meet. The

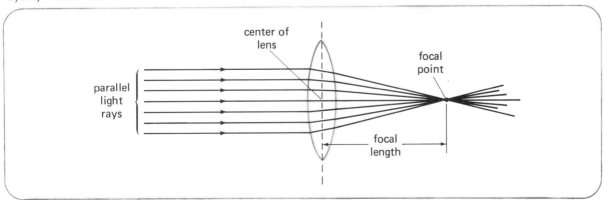

Chapter 8–4 Bending Light Rays

Focal point: The point at which parallel light rays meet after being refracted.

Focal length: The distance from the center of a lens to the focal point.

smoothly rounded surface of a convex lens bends light rays in the same way. However, the lens causes the light rays to meet in a much smaller region. See Fig. 8-28. The point at which parallel light rays are brought together after being refracted by a lens is called the **focal point** of the lens. See Fig. 8-29. The distance from the center of the lens to the *focal point* is called the **focal length**.

Materials
piece of glass
small magnifying glass
piece of white cardboard
paper
pencil

A. Obtain the materials listed in the margin.

B. Lay a piece of flat window glass on top of some printing in your book. While looking through the glass, pick it up and move it slowly away from the printing.

　1. Does the printing appear to get larger, does it get smaller, or does it remain the same size?

C. Lay a small magnifying glass or lens on top of the same printing in your book. While looking through the lens, pick it up and move it slowly away from the printing.

　2. At first, does the printing appear to get larger or does it get smaller?

　3. As you continue to move the lens farther away from the print, does the printing stay clear?

D. Hold a piece of white cardboard upright in front of the windows in your room. See Fig. 8-30.

E. Hold the lens between the

8-30.

Unit II   How Does Energy Affect Matter?

8-31.

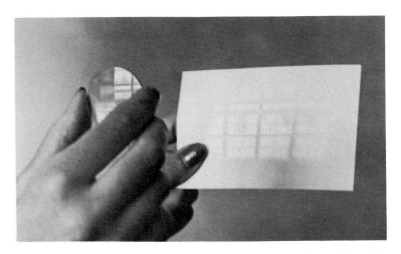

cardboard and the window about 1 cm from the cardboard. Now slowly move the lens away from the cardboard.

F. When an image of the window frames is clear, study it carefully.

4. Is the image larger or smaller than the window itself?

5. Is the image right-side up or is it upside down?

G. Move the lens farther from the cardboard.

6. Does the image stay clear?

H. Now stand on the side of the room opposite the windows.

I. Hold the lens so that the windows are clearly seen on the cardboard. See Fig. 8-31. In this case, the distance from the lens to the cardboard is the *focal length* of the lens.

7. What is the focal length of the lens you are using?

J. Cover 1/2 the lens with a piece of paper or cardboard. See Fig. 8-32. Now try to form an image of the windows.

8. How much of an image

8-32.

Chapter 8–4 Bending Light Rays

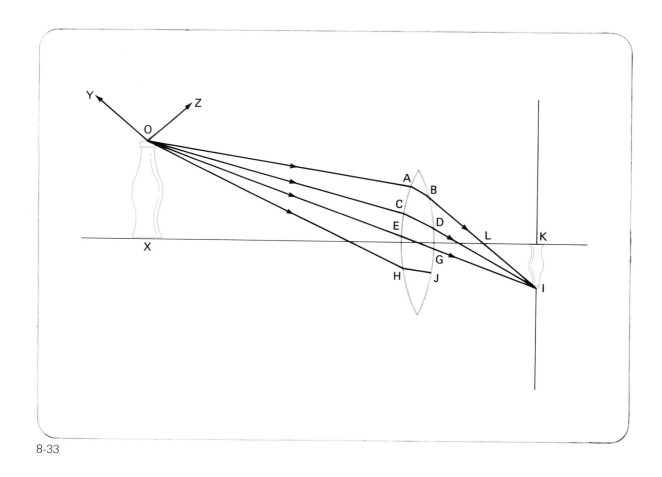

8-33

forms—none, half, or all of it?

The lens behaves as it does because light refracts (changes direction) when going into the lens and coming out of the lens. Figure 8-33 will help you see how this occurs.

K. OX represents a bottle. Rays of light travel in all directions from the top of the bottle. The direction of the rays that hit the lens is changed. These rays come together at I. The image of the top of the bottle forms at I. Other parts of the bottle also send out reflected light rays. These rays form the image of those parts of the bottle.

9. Do the rays OY and OZ help form an image?

L. Notice that all the rays shown change direction in going from the bottle to the image.

10. How many times does each ray change direction in going from O to I?

M. The ray that goes through points O, H, and J is incomplete.

11. To what point must this ray go in order to help form the image?

12. Is the image formed right-side up or upside down?

206 Unit II How Does Energy Affect Matter?

A lens can be used to produce a likeness or image of a scene. You must move a hand lens to make a clear image of close and distant objects. This problem is solved in a camera with a mechanism for moving the lens.

How do your eyes produce clear images of both close and distant objects? The lens of your eye is not hard and rigid. It is soft and flexible. Small muscles change the shape of the lens so that reflected light rays from both near and far objects are refracted properly. See Fig. 8-34.

8-34. *The lens of the eye changes shape to focus on far or near objects.*

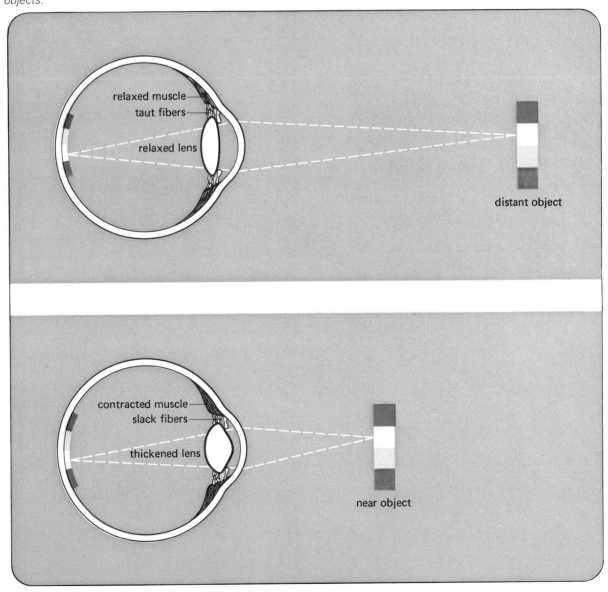

Chapter 8–4 Bending Light Rays

8-35. *Light rays are spread apart by a concave lens.*

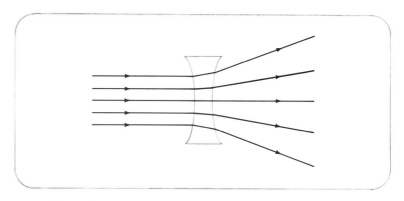

Unlike the eye, most instruments such as telescopes and microscopes have lenses whose shape cannot be changed. These instruments have, instead, combinations of different kinds of lenses. One common type of lens is thicker at the edges than in the middle. This is called a **concave** lens. Light rays are spread apart by a *concave* lens. See Fig. 8-35.

Concave: A lens shape in which the edges are thicker than the center. (The center is *caved* in.)

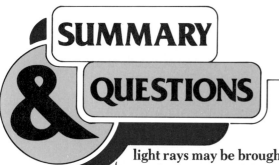

SUMMARY & QUESTIONS

One of the most important parts of a camera is the lens. Lenses refract light passing through them. Depending upon the shape of the lens, light rays may be brought together or spread apart. The lens of the eye can see both near and far objects by changing its shape.

Use the following diagrams which represent a ray of light passing through a convex and a concave lens.

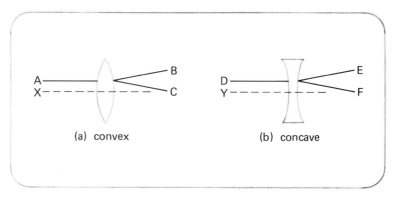

1. When the ray of light labeled A passes through the convex lens, it would be bent as shown by the ray labeled ____.
2. When the ray of light labeled D passes through the concave lens, it would be bent as shown by the ray labeled ____.
3. Explain what happens to a light ray which enters the lens along the dashed line labeled X.
4. Draw a ray diagram to show how a lens forms an image.

VOCABULARY REVIEW

Match the description in Column II with the term it describes in Column I.

Column I
1. electromagnetic waves
2. electromagnetic spectrum
3. visible spectrum
4. image
5. prism
6. mirage
7. light rays
8. focal point
9. lens
10. convex

Column II
a. Straight lines showing the path followed by light.
b. A specially shaped piece of glass which separates white light into colors.
c. A lens shape in which the edges are thinner than the center.
d. A piece of transparent material with curved surfaces which refracts light.
e. A form of energy able to move through empty space at very high speed.
f. The point at which parallel light rays meet after being refracted.
g. The picture formed by a lens.
h. The series of waves with properties similar to light.
i. The band of colors produced when white light is separated into colors.
j. An illusion caused by the refraction of light in which distant objects are seen upside down or floating in the air.

REVIEW QUESTIONS

Choose the letter of the ending which best completes each of the following statements.

1. Light is said to be energy traveling by waves since (a) we need it to see things by (b) experiments show it behaves like other waves (c) it can travel through space (d) it travels at a very fast speed.

2. Light, infrared, gamma rays, and X rays are all parts of the (a) electromagnetic spectrum (b) radio spectrum (c) microwave spectrum (d) wave spectrum.
3. Rays can be used to show the path of light since light (a) is a wave (b) is energy (c) travels in straight lines (d) travels through space.
4. A stick lying partly submerged in water appears bent at the water's edge because of (a) diffraction (b) refraction (c) reflection (d) warping.
5. A mirage is readily explained by the difference in air layers which cause (a) diffraction (b) refraction (c) reflection (d) warping.
6. A glass prism separates white light into a spectrum of colors because (a) each frequency of light is refracted differently (b) it reflects each color differently (c) it absorbs each color differently (d) it speeds up each frequency differently.
7. Different objects may appear different in color because (a) they reflect light differently (b) they refract light differently (c) each frequency of light has its own color (d) all these help to make objects appear colored.
8. The focal length of a lens is the distance from (a) the lens to where an image forms when the light rays are parallel (b) print on a page to the lens when the print is largest (c) print on a page to the lens when the print is in focus (d) print on a page to the lens when the print is smallest.
9. An image formed by a lens can be shown to be the result of (a) reflection (b) diffraction (c) refraction (d) infraction.

REVIEW EXERCISES

Give complete but brief answers to each of the following.

1. Describe an experiment which you could perform to show light has wave properties.
2. Name the six principal parts of the electromagnetic spectrum.
3. Explain why we can point a telescope directly at an object on a distant hill in order to see it.
4. Explain why the water always appears shallow when looking down into a clear water lake.
5. Explain mirages using the idea of refraction.
6. White light is composed of all colors.
   a. Give a reason that this must be so based on common everyday observations.
   b. Describe an experiment which proves that this is so.
7. Assuming white light is a mixture of all colors, explain: (a) why the sky is blue; (b) why a green object looks green in white light and black in red light.
8. Explain what happens to light when it passes through a lens to produce an image.
9. Compare the way a camera focuses faraway and near objects to the way the eye focuses faraway and near objects.

UNIT 3 HOW DO WE MAKE AND USE ENERGY?

CHAPTER

9 ELECTRICITY AND MAGNETISM

1 ELECTRIC AND MAGNETIC FORCES

Have you ever walked across a carpet and then felt a shock when you touched a metal doorknob? Have you ever heard a small crackling sound when you rubbed a cat's fur? Or, did your hair ever crackle and cling as you combed it? If it was dark enough, perhaps you saw tiny sparks jumping between the comb and the hair. We will explore some of these experiences. When you finish this section, you will be able to:

● Describe an *electric charge* and explain what is meant when something is said to be *neutral*.

- Explain what causes *electric* and *magnetic forces* and *fields*.

- Explain the cause of an *electric current*.

- Describe what effects the size of electric and magnetic fields.

▲ Test materials to find if they are insulators or conductors.

Electric charge: The result of something being given an amount of electric energy.

Positive charge: The electric charge given to a glass rod when rubbed with silk cloth.

Negative charge: The electric charge given to a hard rubber rod when rubbed with fur.

Neutral: Describes an object which has neither a positive nor negative charge.

Electric force: The force which causes two like-charged objects to move apart or unlike-charged objects to move toward each other.

The experiences described here are the result of **electric charges**. An *electric charge* results when something is given electric energy. For example, when you walk across a carpet, the friction of your shoes on the carpet may cause your body to be given an electric charge. One way to cause electric charges is to rub two materials together. Rubbing your hands on cat's fur or combing your hair with a nylon comb can cause electric charges.

One of the scientists in the past who studied electricity was Benjamin Franklin. Franklin suggested that the two kinds of electric charge be called **positive** (+) and **negative** (−). The charge on a rubber rod rubbed with fur is a *negative* charge. The charge given to a glass rod rubbed with silk cloth is *positive*. When an object has neither a positive nor negative electric charge, the object is **neutral**. An object which is electrically *neutral* (1) may not have any electric charge, or (2) may have an equal amount of positive and negative charges which cancel each other.

Objects having an electric charge affect each other. Two charged objects may pull toward each other. Two other charged objects may push away from each other. The behavior of objects with electric charges can be described by a simple rule: *Like charges repel each other, while unlike charges attract.*

You have learned that some kind of force is always needed to cause an object to move. When two electrically charged objects cause each other to move, a force must be acting upon them. This force is called the **electric force**. *Electric forces* cause charged objects to move apart or come together according to the kind of charge they have.

Chapter 9–1 Electric and Magnetic Forces

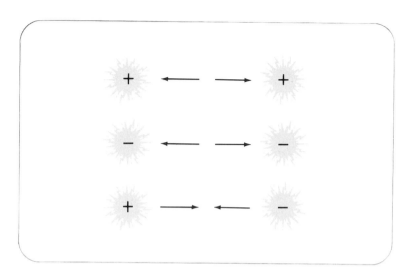

9-1. *Like electric charges repel; unlike charges attract.*

You may be familiar with the behavior of magnets. Two magnets will push or pull each other when their ends are brought together. A force must be acting upon them. This force is called **magnetic force**. All magnets have

Magnetic force: The force that causes two magnets to push or pull each other when their ends are brought together.

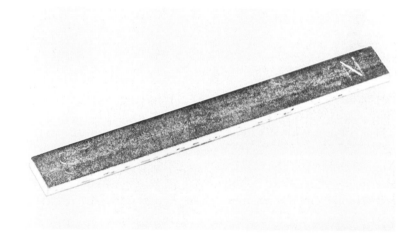

9-2. *A bar magnet.*

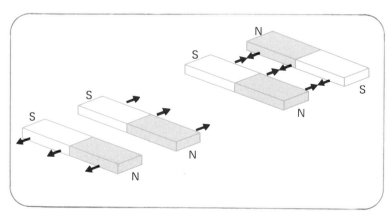

9-3. *Like magnetic poles repel; unlike magnetic poles attract.*

Magnetic pole: The part of a magnet where the magnetic forces are strongest.

places where the *magnetic forces* seem strongest. These places are called **magnetic poles**. Every magnet has at least two poles. *Magnetic poles* are called *north* and *south*. The magnetic force can either attract or repel. In this way, magnetic forces and electric forces are alike.

Two things seem to affect the size of electric forces. First, the more an object is rubbed to give it an electric charge, the stronger the electric force produced. For example, consider a plastic ruler rubbed once with a cloth and held a certain distance from another charged object. The ruler usually produces only a small attraction or repulsion force. Now consider that same ruler rubbed many times. The amount of charge increases and makes the force on the ruler stronger. The strength of the electric force between two objects increases as the amount of electric charge on those objects increases.

Charged objects have a greater effect on each other as they come closer together. If the two charged objects are 100 mm apart, reducing the distance to 50 mm apart causes the electric force to become four times greater. In other words, cutting the distance in half causes the force to become four times greater. In the same way, moving the objects 200 mm apart reduces the force to one quarter of what it was. The relationships between electric force and distance between charged objects are illustrated in Fig. 9-6 on page 216.

9-4. *(left) The rubber rod and the ball have the same type of charge. They repel each other.*

9-5. *(right) The glass rod and the ball have different charges. They attract each other.*

Chapter 9–1 Electric and Magnetic Forces

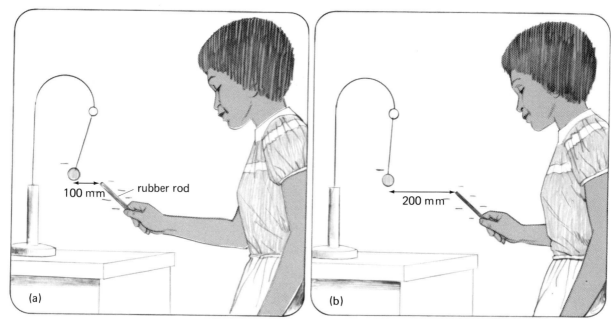

9-6. *This experiment shows that the size of the electric force becomes less with greater distance between the charged objects.*

Electric field: A region of space around an electrically charged object in which electric forces on other charged objects are noticeable.

Magnetic field: A region of space around a magnet in which magnetic forces are noticeable.

Around every charged object there is a space in which the electric force in noticeable. Such a region of space is called an **electric field**. The *electric field* surrounds the charged object in all directions. As you have already seen, the size of the electric force becomes smaller as the distance becomes larger. Finally, at great distances, the electric field grows too weak to be noticeable.

Magnets are also able to affect each other without touching. Magnets are surrounded by **magnetic fields**. A field around a magnet can be seen if small pieces of iron are sprinkled around a magnet. See Fig. 9-7.

To help explain how objects can become electrically charged and how electric charges can move from one place to another, scientists use the model of the *electron*. An electron is a negatively charged particle of matter. The fact that many things can be given an electric charge by being rubbed with another material can be explained by the presence of electrons. Rubbing two materials together can cause electrons to be torn away from one material and added to the other. For example, running a comb through your hair may cause electrons to move from the hair to the comb. In this case, the comb takes on a negative electric charge because of the extra electrons.

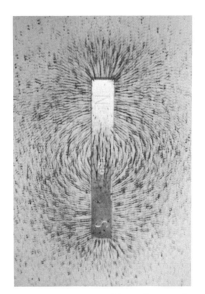

9-7. (left) The iron filings line up in the field of the magnet.

9-8. (right) Some clothes have an electric charge after being dried in a clothes dryer. This charge makes them cling to each other.

Electric current: The result of electrons moving from one place to another.

Materials such as hair do not ordinarily have an electric charge even if electrons are present. Those materials are neutral because they also contain particles similar to electrons but with positive charges. Such particles are actually found in all matter. These particles are called *protons*. A proton is a very small particle of matter with a positive electric charge. Matter, then, is usually neutral because it contains an equal number of protons and electrons. An object, such as a comb, has a negative electric charge only when extra electrons are added.

In some ways, electrons behave like water. Water can flow from one place to another. Usually it is gravity that causes the water to flow downhill. Electrons can also flow from one place to another. It is not gravity, however, that causes electrons to move. Electrons move from a place where there are a greater number of them to a place where there are fewer. This movement of electrons causes an **electric current** to flow. An *electric current* is the result of electrons moving from one place to another.

A common example of an electric current is lightning. Electrons build up in one part of a cloud. These electrons suddenly move from one part of the cloud to another or jump from the cloud to the ground. When this happens, the electrons moving through the air cause it to become

Chapter 9–1 Electric and Magnetic Forces

9-9. *Lightning can jump from one part of a cloud to another.*

very hot. A brilliant light is given off by the heated air. At the same time, the heated air rushes outward to make sound waves that we hear as thunder.

Lightning would not happen if electrons could move easily through the air. Air is an example of an insulator. The electric charge builds up until it is great enough to overcome the resistance of the air. If air were not an insulator, electrons could not build up in a cloud but would flow away through the air.

9-10. *Lightning is caused by the movement of excess electrons.*

Unit III How Do We Make and Use Energy?

Some materials such as metal doorknobs allow electrons to flow freely. These materials are called conductors. Any conductor carries electric currents because electrons can easily move through it.

Materials
Battery
Light bulb
2 wires
Paper clip
Piece of wood
Piece of glass
Piece of plastic
Paint
Chalk
Staple
Eraser
Pencil
Paper

A. Set up a battery, a light bulb, and wires as shown in Fig. 9-11.

When the two free wires touch an object, and the material in it is a good *conductor,* the bulb will light. This means an electric current can move through the material. The bulb will not light if the material is an *insulator.* Electric current cannot flow through an insulator. (You cannot be hurt by the electricity in this equipment.)

B. Test each of the following materials. Record which are conductors and which are insulators: air, paper clip, wood, glass, pencil lead, chalk, staple, paint, eraser, metal holder for eraser on pencil, plastic.

C. Test any other materials you find around you.

1. Which materials are good conductors of electric currents?

2. Which materials are insulators?

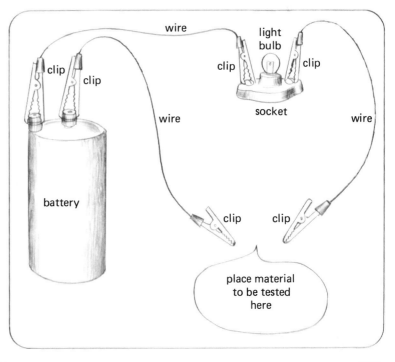

9-11.

Chapter 9-1 Electric and Magnetic Forces

SUMMARY & QUESTIONS

If you build up an electric charge on your body and then touch something, the charge moves from you to the object. Scientists use a model called an electron to help explain how electric charges move. All matter contains both electrons and protons. Objects can become electrically charged when electrons are added or removed. Electrons may move from a place where they are abundant to a place where they are few. The movement of electrons produces an electric current. Some materials allow electrons to pass through freely while other materials do not.

Fill in the blanks with the term described in the clues given for each numbered item. DO NOT WRITE IN THIS BOOK.

```
1.        _ _ E _ _ _ _ _
2.        _ _ _ _ L _ _ _ _
              E
3.        _ _ _ C _
4.    _ _ _ _ _ _ T
              R
5.        _ _ _ _ _ I _ _
              C

              F
6.        _ O _ _ _ _ _ _
7.        _ R _ _ _ _
8.    _ _ _ _ _ C _ _ _
9.        _ E _ _ _ _ _
```

CLUES
1. A negatively charged particle of matter.
2. Does not allow electrons to move easily.
3. The feeling you get when you lose your electric charge.
4. The movement of electrons.
5. A surplus of electrons.

6. A lack of electrons.
7. The positively charged particle of matter.
8. Allows electrons to move easily.
9. Having the same number of protons as electrons.

2 ELECTRIC CIRCUITS

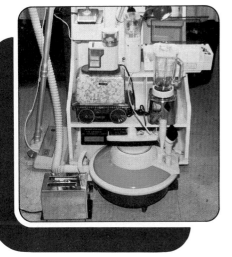

What electric "servants" do you see in the photograph? These common machines provide more servants in a household than a king might have had a few hundred years ago. Each machine does work because electrons flow to it. When you finish section 2, you will be able to:

● Describe how an *electric circuit* can be made in terms of electric current.

● Distinguish between *direct current* and *alternating current* in terms of movement of electrons.

▲ Demonstrate the flow of electric current in an electric circuit.

Switching on a light and turning on a faucet are similar. Opening the faucet lets water flow from a pipe. Turning on an electric light switch permits electrons to flow in a conductor. Water will not flow in a pipe, however, unless a force pushes it along. That force could be gravity causing the water to flow downhill. Maybe a pump could supply the energy needed to move the water. Electrons flowing through a conductor also need something to cause them to move. An ordinary dry cell (flashlight battery) is able to do this. A dry cell works because it causes electrons to build up at one place. This part of the dry cell then has a negative electric charge. On some kinds of cells or batteries this part is marked negative (−). See Fig. 9-13 on page 222. If electrons are to move away from the negative area, a path must be provided. This means that a conductor must be touched to this place on the dry cell. Electrons from the dry cell

9-12. *(a) Turning on a faucet causes water to flow. (b) In a similar way, opening a switch causes an electric current to flow.*

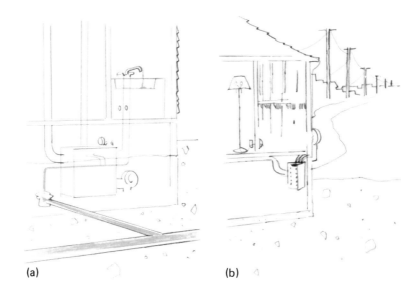

would only pile up on the conductor if they could not move along the conductor to a place which is poor in electrons. A dry cell has such a place which is often marked positive (+). If a conductor such as a wire is touched to both the negative and positive parts of a dry cell, electrons will flow through the conductor. This is called an **electric circuit** (**sir**-cut). An *electric circuit* is a complete path. The circuit allows movement of electrons from a place where there are many electrons to a place where there are few. See Fig. 9-14. The movement of electrons through a conductor in a circuit is called an electric current.

Electric circuit: A complete path which allows electrons to move from a place which is rich in electrons to a place poor in electrons.

9-13. *The terminals on a battery or dry cell are marked + and −.*

222

Unit III   How Do We Make and Use Energy?

9-14. *When the lamp is on, a continuous electric circuit exists between the power station and the lamp.*

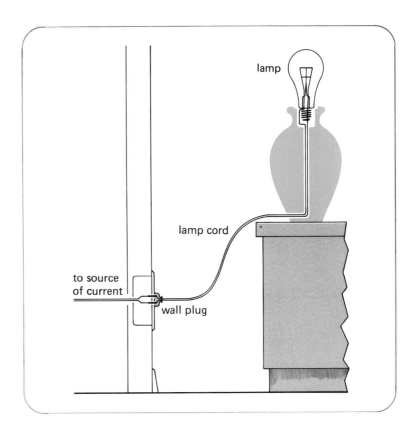

Materials
compass
1 battery
wire
2 alligator clips
pencil
paper

This activity will show you a way to tell that electrons are moving in a wire.

A. Obtain the materials listed in the margin.

B. Lay a compass on a table top.
   1. In what direction does the compass needle point?

C. Connect a battery, light bulb, and wires as shown in Fig. 9-15 on page 224. Place one of the wires on top of the compass needle so that the wire is in the same direction as the compass needle. You can hold it in place with books or a notebook.
   2. Is the compass needle still pointing in the same direction?

Be careful not to bring any iron like the rings in your notebook near the compass. Iron rings or clips may change the direction of the compass needle. If the change in direction isn't great, it won't hurt. Just be certain the wire is along the needle.

D. Watch the light bulb as you connect the wires to complete the circuit so that current flows.

Chapter 9–2 Electric Circuits 223

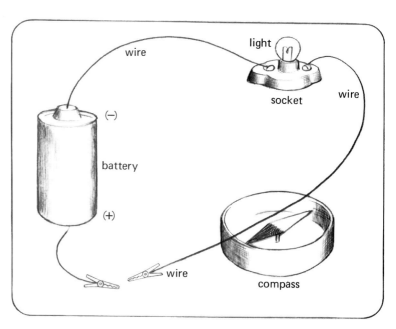

9-15.

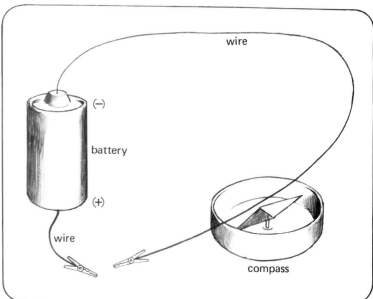

9-16.

3. Does the bulb light?
If the bulb doesn't light, you may not have a complete circuit. Check the connections or check for a dead battery.

4. Does the compass needle move?
The compass needle should react to an electric current in the wire by moving.

E. Now watch the compass needle as you connect the wires to complete the circuit.

F. Take the light bulb out of the circuit and connect the wires as shown in Fig. 9-16. Do not join the final wires yet. Be sure the wire over the compass lies in the same direction as the compass needle. Watch the compass needle as you touch the wires to complete the circuit. Do not leave them connected.

5. When the wires were connected, what evidence did you see that an electric current was present?

When you plug something into an electric outlet, you are making an electric circuit. The two parts of the plug and the attached wires provide a complete circuit for electrons to flow through whatever appliance is plugged in. Some electric circuits in a house are complete only when a switch is turned on. An example are the circuits which operate lights. See Fig. 9-17. While electrons flowing in a circuit are like water flowing in a pipe in many ways, there is also one difference. If you could follow one electron along a wire carrying a current, you would find that the electron itself moved very slowly. Yet when you switch on a light, the electric current takes effect almost instantly. You do not have to wait for the electrons to move along the wires. This is because the

9-17. *The electric circuits controlled by these switches will be complete only when the switches are flipped to "on."*

electrons repel each other. Electrons repel each other because they carry like charges. An electron in a wire repels other electrons in the wire. Electrons all along the wire pass along this movement from one to the next. This effect travels rapidly along the wire. This is what is meant when electrons are said to "flow" along a conductor.

Electrons do not always flow in the same direction in all electric circuits. When a circuit is made with a dry cell, the electrons always move from the negative connection toward the positive connection. This is called **direct current** (DC). In a circuit carrying *direct current*, the electrons always flow in one direction. The current that is most commonly used is not direct current. Electric current supplied by the machines in power stations changes direction many times each second. This is called **alternating** (**all**-ter-nay-ting) **current** (AC) because it continuously alternates or changes its direction of motion. *Alternating current* is used to supply electricity because it can be sent through wires over long distances. Direct current dies out after traveling a long distance through a wire.

Direct current: An electric current that flows in one direction in an electric circuit.

Alternating current: An electric current that changes direction in an electric circuit.

SUMMARY & QUESTIONS

Electrons can flow through a conductor only if there is a complete electric circuit. An electric circuit can be completed with a dry cell by connecting the negative and positive parts of the cell with a conductor. The flow of electrons in a circuit can be controlled by the use of switches. Current from a dry cell is called direct current since it travels in one direction. Current from a house switch is called alternating current since it changes direction rapidly.

1. Describe what happens to the electrons when a circuit is made by your turning on a wall switch, causing an overhead light to go on.
2. Contrast the flow of electrons in an AC circuit with the flow of electrons in a DC circuit.

3. Refer to Fig. 9-15, page 224, to answer the following:
   a. What is necessary in order for a current to flow in the circuit?
   b. When the circuit is completed, do electrons move through it in a clockwise (moving first through the light) or in a counterclockwise (moving first over the compass needle) direction?
   c. When the circuit is completed, what evidence would you see that proves there is a current?

3 MEASURING ELECTRICITY

The diver shown in the photos has something in common with a dry cell. Can you guess how the diver and the dry cell are alike? When she is standing on the diving board, the diver has potential energy. What about the dry cell? Reading section 3 will help you to answer this question. When you finish this section, you will be able to:

● Describe the relationship of *volts, amperes,* and *ohms.*

● Explain the difference between *series* circuits and *parallel* circuits.

Demonstrate the effect of changing the voltage and *resistance* on an electric current.

9-18. *Increasing the number of dry cells increases the voltage.*

Volt: A measure of the amount of work done in moving electrons between two points in an electric circuit.

Ampere (amp): A measure of the number of electrons moving past a point in an electric circuit in 1 sec.

A diver on a diving board has potential energy. She has gained this energy by climbing up to the board. Divers are interested in their potential energy, though they may not use that term, since it determines how hard they will hit the water. The amount of potential energy increases as the height of the board above the water increases. The potential energy of a diver could be found by measuring the height of the board.

A dry cell also has potential energy stored in it. This energy cannot be given off until the dry cell is part of an electric circuit. Then, the flow of electrons will release some of the potential energy. How could we get some idea about the potential energy stored in a dry cell? Some way to measure this energy is needed. Just as the potential energy of a diver is determined by height, the potential energy of electrons from a dry cell is measured in **volts**. A *volt* measures the amount of work that would be done if electrons are moved between two points in an electric circuit. If the flow of electrons is compared to water running down a hill, then voltage is a measure of how high the hill is. An ordinary dry cell, such as the one shown in Fig. 9-18, gives 1.5 volts. This would compare to a diver jumping off a low diving board. The combination of dry cells, also shown, gives 6 volts. This would be like a diver jumping off a higher board. Consider another example. The single dry cell is similar to water flowing down a low hill. The combination of cells is similar to water flowing down a higher hill. The 6-volt combination of cells does four times as much work as the 1.5-volt cell. In other words, the 6-volt combination *pushes* the electrons harder than the 1.5-volt cell.

For most electric circuits, we want to know not only how hard the electrons are pushed, but also how many electrons are flowing. To measure the number of electrons, we use **amperes** (**am**-peers). An *ampere* measures the amount of electrons moving past a point in a circuit in 1 sec. Ampere is often called "amp" for short. Measurement of voltage and amperage describes the behavior of electric currents. For example, a circuit may have high voltage with low amperage. This would be like a very small but swiftly flowing stream. On the other hand, a circuit with high amperage but low voltage would be like a large but slow-moving river.

If you push a book across the top of a desk, the book will probably slide easily. What will happen if you put a rubber eraser between the book and desk top? Much more force will be needed to slide the book. How easily something can be moved changes with different conditions. This is also true for electrons. When electrons move through a material, they meet **resistance** (rhee-**zis**-tance). *Resistance* is the name given to all conditions which limit the flow of electrons in an electric circuit.

The amount of current that flows in a particular electric circuit is determined by the voltage. Think of water flowing through a pipe. The amount of water that will pass through the pipe is determined by the force pushing the water. Suppose that the water meets a narrow place. See Fig. 9-20 on page 230. Less water could then pass through the pipe. The narrow part of the pipe has the same effect as resistance in an electric circuit. If elec-

Resistance: All conditions which limit the flow of electrons in an electric circuit.

9-19. *The fast-flowing stream (top) is like a circuit with high voltage and low amperage. The slow-moving river (bottom) is like a circuit with low voltage and high amperage.*

Chapter 9-3 Measuring Electricity

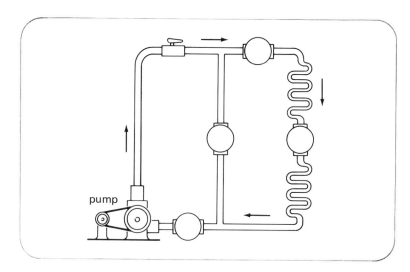

9-20. Water flowing through pipes is similar to an electric current in a circuit. What part of the pipes would have the highest resistance to the flow?

Ohm: A measure of the amount of resistance in an electric circuit.

trons meet a part of the circuit where resistance is high, then the amount of current flowing through the entire circuit is reduced. Resistance is measured in **ohms**. A resistance of one *ohm* allows one ampere of current to flow at one volt.

The voltage, amperage, and resistance in an electric circuit are related to each other by a rule known as Ohm's Law. This relationship was discovered by a German schoolteacher, Georg Ohm, in the early 1800's. The rule is:

$$\text{volts} = \text{amperes} \times \text{ohms}$$

For example, a circuit might have a resistance of 3 ohms and a current of 2 amperes. What is the voltage in the circuit?

Materials
Compass
2 batteries
2 light bulbs
3 wires
Pencil
Paper

A. Obtain the materials listed in the margin.

B. You will need to set up an electric circuit. Look back to section 2, page 224. Set up an electric circuit just like the one in Fig. 9-15.

C. Arrange the compass so the needle points north on the numbered scale. This scale is in degrees.

D. Arrange the wires so that the one over the compass is in the same direction as the needle. Watch the compass needle as you make the final connection.

Unit III   How Do We Make and Use Energy?

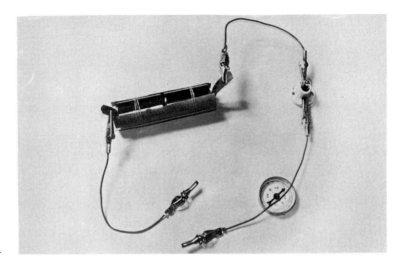

9-21.

E. Read the angle, in degrees, where the compass needle finally settles down. Tap the compass to make sure the needle isn't stuck.

1. When the current is flowing, how many degrees from north does the compass point?

F. Next, increase the number of batteries in the circuit to two. See Fig. 9-21. This will push the electrons more and cause more of them to move.

G. Watch the compass as you make the final connection. Tap the compass to free the needle.

2. Compared with the first circuit, what evidence is there that more current is flowing now?

H. Next, change the circuit, as shown in Fig. 9-22. Notice that the wire over the compass makes a loop and goes over the compass a second time. This is the same

9-22.

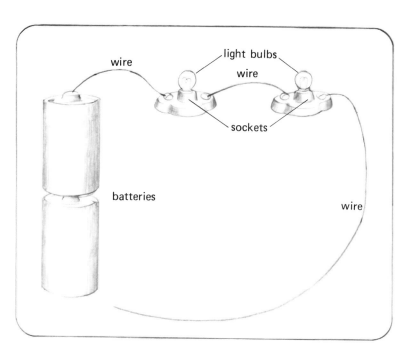

9-23.

as sending more current flowing in the wire over the compass.

I. Watch the compass needle as you make the final connection.

3. Compared with the first circuit, what evidence is there that more current is flowing over the compass?

The effect of increasing the resistance in a circuit can also be shown. A light bulb produces resistance. The brightness of the light bulb in an electric circuit is one way to tell how much current is flowing. The brighter the bulb, the larger the current.

J. Again set up an electric circuit as shown in Fig. 9-21 on page 231. It is the same circuit used in step F but the compass is not needed.

K. Look at the light and notice how bright it is.

L. Now add a second bulb to the circuit as shown in Fig. 9-23. Because each bulb has resistance, the resistance in the circuit has been increased and less current should flow.

4. Compared to a single-bulb circuit, what evidence is there that less current is flowing when there is more resistance?

An electric circuit consists of several parts. There must be a source of the electrons to be moved through the circuit. Conductors, usually wires, are needed to connect all the parts. These parts include switches and the appliance to be operated, a light for instance. These

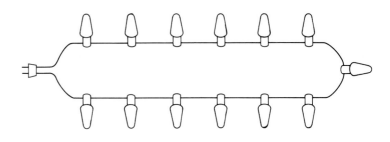

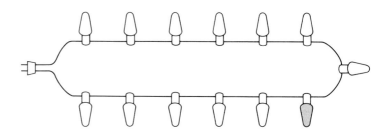

9-24. Many kinds of Christmas tree lights will all go out if one bulb in the string burns out. Why is this true?

Series circuit: A circuit in which all the parts are connected one after the other.

9-25. Can you see why this circuit is called a parallel circuit?

items can be connected one after another. Some types of Christmas tree lights are a good example of such a connection. See Fig. 9-24. This is called a **series circuit**. In a *series circuit*, all parts of an electric circuit are connected one after another. A series circuit can cause some problems. For example, no part of a series circuit can be switched off without turning off everything. If the lights

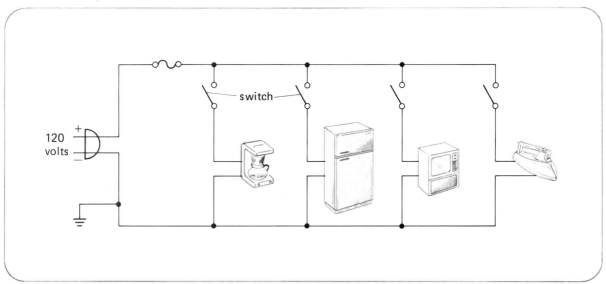

Chapter 9-3 Measuring Electricity

233

in a house were connected this way, they would have to be all on or all off at once.

Another way to connect the parts of a circuit is shown in Fig. 9-25 on page 233. This arrangement is called a **parallel circuit**. In a *parallel circuit*, the various parts are on separate branches. Each branch of a parallel circuit can be switched off without affecting the other branches. This is the way the various circuits in a house are arranged.

Parallel circuit: An electric circuit with the various parts in separate branches.

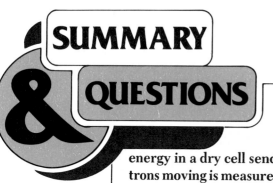

SUMMARY & QUESTIONS

Can you see how a diver on a board is like a dry cell? Both have potential energy. For the dry cell, potential energy is measured in volts. If the energy in a dry cell sends electrons through a circuit, the number of electrons moving is measured in amperes. Whatever resistance those electrons meet is measured in ohms. Circuits may be arranged so that the parts are connected either one after another or with branches.

Match the description with the electrical term which it properly describes.

Electrical Term
1. ohm
2. ampere
3. volt
4. parallel circuit
5. series circuit
6. resistance

Description
a. The measure of potential energy of a dry cell.
b. The way electric lights are wired in a house.
c. Unit used in measuring the amount of electrons flowing in a circuit.
d. The kind of circuit in which current stops flowing when any part of the circuit is turned off.
e. Any condition which limits the flow of electrons through a circuit.
f. A unit used in measuring resistance.

VOCABULARY REVIEW

Match the description in Column II with the term it describes in Column I.

Column I
1. alternating current
2. electric charge
3. electric circuit
4. electric field
5. insulator
6. negative
7. neutral
8. magnetic field
9. magnetic poles

Column II
a. The space in which an electric force exists.
b. Having equal numbers of protons and electrons.
c. The parts of a magnet where magnetic forces are strongest.
d. The charge given to a rubber rod when rubbed with wool.
e. Electrons continuously reversing their direction of flow.
f. Having a given amount of electric energy.
g. An arrangement which allows the flow of electrons.
h. A region of space around a magnet in which magnetic forces are noticeable.
i. A material which does not allow easy movement of electrons.

REVIEW QUESTIONS

Choose the letter of the ending which best completes each of the following statements.

1. An object which has an electric charge will cause another charged object to (a) be attracted (b) be repelled (c) be attracted or repelled (d) remain unchanged.
2. An object which has the same amount of positive charge as it has negative charge is (a) neutral (b) opposite (c) like charged (d) unlike charged.
3. If the size of the electric charge on two objects were increased, the electric force would (a) increase (b) decrease (c) remain the same.
4. An electric current is the result of (a) electrons moving from a place where protons are to a place lacking protons (b) protons moving from a place where electrons are to a place lacking electrons (c) electrons moving from a place where there are many of them to a place where there are few (d) rubbing an insulator with a conductor.
5. Electrons move easily through most objects made of metal. These metals are called (a) insulators (b) transmitters (c) conductors (d) transformers.

CAREERS IN ELECTRICITY

Electric lighting in the form of arc lights first appeared on city streets in Europe in the late 1800's. In America, Thomas Edison took out a patent for an electric light bulb in 1880. More important, Edison developed a system for distributing electricity all over the country.

Today, most people take electricity for granted. In our homes, electricity provides light, heat, and refrigeration and can also cook our food. Power lines and telephone cables stretch across the continent. The number of careers involving a knowledge of electricity is almost limitless.

(below) **One of the most imporant inventions of the 20th century is the electronic computer.** Many parts of our daily lives, from bank accounts to telephone bills, are computerized. The computer field has opened up a wide range of careers for people with a knowledge of and interest in the uses of electricity.

(above) **Television has become one of the most popular electrical appliances. Almost everyone owns a TV set. But not many people understand the inner workings of a television or how to repair one. They take their broken television sets to a TV repairperson. The wires and tubes in a TV carry high-voltage electricity. A thorough knowledge of electricity and wiring is necessary for repairs to be made safely.**

(right) **A telephone lineperson maintains and repairs the cables and wires that form our communications system. This job can involve climbing telephone poles, as shown here, or working underground or even under water. A lineperson must also be able to operate the necessary electrical power equipment.**

6. A magnet is similar to an object which has an electric charge in that (a) both may attract or repel (b) their forces act through a distance (c) their forces increase with a decrease in distance (d) they are similar in all of the above ways.
7. The part of a magnet where the magnetic forces are the strongest is called a magnetic (a) field (b) pole (c) attraction (d) repulsion.
8. The number of electrons moving past a point in a circuit per second is measured in (a) volts (b) ohms (c) watts (d) amperes.
9. The kind of circuit in which a branch may be shut off without affecting the other branches is called a (a) parallel circuit (b) series circuit (c) direct circuit (d) alternating circuit.

REVIEW EXERCISES

Give complete but brief answers to each of the following exercises.

1. Describe one way you could convert the energy of motion into electric energy and become electrically charged.
2. Predict whether the electric force acting between two charged objects would increase or decrease when the following changes are made:
   a. one object is given more charge
   b. some charge is taken from both objects
   c. the objects are moved closer together
   d. the objects are moved farther apart
3. If matter is composed of electrons and protons, explain how it can become either positive, negative, or neutral.
4. Select from the following list those objects which would usually be classed as insulators: rug, doorknob, cat's fur, penny, plastic bag, kitchen fork, wooden spoon, pencil lead.
5. Describe, in volts and amperes, the current in a circuit if a large number of electrons are passing through it, but the electrons have very little push behind them.
6. Explain what happens in a circuit when the resistance is greatly increased.
7. Describe the two ways a magnet may behave when brought near another magnet.
8. Explain how you would determine whether a piece of metal is a magnet.
9. Give two bits of evidence which show that magnetism and electricity are related.
10. Describe three ways an electromagnet differs from a permanent magnet.

CHAPTER 10
ENERGY FOR EVERYDAY USE

1 ELECTRIC POWER

This photograph was taken during the New York City blackout in July, 1977. Parts of the city were without electricity for as long as 26 hours. How would your life change if the electricity went off? When you finish this section, you will be able to:

- Explain why electricity is such a commonly used form of energy.
- Describe how electric circuits are used to make electricity available in a home.
- Indicate what is measured by an electric meter in a home.
- ▲ Make and observe the action of an electric *fuse*.

10-1. *We use many electric appliances every day.*

Transformer: A part of an electric circuit which changes the voltage.

10-2. *A transformer lowers the voltage for use in homes.*

Electric energy is used in so many different ways for several reasons. Most important is the fact that it can be changed into almost any other form of energy. Electric motors change electricity into motion. Lamps change electricity into light. The temperature inside buildings is controlled by air conditioners and heaters run by electricity. Electric energy also can be sent over long distances through wires. Because of this, factories can be located far from a source of power. The same wires allow each building to be connected to a source of energy.

Buildings are connected to power lines by at least two wires in order to make a complete circuit. The voltage sent out through the main power lines is very high. This high voltage must be reduced to a lower voltage before being sent into homes. Voltage can be lowered by use of a **transformer**. A *transformer* is used to change the voltage into the 110 volts needed in homes. Buildings often have three wires coming from the power lines. When all three wires are connected to a circuit, the voltage is doubled to give 220 volts. Electric stoves and clothes driers need 220 volts.

Wires bringing electricity into a home are connected with parallel circuits in the home. To use the electric energy, the appliance is made part of one of the circuits. For example, a toaster is plugged into a wall socket. The plug and the cord on the toaster allow it to become a part of the circuit. Electrons flow through special wires in the toaster that have high resistance. When electrons meet resistance, they are robbed of some of their energy. The energy lost by the electrons is changed into heat. The heat toasts the bread.

Many other common appliances such as electric stoves and irons work because electric energy is changed into heat energy. When the current flowing through an appliance meets high resistance, the electric energy is changed into heat. A common electric light bulb also works this way. The bulb contains a thin wire made of special metal. Because of its high resistance, this thin wire becomes hot enough to give off a bright light.

Too much current flowing in one of the house circuits is dangerous. For example, if you attach a toaster, a steam iron, and an electric coffee maker to the same circuit, more than 20 amperes of current would be needed.

Unit III   How Do We Make and Use Energy?

10-3. (above) The wire in this bulb becomes hot because of its high resistance. The wire becomes hot enough to glow.

10-4. (right) Common types of fuses and circuit breakers used in household circuits.

Fuse: A part of an electric circuit which prevents too much current from flowing.

The wires of the circuit could become overheated and start a fire. To prevent this overheating, each circuit in a house is provided with a **fuse**. A *fuse* is a part of a circuit which prevents too much current from flowing. Some fuses contain a part that melts and breaks the circuit if there is too much current. A more common arrangement is a special switch, called a *circuit breaker*. The circuit breaker opens when too much current flows. All home circuits must have some kind of overload protection to prevent a serious fire threat.

Materials
battery
light bulb
3 wires
fuse wire
paper
pencil

A. Obtain the materials listed in the margin.

B. Set up a circuit which includes a *fuse*. See Fig. 10-5.

Watch the light as you connect the last wire to the button on the end of the light bulb.
 1. Does the bulb light up?
 2. Does it remain lighted?

Chapter 10–1 Electric Power

C. When the bulb is lighted, connect a wire for a very short time between the metal cap of the bulb and the metal button on the end of the bulb.

3. What effect does making this connection have on the bulb?

4. Does the circuit return to normal when you disconnect this last wire?

D. Draw a circuit similar to Fig. 10-6 on a sheet of paper.

5. What connection could be made that would cause the bulb to light? Draw this connection in your diagram. Label it with the number 1.

6. Where could you add a wire that would form a circuit that would burn out the fuse? Draw the wire connection and label it with the number 2.

7. What is the purpose of a fuse in an electric circuit?

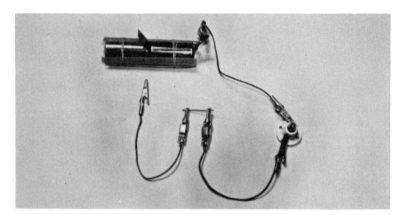

10-5.

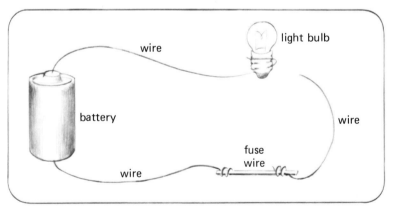

10-6.

Watt: The amount of work done by 1 ampere of current through 1 volt of force.

You have probably noticed that many electric appliances are marked to show how much electric energy they use. For example, a light bulb might be marked as "100 watts." A **watt** measures the speed at which electric energy is changed to other forms. *Watts* are useful in measuring how much electric energy is consumed. The amount of electricity used by a light bulb, for example, takes into account two things: (1) how fast the bulb used the electricity by changing it into another form of energy; and (2) how long the bulb was turned on. A 100-watt light bulb burning for one hour could be said to use 100 *watt-hours* of electricity. A watt-hour is a small amount of electric energy. It is more common to use a **kilowatt-hour** which is a thousand times larger. A *kilowatt-hour* is the amount of energy supplied in one hour by one kilowatt of power. Homes have a meter attached to their lead-in wires. The meter measures how much electricity is consumed in kilowatt-hours. The electric company then charges a certain amount for each kilowatt-hour used.

Kilowatt-hour: The amount of energy supplied in one hour by one kilowatt of power. Used to measure how much electric energy is consumed.

How would your life be different without electricity? If you think of all the ways your life would change if the electricity went off, you will see why electricity is one of the most commonly used forms of energy. Electricity is useful mainly because it can be changed into other forms of energy so easily. Electricity can also be sent over long distances. Electricity is supplied to a home by several low-voltage circuits which are connected to the power lines. Each circuit must be protected from overloads. Electric energy consumed in a home is measured in kilowatt-hours.

1. Give at least two reasons why electricity is such a commonly used form of energy.
2. Describe what you must do and what happens in the circuit when you use an electric toaster.
3. Why are fuses made a part of electric circuits in a home?
4. What does an electric meter in a home circuit measure?
5. Using a drawing of a circuit such as Fig. 10-6 on page 242, explain how a fuse operates in the circuit.

2 MAKING ELECTRICITY

When you plug in a radio or stereo at home, where does the electricity come from? You could find out by following the electric power lines all the way back to the power plant. Inside the power plant you would probably find giant boilers making steam by burning coal or oil. You would also find large generators in action. How does this machinery make electricity? You can understand the generation of electricity by knowing the complete relationship between electricity and magnetism. Magnetism can be used to make electricity. When you finish section 2, you will be able to:

- Predict what happens when a wire moves in a magnetic field.

- Define *electromagnetic induction.*

- Explain how an electric generator produces electricity.

▲ Demonstrate what happens when a wire moves in a magnetic field.

Electricity can be changed into magnetism. Can magnetism also be changed into electricity?

Materials
Bar magnet
Wire, 1.5 m
Pencil
Compass
Paper

In this activity, you will try to generate an electric current in a wire. You will move a magnetic field past the wire.
A. Obtain the materials listed in the margin.

B. Wrap one end of the wire around a compass so that about five loops go over the top and underneath the compass needle. See Fig. 10-7 on page 245.

Unit III How Do We Make and Use Energy?

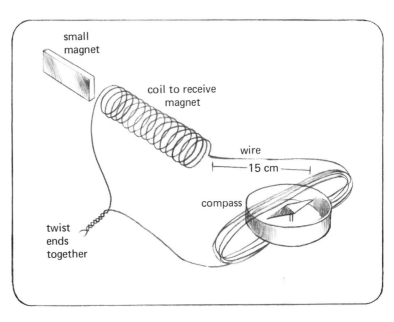

10-7.

C. At least 15 cm from the compass, wrap the wire into another coil around a pencil. Make about 10 to 15 coils. Then remove the pencil.

D. Bring the loose end of the wire back and join it to the end of the wire near the compass. This makes a complete circuit. See Fig. 10-7. The coils should be at least 15 cm apart.

The compass will be used to detect any current that flows in the wire.

E. Arrange the compass and its coil so that the compass needle is pointing in the same direction as the wires in the coil around it. Be sure the compass needle is free to move. You may need to tap the compass to be sure the needle is free.

F. Insert the magnet in the coil you made with a pencil.

G. Watch the compass. Now quickly pull the magnet from the coil.

1. Does the compass needle move?

If the coils are 15 cm apart, any movement of the needle is due to an electric current.

2. Can you produce an electric current with a magnet?

When a wire is moved in a magnetic field, an electric current flows in the wire. This generation of current is an example of what is called **electromagnetic induction** (e-leck-tro-mag-**net**-ick in-**duck**-shun). An electric current produced by *electromagnetic induction* is always the result of motion in a magnetic field. When a wire moves

Electromagnetic induction: Production of an electric current by motion in a magnetic field.

Chapter 10–2 Making Electricity

10-8. *An electric current is produced in a wire which is moved through a magnetic field. The current can be measured on a galvanometer. The galvanometer contains a small electromagnet which is activated by the current produced. The electromagnet moves the dial on the meter.*

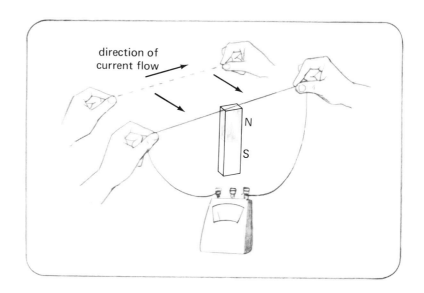

past a magnet, a current will flow in the wire. If the wire is held still and the magnet is moved past, a current flows in the wire. The only requirement for producing a current in this way is to cause motion within the magnetic field. Experimenting with a wire moved past a magnet proves that the current flows in a certain direction. See Fig. 10-8. If the wire is moved in the opposite direction, the direction of current flow is reversed.

A knowledge of electromagnetic induction can be used to build a machine to produce electricity. A loop of wire

10-9. *A diagram of an electric generator.*

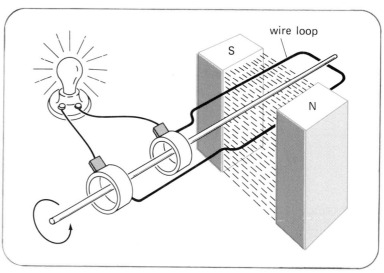

Unit III   How Do We Make and Use Energy?

10-10. *An alternating current electric generator.*

is put into a magnetic field. See Fig. 10-9 on page 246. As the wire loop is turned, an electric current is produced. Spinning the loop can produce a current which is fed into wires.

Remember that the direction of current flow changes with the direction of movement of a wire in a magnetic field. When one side of a wire loop moves down in a magnetic field, the current produced will flow in one direction. The same side moving up during the other half of its turn causes the current to flow in the opposite direction. See Fig. 10-10. As a result, the current flows first in one direction and then the other. This change of direction is alternating current.

The big generators in power plants have many loops of

MAGNETS AND ELECTROMAGNETISM

The interaction between magnetism and electricity is one of the basic forces in the universe. This interaction was first discovered by Hans Christian Oersted in 1820. He showed that an electric current in a wire can produce magnetic effects. In 1831, Michael Faraday, an English scientist, demonstrated the opposite effect. Faraday proved that when a magnet is moved near a wire, a current is produced in the wire. This effect is called electromagnetic induction. The discovery of electromagnetic induction led to the invention of generators and electric motors.

ELECTRICAL ORE FINDER.

(right) **Hans Christian Oersted was a Danish scientist of the 19th century.** As early as 1813, Oersted predicted that there was a relationship between electricity and magnetism. This picture shows Oersted's experiment, in 1820, which proved the existence of such a relationship. Oersted lined up a conducting wire along the north-south direction of the earth's magnetic field. Beneath this wire, he placed a compass. The compass was also lined up in a north-south direction. Oersted then connected the wire to the terminals of a battery. When he did, the compass needle swung around to an east-west direction. This result proved that the flow of current in a wire has an effect on a magnetic compass needle.

(far left) **Oersted's discovery of electromagnetism led to some odd uses.**

(left) **Electromagnet forces may someday be used to launch payloads into space instead of rockets. This picture shows such a launcher on the moon. It is launching packets of minerals towards a space station being constructed in orbit.**

(below) **Electromagnetic forces lift the test train below about 2–3 cm off the track. This is called electromagnetic levitation (lev-uh-tay-shun). Other electromagnets on the track create an alternating current in wire coils on the train. This current moves the train at speeds over 200 km/hr. Trains like this are being tested in Europe and Japan. These experiments may lead to a new transportation system in the near future.**

wire spinning inside large electromagnets. The speed of the generators is carefully controlled. The direction of current flow reverses 60 times each second. This change of direction produces the form of electricity we use. This form of electricity is alternating current with a frequency of 60 Hz.

At first glance, it may seem that the electric energy in the wire is coming from nowhere. You know this is impossible. The Law of Conservation of Energy says that energy cannot be created. The electric energy really comes from the energy of the moving wires. If the wires stop moving, the electric current stops. An electric generator needs a source of energy to cause motion. Falling water can be used to turn the generator, so can steam from water boiled by heat from a fuel or from nuclear energy. In short, chemical or nuclear energy is changed to heat energy. The heat energy is then changed to energy of motion. The energy of motion is transformed into electrical energy.

SUMMARY & QUESTIONS

The next time you use an electrical appliance, think of where the energy comes from. Electricity is generated by electromagnetic induction. This involves the motion of wires in a magnetic field. Electric generators usually produce an alternating current. The electrical energy is really other types of energy converted into electricity by a generator.

1. An electric current is generated in a circuit when a wire is moved through a magnetic field. Predict the effect on this current when (a) the wire is moved at a faster rate (b) the wire is moved first in one direction then back in the reverse direction.
2. Use Fig. 10-9, page 246, and describe how a generator produces electricity.
3. Describe how you can detect the current in a circuit caused when a magnet is moved in and out of a coil of wire.

3 HOME HEATING

The unusual picture above is not a photograph of a haunted house. It is an infrared picture of heat escaping from someone's home. The picture was taken with a machine called a heat sensor. The various colors show different temperatures. From warmest to coolest, the colors are white, red, orange, green, blue, purple, and black. The white parts of the house are areas where the greatest amounts of heat are escaping through windows and doors. Other colors show areas that are poorly insulated. What would a heat sensor picture show about your home? When you finish this section, you will be able to:

- Name the major sources of home heating energy.
- Compare and contrast the major types of home heating systems.
- ▲ Build a device to detect drafts.

In the colder parts of our country, people use a lot of **fuel** to keep warm in the winter. The cost of *fuels* for heating systems has more than tripled in recent years. People have become more and more concerned with

Fuel: Any material used as a source of chemical energy.

10-11. *Keeping homes warm in winter requires a heating system and insulation.*

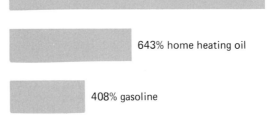

10-12. *This graph shows the average percentage increase in energy cost from 1950 to 1980.*

conserving their fuel as well as their money. This means they also have become more concerned with not allowing heat to escape from their homes. For some people, these problems are a matter of survival in extremely cold weather.

Heating problems can be grouped into two main areas: using the most efficient fuel and keeping the heat in your home. Since the first fireplaces were used for light and heat, people have sought better fuels and better heating systems. Wood, which was the first fuel, was gradually replaced by coal. Coal burns at a higher temperature; therefore, it produces more heat than an equal amount of wood. However, burning coal also produces soot and ashes as well as gases that pollute the air and can be harmful. In this century, coal was replaced by oil and natural gas. These two fuels burn with less pollution than coal and also give more heat per kilogram. Until recently, oil and natural gas were also cheaper to use than coal and were thought to be plentiful. Many newer homes are heated electrically. However, most electricity comes from coal-, oil-, or natural gas-powered generating stations.

Since the mid-1970's, we have faced a growing crisis over oil and natural gas. People were using these fuels faster than had been expected. Oil-producing countries began to force the price of oil and gas to rise and to limit the amounts available. People became concerned about how long the world's supplies of these fuels would last. New discoveries of oil and natural gas only relieved the problem for a while. As a result, more and more people have been returning to coal or wood to help heat their homes. Others are using the sun's energy for heating the air and water in their homes. People are also

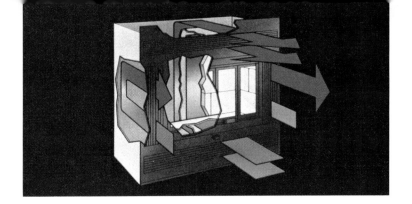

10-13. *All fireplaces can transfer heat by radiation. This fireplace also uses a blower to transfer heat by convection.*

learning to use less fuel through conservation practices. These measures are helpful and necessary. Exactly how much energy can be saved remains to be seen.

How is energy in a fuel used to heat a home? Heat released from a burning fuel is transferred in three ways. Some heat passes by rays into the surrounding space. You can feel this *radiation* on your hands and face from a fireplace or electric heater. Heat is also carried through metal objects. If one end of a nail is held in a candle flame, the other end soon becomes hot. This is called *conduction*. Finally, heat is carried by moving air or water. This is called *convection*. All heating systems use some or all of these transfer methods.

Some heating systems burn a fuel such as coal, oil, natural gas, or wood in a furnace. The chamber in which the burning takes place is called the firebox. The fire heats up the walls of the firebox. Smoke, fumes, some heat, and unburned fuel are carried up the chimney by convection currents. The rest of the heat is transferred through the firebox walls by conduction. It then warms either air or water by conduction and radiation. The heated air or water will then carry the heat to the rest of the home.

Electric heaters and coal or wood stoves heat the air in the room directly by conduction and radiation. Convection then transfers the heat through the room. Solar heating may be *active* or *passive*. Passive systems allow solar energy to warm interior walls and floors during the day. This heat radiates back into the room at night. Active systems use air or fluids in collectors to trap heat. The heated air or fluid then passes through a storage tank. Here water or rocks absorb the heat from the transfer fluid and store it for use later. The most common home heating systems are described on the next two pages.

Chapter 10-3 Home Heating

Home Heating Systems

ELECTRIC HEAT

Heating units are usually located along the baseboard. Electric current passing through special wires causes these wires to become very hot. Cool air flows into the bottom of the unit, is warmed by the hot wires, and rises. This creates a convection current that warms the room. In homes with electric heat, thermostats can keep different rooms at different temperatures to save energy.

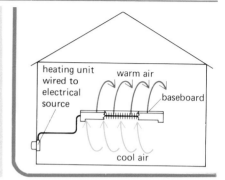

FORCED HOT AIR

Cool air is brought in contact with the hot furnace walls. The air is warmed and rises into the rooms above through a system of ducts. The air enters the room through an opening called a *register.* It circulates in a convection current, warming the room. The cool air is drawn back into another register and returned to the furnace. Fans within the system are used to speed the air flow.

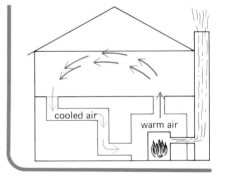

PASSIVE SOLAR HEATING

Passive systems are so varied, only the general concepts can be shown here. In a passive system, the home and its furnishings are heat storage units. Large, south-facing windows, greenhouses, and skylights admit the winter sunlight. As sunlight falls on walls, floors, and furniture, they heat up. In turn, these objects warm the air in the home. Convection currents, possibly aided by circulating fans, carry heat to all parts of the home. Heat absorbed during the day is radiated into the rooms at night.

In summer, roof overhangs, awnings, and shades keep the sun's rays out. Open roof vents allow warm air to escape. The house can be shaded by leafy trees in summer. In winter, the bare branches will not block the sunlight. Evergreens on the north and west sides can block winter winds. Solar homes in most of the country may need backup heating systems for long cloudy periods.

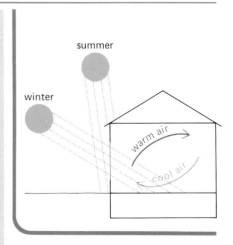

STEAM HEAT

The furnace is surrounded by a hollow space partly filled with water called a boiler. Steam from the boiling water rises through pipes to room radiators. These metal radiators transfer the heat to the surrounding air by conduction and radiation. The air warms the room by convection. As the steam loses heat, it changes back to water. The water drains back to the boiler through steam pipes.

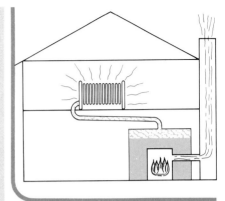

FORCED HOT WATER

This system needs two sets of pipes that are full of water. Hot water from the boiler rises through one set of pipes to the radiators by convection. The cooler water returns through the other set. This system needs a tank in the attic into which the hot water can expand. Without this tank, the expanding hot water could break the pipes.

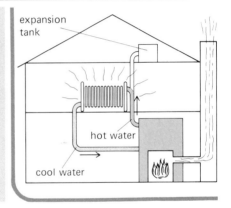

ACTIVE SOLAR HEATING

Solar collectors are mounted on a south-facing roof or in the yard. They are angled to receive the most solar energy possible. The collector is an insulated box with a glass cover. The interior is painted black to absorb the most energy. A heat transferring fluid is heated as it circulates through pipes or channels in the collector.

In a water transfer system, the fluid is piped to a large insulated storage tank. The transfer fluid heats the water in the storage tank, then is pumped back to the collector. The heated water in the storage tank is piped through the radiators to heat the home.

In an air system, air heated in the collector can be blown directly into the rooms or into a storage tank containing rocks. The rocks absorb and store heat from the air. Later, air from the room is circulated over the rocks to be warmed. Both systems need electrical energy to run the various pumps, fans, and controls.

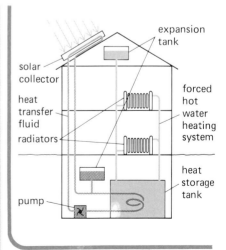

Materials
Pencil or wooden dowel,
 15 cm long
Sheet of plastic wrap,
 13 × 25 cm
Stapler or tape
Quarter
Pencil
Paper

A. Obtain the materials listed in the margin.

B. Staple or tape the short end of the plastic wrap to the pencil. Roll the plastic on the pencil until 20 cm hangs free.

C. The free end of the plastic wrap should move in even a slight breeze. Blow gently at the draftometer from arm's length to test it. Be aware that the draftometer will respond to air movement caused by people walking by.

D. Make a simple floor plan of your home. Include windows, entry doors, fireplaces, attic openings, and outlets.

E. On a day when the outside temperature is at least 5°C lower than the inside temperature, check each of these locations for drafts. Mark the drafts on the floor plan.

F. Try to slide the quarter under closed outside doors. If it fits easily, weather-stripping is needed there.

G. Suggest ways to prevent heat-stealing leaks.

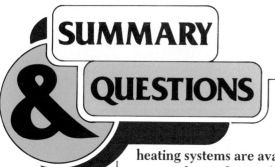

Most people use a heating system for at least part of the year. To conserve fuel, insulation to prevent heat loss is necessary. Several types of heating systems are available. These systems use fuels such as oil, natural gas, coal, wood, or solar energy.

1. List several sources of home heating energy. Which of these sources can be described as fuels?
2. Why are oil and natural gas considered better fuels than coal? Why then are many people switching back to coal and wood as fuels?
3. Compare and contrast active and passive solar heating systems.
4. What types of fuels are used to power most of our electrical generating plants?
5. Compare and contrast a steam heat system and a forced hot water system.

4 ENERGY FOR EVERYDAY USE

Bills! Bills! Bills! We get bills for clothing, food, rent, for everything we need. Did you ever stop to think about how much of a family's income goes to pay bills that are related to energy? As the price of just about everything continues to rise, people are looking for ways to save. How could your family reduce the amount you spend on energy? When you finish this section, you will be able to:

- Explain how energy is used in the United States.
- Define two ways of measuring energy.
- List several ways to reduce energy use.
- ▲ Calculate the electrical energy used in a home.

Our society depends heavily on energy to provide the things we want and need. Until now that energy has been rather cheap. But in recent years the cost of energy has risen sharply. Some of these costs are obvious in the bills we pay. Gasoline bills, electric bills, heating oil bills, and natural gas bills are all direct payments for energy. But what about indirect energy bills? Every time we buy clothing, or furniture, or any of the many objects we have available to us, we are paying for energy. The energy used to make these objects is part of the price we pay. So is the energy used to transport these materials from the factories to us. These are indirect energy payments. Even the cost of the food we eat contains energy payments. The production, processing, and transportation of food uses energy for which the consumer must pay. It seems that everything we use or everything we do consumes energy.

Americans use far more energy than any other people on earth. The United States has less than 5% of the world's population but uses more than 25% of all the energy produced each year. Until the mid-1970's, the rate at which we used energy increased steadily each year. One reason for this was because our population also increased steadily. Another reason was that each person used more energy. Between 1960 and 1979, the population of the United States increased by about 22%. Our energy use increased by almost 80%.

During the mid-1970's, the cost of energy began to rise rapidly. Our rate of use continued to increase but not as quickly as before. In the late 1970's, energy conservation became more important as people tried to reduce energy bills at home and at work. It is estimated that during the 1980's, our energy use will continue to level off. Some of this will be due to conservation efforts, but most will be because of increased efficiency in industry, transportation, and in our homes. Right now experts feel that between 40% and 50% of the energy produced each year is "lost" as heat or wasted by inefficient

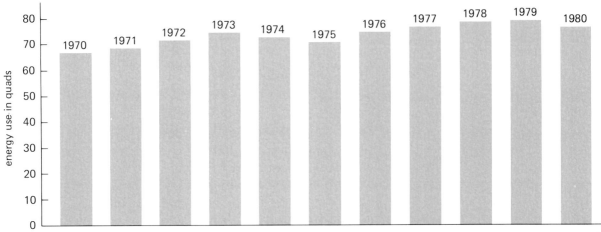

10–14. *Total United States energy consumption from 1970 to 1980.*

BTU (British Thermal Unit): The amount of energy needed to raise the temperature of 250 g of water 1°C.

machines or processes.

How do we measure energy use? In section 1, you learned that electrical energy is measured in *kilowatt-hours (Kwh's)*. The heat energy from the fuels we use is usually measured in **British Thermal Units (BTU's)**. A *BTU* is the amount of energy needed to raise the temperature of 250 g of water one degree Celsius, or about 250 calories. One liter of crude oil will give off about 37,800 BTU's when burned completely. A kilogram of coal releases about 29,700 BTU's and a cubic meter of natural gas gives off about 28,000 BTU's. Kwh's can also be converted to BTU's by multiplying the Kwh's by 3.413.

In 1980, the United States used about 76,350,000,-000,000,000 BTU's of energy. Numbers this large are difficult to understand and to use. For this reason the term *quad* is used. A quad is one quadrillion BTU's (1 with 15 zeros). The number above equals 76.35 quads. Fig. 10-14 shows our energy use in quads from 1970 to 1980. Note that the total decreased between 1973 and 1975. This was due to the oil embargo and economic recession. Another decrease occurred in 1980. This seems to be due mostly to conservation and higher efficiency.

Energy users fall into three groups: homes and businesses, industry, and transportation. Fig. 10-15 shows energy use by groups for 1980. In 1978, transportation used 20.6 quads or 26.4% of all the energy consumed in the United States. Conservation and more efficient engines reduced that amount to 18.6 quads or 24.4% in 1980. Cars that get better mileage, mass transportation, car pools, and a general decrease in unnecessary travel can reduce consumption even further.

Energy use in businesses and homes has also dropped

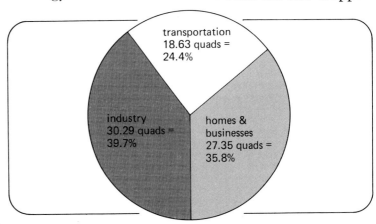

10-15. *This graph shows the total consumption of energy in the United States by each group in 1980.*

Chapter 10-4 Energy for Everyday Use

Reading Meters

READING ELECTRIC METERS

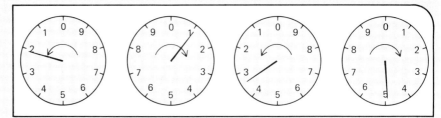

Your electric meter may have four or five dials. The dial on the right shows units of kilowatt-hours of electricity used. The next dial indicates tens of Kwh, the next hundreds, and so on. Note that some of the dials turn counterclockwise and others clockwise. On some meters, all turn clockwise. To read the meter, record the number the pointer has just passed. The meter above reads 2,134 Kwh. Meters are read each month. Your bill is based on the difference between each reading.

READING GAS METERS

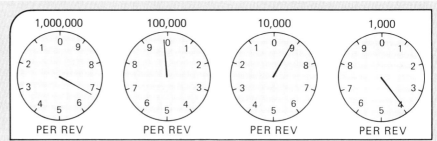

Reading a gas meter is very much like reading an electric meter. First, check the value of each dial. On some meters, the right hand dial measures every 100 cubic feet of gas. On others it measures every 1,000 cubic feet. On each dial, record the number the pointer has just passed. Add two zeros if this dial measures in hundreds and three zeros if it measures in thousands. The meter reading above is 6,993 cubic feet.

from 28.2 quads in 1978 to 27.4 quads in 1980. Some of this energy is used for heating and lighting. At home, we also use energy for preparing and storing food, cleaning, and entertainment. Do you know how much energy your home uses in a day? in a month? in a year?

You can get an idea of these amounts by learning to read electric and gas meters. This is a skill you should develop. You may someday have to read a meter. The feature at the top of this page will help you learn this skill.

As the number of electrical appliances we use increases, so does the amount of energy we use. Some of these appliances are necessary and some are not. Each person should make careful decisions as to what appliances to buy and how much to use them. Table 10-1 lists

TABLE 10-1
What It Costs to Run Electric Appliances

		Estimated at 15.2¢ per kwh			Estimated at 15.2¢ per kwh
Refrigerator (single door)			**Television**		
12 cu. ft., manual defrost	cents per day	32	Color	cents per hour	5.1
Refrigerator/Freezer (two door)			B&W	cents per hour	3.8
14 cu ft., cycle			**Light Bulb**		
manual defrost	cents per day	56	100-watt	cents per hour	1.5
14 cu. ft., frost-free	cents per day	65	60-watt	cents per hour	0.9
18 cu. ft., frost-free	cents per day	83	25-watt	cents per hour	0.4
Freezer			**Room Air Conditioner***	cents per hour	17.4
15 cu. ft. chest, manual defrost	cents per day	60	**Electric Blanket**	cents per hour	1.2
16 cu. ft. upright, manual defrost	cents per day	78	**Clock**	cents per month	30.4
16 cu. ft. upright frost-free	cents per day	106	**Vacuum Cleaner**	cents per hour	10.0
Dishwasher	cents per use	9.1	**Washing Machine**	cents per use	2.9
Toaster	cents per slice	0.4	**Dryer**	cents per load	45.5
Coffee Maker			**Iron**	cents per hour	8.7
Brew	cents per pot	2.3	**Fan**	cents per hour	1.3-3.0
Keep Warm	cents per hour	1.2			

*10,000-BTU unit with Energy Efficiency Rating (EER) of 8.7.

many of the electrical appliances that are found in the home. Make a list of those you have. Estimate the amount of time you use each appliance in one year. Add up the approximate total cost of using these appliances for one year. Which appliances could you do without or use less often if you wanted to reduce your electric bill by about 25%?

Several methods of conserving energy and natural resources are discussed on pages 262 and 263. The easiest way to conserve fuel energy at home is to lower the thermostat. A setting around 20°C is recommended for daytime. At night or when no one will be home for a period of time, the thermostat can be lowered even more. New "set-back" thermostats can do this automatically. Closing doors to unused rooms and turning off their radiators also conserves fuel. So does reducing the amount of hot water used by your family. This can easily be done by lowering the temperature setting on the water tank and by using a flow restrictor inserted in the shower head. Taking a shower uses about one half the hot water taking a bath does.

For many people, keeping heat in their homes is a big problem. Insulation is one solution. Insulation is made

INSULATION

Insulation is an excellent energy saving idea. Ceilings under attics, walls, and floors over unheated garages, basements, and crawl spaces should be well insulated. Insulation comes in rolls, pieces called batts, and as loose material to be poured or blown into place. A vapor barrier between the house and the insulation will keep moisture out of the insulation. Good insulation also keeps a house cool in summer.

DRAFTS

Windows should have double panes or be covered with storm windows. Inexpensive plastic sheets can be used to cover windows. Storm doors and weatherstripping around all windows and doors will stop drafts. Seal openings around window frames and pipes with caulking compound. Use plastic shields to stop drafts through electrical outlets. Insulated window drapes and shades are good energy savers.

THERMOSTATS

Turn thermostats as low as is comfortable in winter. Lower the thermostat even more at night. A 5°C set-back at night can save much fuel. A set-back thermostat will automatically lower the temperature at night and raise it before you wake in the morning. Thermostats should be on interior walls away from drafts. In summer, set thermostats on air conditioners as high as is comfortable.

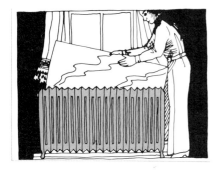

RADIATORS

Turn off radiators in unused rooms. Keep the doors to these rooms closed. Radiators should be kept clean. Make sure air circulation to and from radiators is not blocked by drapes or furniture. When refinishing radiators, use a flat paint. If possible, use black paint since it transfers heat better. Metal reflectors can be placed behind radiators to reflect more heat into the room.

Energy Conservation Ideas

HOT WATER SAVERS

Fix leaky faucets immediately. Taking showers uses less water than baths. Use flow restrictors and aerators (ar-ate-ur) in water faucets to reduce use of water. Most clothes washers use 85 liters of hot water per wash. You can use cold water and cold water detergents instead. Insulate hot water heaters and pipes well, but do not cover vents, switches, and drains. Lower the temperature setting to 45° or 50°C.

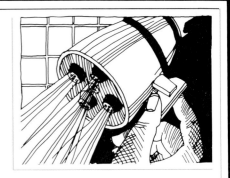

FIREPLACE HINTS

Keep the damper tightly closed when not in use. Glass doors on the fireplace prevent loss of some room heat while the damper is open. Various types of heat exchangers can be placed in the fireplace to circulate heat that would have otherwise escaped up the chimney. Lower the house thermostat when using the fireplace. If possible, close doors to room with fireplace but open a window slightly to provide ventilation.

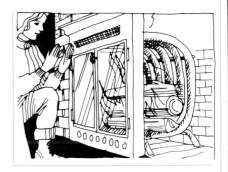

LIGHTING

Lighting is another area where we waste energy. Turn off lights when leaving rooms. Provide "task lighting" over work areas instead of lighting the whole room. Use lower wattage bulbs where less light is needed. Use lighter colors for walls, carpets, and furniture. Fluorescent lights are more efficient than incandescent lamps. Keep lamps, shades, and reflectors clean for more light.

SAVING ELECTRICITY

In addition to lights, turning off other appliances can save energy. Some TV sets called "instant-on" types use electricity even when the picture is turned off. Unplug these sets at night or during vacations. Air dry clothes and dishes instead of using dry cycles on washers. Towel or air dry your hair. Check seals on refrigerator and freezer doors to make sure air doesn't leak out. Keep coils under refrigerator clean. Don't open the refrigerator door unnecessarily.

10-16. *A well-insulated home wastes a minimum of heat.*

from materials that do not transfer heat well. It is available in rolls, pieces called "batts," or as a loose material that is blown or poured into place. How much insulation a home needs depends on the average winter temperatures of the area in which the home is located.

Another serious problem is the loss of heat through leaks. Much heat is lost through doors and windows that are not covered with storm windows or sealed with weatherstripping. Estimates show that up to 50% of the heat lost escapes through windows. Use of window shades and insulated drapes can reduce this amount.

Drafts of cold air that enter your home can also cause your use of energy to increase. Even small air leaks around electrical outlets account for up to 20% of the leaks in an average home. Fortunately, these leaks can be reduced by the use of nonflammable shields.

Local utilities and conservation groups can give you many ideas for saving energy that are easy to follow. For example, many of these groups will do an inexpensive survey of your home to see where you are losing heat or using unnecessary energy. Often they will use a heat sensor to check for heat losses. These checks are called Energy Audits.

Materials
Paper
Pencil

A. Obtain the materials listed in the margin.

B. Copy columns II, III, IV of Table 10-2 onto your paper.

C. Column I of Table 10-2 shows the dials of an electric meter at the same time of day for 6 days. Record the meter reading for each day in column III.

D. Complete column IV by subtracting the reading for day 1 from day 2, day 2 from day 3, and so on.

   1. What is the total usage for 5 days?
   2. What is the average use per day?
   3. If that is the daily average, what would the usage be per month (30 days)?
   4. What would it be for a year (12 months)?

E. The billing rate shown on this person's bill was 15.2¢ per Kwh.

   5. What would this person's monthly bill be?
   6. How much would this person pay for the year?
   7. Do you think a home uses the same amount of energy each month? Explain.
   8. Using the answer to question 3, calculate how many BTU's of energy this home uses per month.

TABLE 10-2

Column I Meter Dials				Column II Day	Column III Meter Readings in Kwh	Column IV Kwh used since last reading
				1		
				2		
				3		
				4		
				5		
				6		

SUMMARY & QUESTIONS

Much of the energy we use could be saved by better conservation practices and improved efficiencies. Amounts of energy used can be described by several measurement units. Such comparisons help us to keep track of and conserve energy.

1. List two causes for the rapid increase in energy use in America in the past thirty years.
2. Name two factors that are expected to help reduce the amount of energy we use.
3. Explain how the energy used in 1980 was distributed among energy users.
4. What is a BTU? What is a quad? How many quads of energy did Americans use in 1978?
5. List several ways to reduce heat loss from a home.

Chapter 10-4 Energy for Everyday Use

VOCABULARY REVIEW

Match the description in Column II with the term it describes in Column I.

Column I
1. fuse
2. transformer
3. watt
4. kilowatt-hour
5. electromagnetic induction
6. fuel
7. BTU

Column II
a. A part of a circuit that changes the voltage.
b. Measures how fast electrical energy is changed into another form.
c. Prevents too much current from flowing through an electrical circuit.
d. A measure of the amount of heat energy in a fuel.
e. Any substance used as a source of chemical energy.
f. Produces electricity by motion in a magnetic field.
g. Used to measure the consumption of electric energy.

REVIEW QUESTIONS

Choose the letter of the phrase that best completes each of the following statements.

1. The most important reason electric energy is commonly used is that it (a) is cheap (b) is clean (c) can be changed into most other forms of energy (d) can be made without polluting.
2. The electric meter in your house measures (a) volts (b) watts (c) amps. (d) kilowatt-hours.
3. Electromagnetic induction is the process that produces (a) a magnetic field from a current in a wire (b) an electromagnet using electricity in coils of wire and an iron core (c) an electric current from motion in a magnetic field (d) a permanent magnet from a piece of iron.
4. The electric current supplied to most homes is called (a) DC (b) AC (c) low voltage (d) high voltage.
5. An electric generator is a machine that converts (a) nuclear energy into electric energy (b) magnetism into electricity (c) the energy of burning coal into electric energy (d) some form of energy into electricity.

6. Solar heating systems are classified as (a) positive and negative (b) active and passive (c) AC and DC (d) active and indirect.
7. Which of the following groups consumes the greatest percentage of energy in America? (a) business and homes (b) transportation (c) industry (d) electric companies.
8. Which of the following heating systems does not need a furnace? (a) steam (b) electric (c) hot water (d) hot air.
9. Furnaces are usually fueled by (a) oil (b) natural gas (c) coal (d) any of these.
10. The voltage that comes into a home is usually (a) 110 volts DC (b) 110 or 220 volts AC (c) 110 or 220 volts DC (d) 110 volts AC/DC.

REVIEW EXERCISES

Give complete but brief answers to each of the following.

1. Explain how an electric motor works.
2. Predict what would happen in an electric circuit if a magnet is rapidly moved into and out of a coil of wire.
3. Explain how an electric generator is able to produce electricity.
4. What is a 60 hertz alternating current?
5. Why are oil and gas considered more efficient fuels than wood?
6. Explain how the three types of heat transfer (conduction, convection, radiation) are used in a forced hot water heating system.
7. List five different ways you can reduce the amount of energy you use.
8. How can a home be built to use passive solar heat?
9. How does a forced hot water heating system differ from a forced hot air system?
10. Explain how an active solar hot water heating system might work.

CHAPTER 11 SOURCES OF ENERGY

1 FOSSIL FUELS

Over a hundred years ago, thousands of people headed west in search of quick fortunes. They were drawn by the newly discovered gold fields of the western states. Overnight, small towns "boomed" as miners poured in. In time, the gold that was left became too expensive to mine. Most people moved on. Many towns became ghost towns and fell into ruin. Today there is another gold rush to the western states. More "boom towns" are developing. This time, however, the search is not for "yellow gold" but for what is called "black gold"—coal. When you finish this section, you will be able to:

Unit III   How Do We Make and Use Energy?

- Explain how fossil fuels are formed.
- Name the three major energy sources of today.
- Describe the known fossil fuel reserves of the United States.
- Interpret graphs on energy data.

Long ago, North America looked very different from what it does today. Vast, swampy areas existed between what are now the Appalachian and Rocky mountains. For long periods of time, the climate in these areas was favorable for the growth of plants and trees in the warm, marshy lowlands. As they died, the remains of these plants and trees became part of the thick layers of partly decayed matter on which they grew. This brown, partly decayed plant matter is the first step in the formation of coal. It is called *peat*. In places, the peat deposits were more than a thousand meters thick.

Gradually, the peat was buried under sand and other material from rivers and shallow seas that sometimes covered these areas. The layers of sediment became thicker. Over many years, the increasing weight of the sediment squeezed the peat more and more. Temperatures in the peat rose. This pressure and rising temperature caused chemical changes in the peat. Much of the water was squeezed out of the deposit. The plant remains were slowly changed to a type of coal called *lignite* (**lig**-nite). Lignite is brownish-black in color. It is sometimes called brown coal. Sometimes the original plant material can still be identified. Lignite catches fire easily but burns with a smoky flame. It gives off more heat per unit of mass than peat or wood.

In many places, the sediments continued to pile up. The lignite deposits were subjected to even higher pressures and temperatures. More chemical changes took place. Most of the remaining water was driven off. It carried with it many elements in solution. What was left behind became a dark brown or black material with a high carbon content. This was *bituminous* (buh-**too**-muh-nus)

11-1. Coal was formed from the remains of plants like these that lived in marshes long ago.

11-2. Chemical changes convert peat into lignite, then to bituminous, and finally to anthracite coal.

Petroleum: A black, oily liquid formed from the remains of ancient sea life.

Natural gas: A colorless, odorless gas formed from the remains of ancient sea life.

Hydrocarbons: Substances made up only of carbon and hydrogen.

11-3. Oil and natural gas are often found together in underground traps or pools.

or soft coal. It takes about 10 meters of peat to make 1 meter of bituminous coal. Bituminous burns with a somewhat smoky but hotter flame than lignite. Most of the coal in the United States is bituminous.

In a few places, the formations of bituminous were squeezed and compressed even more. This happened mostly in areas where mountains were forming. The pressures and temperatures were very high. Under these conditions, bituminous becomes *anthracite* (**an**-thruh-site). Anthracite is a hard, black, and brittle form of coal. It burns with a blue smokeless flame and gives off much heat. Anthracite makes up only about 2% of the coal reserves in the United States. Most of this is located in the Appalachian Mountains.

The other two major fossil fuels are **petroleum** and **natural gas**. The formation of these two fuels is not yet fully understood. Both are thought to form from the decay of organisms that once lived in the sea. The remains of these animals and plants settled to the sea floor and were gradually covered by sediments. After millions of years, the remains were slowly changed by heat and pressure into a substance made mostly of hydrogen and carbon. Such substances are called **hydrocarbons**. *Petroleum*, which is often called oil, and *natural gas* are mixtures of many different *hydrocarbons*.

The tremendous pressure deep underground forces the oil and natural gas out of the rock material in which they usually form. These substances begin to move upward through spaces or fractures in the rocks. Often they meet a rock type or a formation they cannot pass through. In this way, they are trapped and form *oil pools* and *gas pockets*. Water, also trapped in the rocks, often forces the

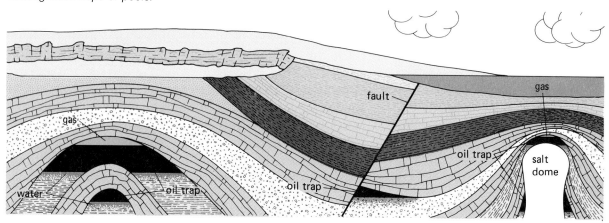

Unit III How Do We Make and Use Energy?

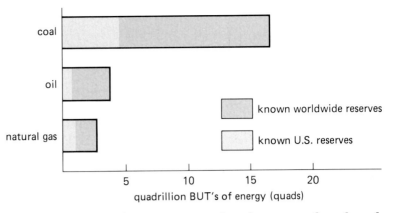

11-4. *The shaded portions of this bar graph show that the United States has only a small part of the world's proven oil and natural gas reserves.*

less dense oil and gas to rise. Oil and gas are often found together. However, since gas can move through the rock spaces more easily than oil can, there are many gas deposits without oil.

The supplies of fossil fuels in the earth's crust are limited. It has taken millions of years for them to form. In addition, they are not evenly distributed throughout the world. The United States has only 4% of the known oil reserves and 8% of the known natural gas reserves in the world. However, 28% of the world's known coal deposits are found in the United States.

At the current rate of use in the United States, the known supplies of oil and gas will last less than fifty years. Only coal, which now provides less than 30% of the world's energy, could last any great length of time. At current rates, coal would last at least 200 years more. These figures deal only with known resources of common sources of fossil fuels. Several uncommon sources will be discussed in later sections.

11-5. *Coal fields of the United States.*

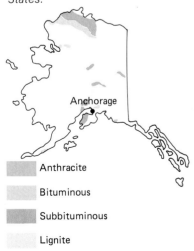

Chapter 11-1 Fossil Fuels

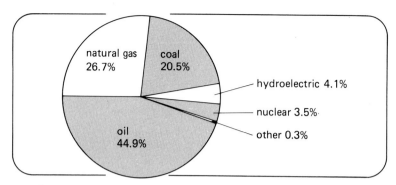

11-6. *This graph shows the sources of all of the energy consumed in the United States in 1980. Notice that fossil fuels provided over 90% of all the energy used.*

There are probably greater amounts of undiscovered coal remaining within the earth than petroleum. Coal comes from areas that were under shallow seas that once covered large parts of the continents. Because the continents are very old, enough time has passed for large deposits to be formed. Petroleum, on the other hand, is found in rocks formed along the edges of the continents. Many of these rocks have been destroyed by mountain building and erosion. As a result, petroleum and natural gas deposits are younger and less common. Some geologists feel that the undiscovered deposits of oil, natural gas, and coal may be equal to our known reserves.

Over 90% of all the energy that is used at present comes from the burning of fossil fuels. For the past hundred years, the increasing need for energy has been met by these fuels. The supply is no longer able to meet current demands. In the future, other sources of energy may be able to replace fossil fuels. However, the best answer to the present energy crisis may be to carefully conserve the energy supply now available.

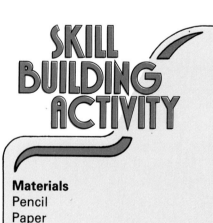

SKILL BUILDING ACTIVITY

Materials
Pencil
Paper

READING GRAPHS

A. Obtain the materials listed in the margin.

B. You may wish to review the section on "Reading Illustrations" in the Resource Book.

C. Graphs are often used to present data in an easy to understand manner. There are three types of graphs used in this text: circle graphs, bar graphs, and line graphs.

D. Look at Fig. 11-4 on page 271.
   1. Which type of graph is used here?
   2. What information is being presented in this graph?
   3. What does the number scale along the bottom stand for?
   4. Which types of fossil fuels are discussed?
   5. Which type of fossil

fuel is most abundant in the world?

6. Which type is the least abundant?

7. What percentage of the world's coal reserves are found in the U.S.?

8. How many quads of energy does the U.S. supply represent?

9. In 1980, the U.S. used about 17.5 quads of energy from its reserves. At that rate, how many years will these reserves last?

10. What percentage of the world's reserves of oil and natural gas are found in the U.S.?

E. Look at Fig. 11-6 on page 272.

11. Which type of graph is used here?

12. What information is being presented here?

13. Which energy sources are discussed here?

14. Which source provided the greatest portion of our energy in 1980?

15. Approximately what percentage was that?

16. List each of the sources and the percentage of our energy they provided in 1980.

17. What total percentage of our energy was provided by fossil fuels in 1980?

18. What sources might be included in the column labeled *Other*?

SUMMARY & QUESTIONS

For over a century, people have depended on fossil fuels to supply the bulk of the energy needed for society. However, the supplies of fossil fuels are limited. The rate at which we are currently using these fuels is increasing. Known resources of these fuels are fast being used up.

1. Briefly describe how the four stages of coal are formed.
2. Describe the formation of hydrocarbons.
3. Name the three major sources of energy today.
4. Why do scientists feel there is more undiscovered coal in the world than petroleum?

2 ENERGY TODAY AND TOMORROW

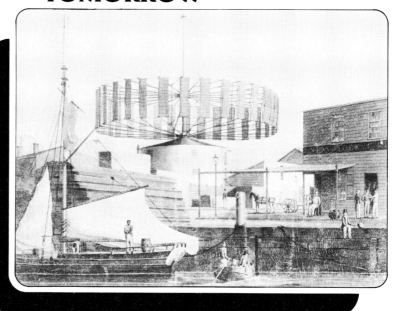

11-7. *The first windmills were developed about* A.D. *750. Wind power is still an important resource.*

Flying kites, moving ships, turning grinding wheels, pumping water—the wind has long been used as a source of energy. Now scientists are taking a closer look at the wind as well as other "old-fashioned" energy sources. They are looking for new ways these old friends can help us in our constant search for energy. When you finish this section, you will be able to:

● Describe the changing nature of our energy sources.

● Describe two types of nuclear reactors that may be used as energy sources in the future.

● Discuss the future role of today's energy sources.

▲ Construct a graph of energy use from data tables.

In the endless search for more energy to do work, humans first relied on their own muscles. Later the load was shifted to the muscles of animals such as oxen, camels, and horses. Centuries ago, the wind was har-

Unit III How Do We Make and Use Energy?

11-8. Water power can be used to turn machinery or generate hydroelectricity. This is part of a generating plant.

Hydroelectric power: Energy obtained from water moving in rivers and streams.

Nuclear power: Energy obtained by splitting atoms in a nuclear reactor.

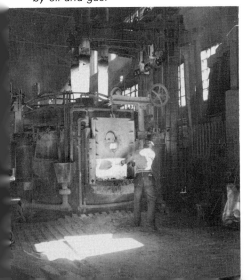

11-9. Coal replaced wood as a fuel because it was more efficient. In turn, it was replaced by oil and gas.

nessed to do useful work. Windmills capture the kinetic energy of the moving wind to drive a series of turning shafts and gears. This energy can be made to pump water and turn grinding wheels or other machines. Holland has long been famous for its windmills. By connecting the windmill to a generator, wind can be used to create an electrical current. In the early part of this century, thousands of American farms used wind power. Estimates say over 150,000 windmills are still at work in the United States today.

The energy of falling water was also harnessed long ago. At first, it was used to turn water wheels. These wheels could also turn grinding wheels or be connected by pulleys, ropes, or belts to drive machines. As with wind power, the energy of falling water was later used to turn electrical generators. This type of power is called **hydroelectric power.** Very large dams such as the Hoover or Glen Canyon dam in Arizona store large reservoirs of water. Some of this water is moved through shafts to spin the generators.

When James Watt developed the improved steam engine, a new source of power was introduced. Heat from a burning fuel was used to produce steam. This steam can run machines or can be used to turn electrical generators. Early steam engines used wood as a fuel. Wood was soon replaced by coal because coal provided more heat per unit of mass. It was also easier to work with and was plentiful. By the late 1940's, coal had been replaced by oil and natural gas as the major fuels for American industry. Again the reasons were more heat, easier handling, cleaner burning, and less cost. At the present time, oil provides about 45% and natural gas about 26% of all the energy used in the United States. Coal supplies about 20%.

During the 1960's, **nuclear power** was added to the nation's energy pool. In an atomic power plant, *chain reactions* in the fuel are controlled by the control rods that absorb some of the nuclear particles. The tremendous energy that is released as heat can be used to create steam. The steam in turn can power an electric generator. The fission of one gram of uranium produces as much heat as the burning of three tons of coal. In 1980, there were 75 nuclear power plants operating in the United States. Together they produced 3.5% of all the energy we used. This amount is misleading. Most of the reactors

11-10. *Wind generators like this may contribute to our future energy supply.*

are located in the northeast and midwest. *Nuclear power* supplies 80% of New Hampshire's electricity, 60% of Maine's, and 50% of Connecticut's and Nebraska's.

What role will these sources play in our energy future? Their present and future roles are controlled by several factors. These include technical and environmental problems, availability, and cost. For example, the wind is not a dependable source of power. There are not many locations where the wind is steady enough to produce a constant flow of electric current. The average wind speed in the United States is 12 km/hr. A wind generator with 30 m blades needs an 18 km/hr wind to produce 200 kilowatts of electricity. It takes a lot of wind generators to produce enough electricity for even a small town. Some type of storage system is needed for calm periods. These problems make wind generated electricity costly at the present time.

There is also an environmental problem with the high level of noise large generators make. However, wind power has been successfully used to produce part of the energy needed for farms or single-family homes. So, wind generating systems can contribute some energy to our overall needs in certain situations. Much research is being done on new designs for wind-powered generators.

Hydroelectric (hie-droe-ih-**lek**-trik) *power*, like the wind, is a *self-renewing* energy source. The supply of water is constantly being renewed every time it rains upstream. However, there are only a limited number of locations where large dams like the Glen Canyon can be built. Most of these sites have already been used. Other proposed locations have been dropped because of public concern over scenic areas and plant and animal communities such large reservoirs would alter or destroy. There are, however, two sources of hydroelectric power that may play an increasing role in future energy plans.

The first of these is called *pumped storage*. Many large fossil fuel and nuclear plants run at a steady pace, no matter what the demand for electrical power is. These plants can use the extra power during the night when demand is low to pump water to storage areas on top of nearby hills. Then, when demand is high, the water is released through a hydroelectric generator to generate more power. The second option is a small local hydro plant. Hundreds of millponds and old, small water stations exist that may be useful for generating small amounts of electricity for local use. Newer hydrogenerators can generate electricity from the normal flow of a river without dams.

We are all aware of both the pluses and the minuses of nuclear energy. Accidents at nuclear generating plants have called public attention to the complex problems of reactor safety. There is also concern over disposal of radioactive wastes and the environmental problems such plants may cause. Even though doubts raised by these questions make the future of nuclear energy uncertain, the amount of energy we get from this source will continue to increase in the near future. The number of new plants under construction is equal to those in operation. With these new plants in operation, the government projects that the electricity from nuclear power will triple by the year 2000.

Two new types of reactors should be mentioned here. Uranium is needed to operate the nuclear reactors now in use. The United States has 29% of the known uranium resources, yet even that amount will not last forever. A new technology being used in four countries is the

11-11. *Water is pumped to storage during off-peak hours and released later to generate hydroelectric power.*

Breeder reactor: A nuclear reactor that creates more nuclear fuel than it uses.

Nuclear fusion: A nuclear reaction in which two small nuclei join to form one large nucleus.

breeder reactor. This reactor actually makes more fuel than it uses. This can extend our fuel supply as much as 70 times what is available now. However, the new fuel is plutonium. It only takes a few kilograms of plutonium to make an atomic bomb. Any country with a *breeder reactor* can make nuclear weapons. This raises serious questions in many people's minds about threats to world peace. These questions are affecting the future of the breeder reactor program in the United States.

Many scientists think the development of a **nuclear fusion** reactor will be the solution to the energy crisis. Fusion reactors do not have meltdowns or high levels of radioactivity in their wastes. Hydrogen isotopes are the fuel. These are so abundant in the oceans that we would never fear shortages.

However, a nuclear fusion reactor is not yet possible. Fusion requires maintaining temperatures of over one million degrees Celsius for the nuclei to collide and fuse. Experimental devices using electromagnetic forces and laser beams are now being tested. Research on fusion reactors has been going on for twenty-five years, but it will probably be another forty years before it pays off. If the technology problems can be solved, *nuclear fusion* may well be an answer to our long range needs.

The very near future concerns us now, however. What role will fossil fuels play? We have already said that the easily recovered resources of oil and gas are running out quickly. It becomes more and more expensive to recover the deeper and harder to reach known resources. Coal will last longer but increased rates of use will cause increased problems. Coal contains a number of impurities. When coal is burned, ashes, soot, and such gases as carbon dioxide and sulfur dioxide are released. Unless these are removed by expensive processes, serious environmental problems may result.

11–12. *Land destroyed by strip mines such as this one can be reclaimed for useful purposes.*

Skill Building Activity

CONSTRUCTING A GRAPH

Materials
Graph paper, quarter-inch rule
Ruler
Lead pencil
Colored pencils, 5

A. Obtain the materials listed in the margin. Use Fig. 11-13 to help you understand the directions.

B. Table 11-1 shows the use of several types of energy in the U.S. every 5 years from 1950 to 1980. You will plot this information on a line graph.

C. Hold your graph paper so that the short edge is along the bottom. Find the point where the third line up from the bottom of the paper and the third line in from the left edge of the paper cross each other. Draw a dot on that point.

D. Use your pencil to darken the horizontal line which runs from your dot to the right hand edge of the paper. This is the X axis. Now darken the line which runs from your dot to the top of the page. This is the Y axis.

E. The years will go on the X axis. Each block represents one year. The quads of energy used each year will go on the Y axis. Each block represents one quad. Mark off both scales.

F. Begin by plotting the use of coal during these years. Use a colored pencil. In 1950, we used 12.9 quads of energy from burning coal. Find the line marked 1950. Come up this line until you meet the line for 12 quads. You will have to estimate the position for the .9 quads. It would place your dot just beneath the 13 quad line.

G. Plot the rest of the positions given for the use of

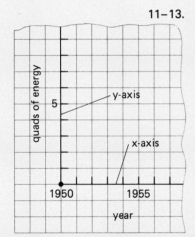

11-13.

TABLE 11-1
Consumption of Energy by Type, 1950–1980 in Quads

Year	Coal	Natural Gas	Oil	Hydropower	Nuclear Power
1950	12.9	6.0	13.3	1.4	0
1955	11.5	9.0	17.3	1.4	0
1960	10.1	12.4	19.9	1.7	0.01
1965	11.9	15.8	23.3	2.1	0.04
1970	12.7	21.8	29.5	2.7	0.24
1975	12.8	20.0	32.7	3.2	1.9
1980	15.7	20.4	34.3	3.1	2.7

TABLE 11-2
Oil and Natural Gas Consumption 1970–1980 in Quads

Year	Oil	Natural Gas
1970	29.5	21.8
1971	30.6	22.5
1972	33.0	22.7
1973	34.8	22.5
1974	33.5	21.7
1975	32.7	20.0
1976	35.2	20.4
1977	37.1	19.9
1978	38.0	20.0
1979	37.1	20.7
1980	34.3	20.4

Chapter 11-2 Energy Today and Tomorrow

coal. Connect the dots and label the line "coal."

H. Repeat steps F and G for each energy type given. Use a different color for each. Label each line.

 1. What were the first and second most important fuel sources in 1950? in 1960?
 2. What happened to coal use between 1950 and 1980?
 3. Between what years did nuclear power begin to make an important contribution to our energy use?
 4. Which type of energy changed least during this period? Why?

I. Table 11-2 lists the use of oil and natural gas each year between 1970 and 1980. Add these points to the graph and connect them.

 5. What caused the drop in oil use between 1973 and 1975?
 6. In which year did oil use reach its highest point?
 7. What may have caused the drop between 1978 and 1980?
 8. In which year did natural gas use reach its peak?
 9. Did the 5 year points show what actually happened between 1970 and 1980?

SUMMARY & QUESTIONS

As our need for energy has steadily increased, our sources have gradually changed. In the past, such changes were made by choice rather than out of need. Today we are making changes because our current energy sources are running out.

1. Describe how the wind and running water can be used as energy sources.
2. Describe what a breeder reactor does.
3. How does a fusion reactor produce energy? What are the advantages of fusion reactors over fission reactors?
4. Explain some of the problems associated with using coal as an energy source.
5. What is hydroelectric power? Describe two possible new sources of hydroelectric power.

3 ENERGY ALTERNATIVES

Early settlers in parts of Colorado built their fireplaces from stones they found in the area. Imagine their surprise when the heat from their fires caused these same stones to burst into flames. They later realized the stones contained a black oil that burned. Today, millions of kilograms of "the stone that burns" are picked up with every bite of the giant power shovels that work the area. Modern scientific "pioneers" are searching for ways to "cook" the oil from the stones. As you can imagine, it is not an easy job. When you finish this section, you will be able to:

● Describe several unconventional fossil fuel sources for future energy supplies.

● Describe several other possible sources of energy for the future.

● Explain the problems associated with several future energy sources.

▲ Determine the effect of colors on solar heating.

We have heard a lot about unconventional sources of fossil fuels. These are fuels that are not able to be obtained in the usual ways. Examples of these are the *oil shale* and *tar sands* deposits of the western states. These contain many times more oil than typical sources. Formations called *tight sands* (tightly packed sandstones) and *geopressured zones* (deeply buried rocks) contain many times the supply of natural gas currently available. Yet all of these resources face the same problems. They will be very difficult and very expensive to obtain.

Removing the oil from shale and tar sands requires moving vast quantities of usable rock and getting rid of even more waste rock. Very expensive and complicated processing plants are needed to extract the oil. Turning coal into a gas or a liquid is another possibility for the

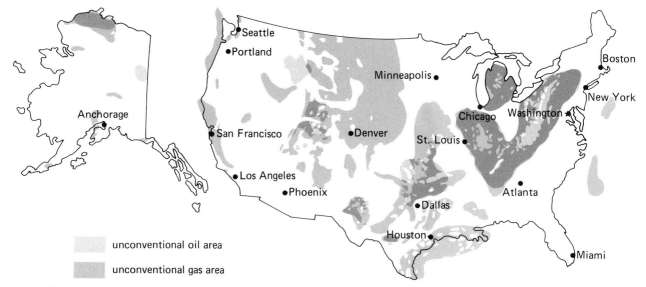

11-14. *Unconventional sources include tar sands, oil shale, tight sands, and geopressured zones.*

future. These processes are also difficult, expensive, and possibly polluting. Fuels made this way are called synthetic (sin-**thet**-ik) or *synfuels* for short.

There are many who feel that the ecological problems that go with these resources make them impractical. Some methods of getting the oil and gas from the rocks are difficult. They require surface mining and the use of large amounts of already scarce water. These steps may cause increasing air pollution by dust and particles from the refining process. They may also cause disruption of wildlife and of scenic areas. There could also be social problems caused by the rapidly increasing populations of the mining towns.

Unless new sources are found or breakthroughs in technology occur, fossil fuels will continue to play the major role in energy production at least into the twenty-first century.

Why is all this concern about energy necessary? It has been estimated that in the next twenty-five years we will use an amount of energy equal to all the energy used

11-15. *Mining oil shale and tar sands may cause serious environmental problems.*

in all recorded history. It is clear that our supplies of fossil fuels will not last forever. There is also only a limited amount of uranium available for fission plants. But, as one scientist put it, "We are not running out of energy, only cheap forms of energy." We must look to other sources of energy in the future.

The sun is one of these sources. We have been using solar energy indirectly for centuries. Fossil fuels formed from plants and animals that stored up solar energy when they lived along ago. But these fuels are limited. One day they will be gone.

The sun is also the source of energy for wind and hydropower. The wind blows because the sun heats the earth unevenly. Therefore, when we use wind power we are using solar power indirectly. Sunlight also causes water to evaporate from the oceans. Later, it falls as rain on the land. As this water returns to the sea, it can be used to generate hydropower. Other ways of producing power from moving water are being studied. (See page 284.)

Another form of solar power uses fast growing plants. Cottonwood trees and other plants grow quickly. They can be burned as fuel in a nearby power plant. Other plants, such as corn and sugar beets, can be converted into alcohol and used as an automobile fuel. Gasohol is a mixture of gasoline and alcohol. These types of fuels, made from living plants, are called **biomass fuels.** Even the wood we burn in fireplaces or stoves is a *biomass fuel*.

The best way to make use of solar energy would be the direct use of sunlight. This use of solar energy changes the incoming radiant energy from the sun directly into a usable form. For example, solar energy can be changed into heat for a home. Heating hot water and direct passive and active space heating were discussed in Chapter 10.

A second direct use of solar energy is the use of **solar cells.** Sunlight falling on a *solar cell* is converted into electricity. However, each cell produces only a small amount of electricity. Solar cells are also very expensive to make. At the present time their use is limited to spacecraft and other areas where no other source of electricity can be used. Much research is now directed at reducing the cost of making solar cells.

In the future, there may be another way of using solar energy. Central solar power stations may be built in sunny places such as a desert. At the stations, large

Biomass fuel: Any plant or animal matter that is used as a source of energy.

Solar cell: A device that changes sunlight directly into electricity.

11–16. *Biomass fuels may become more important in our energy future.*

Chapter 11–3 Energy Alternatives

Alternate Energy Sources

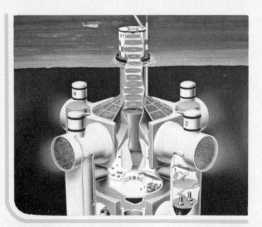

In tropical waters, the surface layers are hot enough to make liquid ammonia change into a gas. Floating power plants like the one in this photo could use the gas to spin generators. The ammonia would be turned back into a liquid by cooling it with cold water from the deep ocean. The electricity could be used by "floating factories" to remove minerals, chemicals, and hydrogen for fuel from the ocean water. A test project is now underway off Hawaii.

Tidal power might be used in a few special locations. A dam must be built across the mouth of a bay. Water is then trapped behind the dam at high tide. At low tide, the water is allowed to run out through the dam and used to turn electrical generators. Only a few places along the coast can be used for this purpose. In the United States, there are only two locations where the tides are high enough. Both are far from where the energy is needed. The picture shows a tidal power plant in France.

The earth's heat can be used in many ways. Geo-pressured areas contain natural gas dissolved in hot water. When tapped, the gas can be separated and fed into pipelines. The hot water can generate electricity.

Natural hot springs can be tapped to provide energy. This is now done in Italy, Iceland, and California. In areas with hot rocks but little ground water, water can be pumped in to be heated. This water is then drawn back up through other pipes and used to generate power. The photo at the left shows a geothermal power plant in California.

11-17. *Practical and affordable ways to use solar energy will help solve our energy problems.*

Geothermal energy: Energy obtained from heat within the earth's crust.

numbers of solar energy collectors would produce heat. This would generate steam to run a central power plant. In effect, solar energy would take the place of fossil fuels or nuclear power sources.

In some places, it is possible to use **geothermal (jee-oe-thur-mul) energy.** *Geothermal* energy comes from heat within the earth's crust. The heat causes ground water or water pumped into these areas to become hot. The hot water can be used for heating. Steam from such sources can be used to generate electricity. Large amounts of natural gas dissolved in hot water have been found deep underground in many areas. This gas is under great pressure. It is hoped that the gas in these *geopressured* areas can be removed. If so, it will extend our usable gas resources many times.

There are only a few areas where geothermal energy is now being used. One is outside San Francisco. Half of the electrical power for the city is generated there. More locations are being investigated now. Geothermal power may provide an important share of future energy needs.

What does our energy future look like? Our future energy needs will be filled by an even greater combination of energy sources than we have now. In the near future, no single source will provide a majority of our energy as in the past. Conservation will continue to play an important role. We must also increase the efficiency of the machinery we use. Possibly, somewhere in the future, new technology will solve the energy problem. It will not be easy, and it will not be cheap.

A. Obtain the materials listed in the margin on page 286.

B. With your pencil, punch a hole in the bottom center of both cups. Make each hole just large enough to slide a thermometer through.

C. Place the black cup upside down over the black can to form a tight lid. Insert a thermometer through the hole. Do the same with the unpainted can and cup. **CAUTION:** DO NOT FORCE THE THERMOMETER. IF IT SHOULD BREAK DO NOT TOUCH IT. CALL YOUR TEACHER.

D. If possible, place both cans and the third thermometer in bright

Chapter 11-3 Energy Alternatives

Materials
2 soup cans, with labels removed, 1 painted black
2 styrofoam cups, 1 painted black
3 laboratory thermometers
Light source, 200 watts or more (or use sunlight)
Pencil
Paper

sunlight. If not, set the two cans and thermometer an equal distance from a bright light source (not more than 30 cm). Arrange it so that the thermometer bulb is shaded from direct light.

E. Record the beginning temperature of each thermometer before placing it in the light.

F. Record the temperature of all three thermometers every 2 min. Do this for a total of 10 min. Remove the light. Record the temperatures every 2 min. for another 10 min. If using sunlight, record for 15 min. in the sun and 15 min. in shade.

1. Which thermometer showed the greatest increase in temperature over time?
2. Which thermometer showed the least increase?
3. Why was the thermometer without the can used?
4. Which can showed the greatest drop in temperature over the second time span?
5. Based on your results, which color clothing would be best for wear during the summer? Why? During the winter? Why?
6. What color would you paint the inside of the box and the pipes in a solar hot water heater? Why?

SUMMARY & QUESTIONS

"We are not running out of energy sources. We are running out of cheap energy sources." This statement summarizes the future of our energy resources. There are many unconventional sources of energy being explored today. Each has its promises and problems. The solutions to the energy problem will not be easy, and they will not be cheap.

1. Describe three unconventional sources of oil and natural gas.
2. What are some of the technical problems involved with utilizing these sources?
3. What are some of the ecological problems involved?
4. Describe three proposals for using indirect solar energy.
5. Describe three direct uses of solar energy.
6. How can heat within the earth be used as an energy source?

VOCABULARY REVIEW

Match the description in Column II with the term it describes in Column I.

Column I
1. nuclear fusion
2. natural gas
3. biomass fuel
4. breeder reactor
5. solar cell
6. petroleum
7. acid rain
8. geothermal energy
9. hydrocarbon
10. hydroelectric power

Column II
a. A reactor that creates more nuclear fuel than it uses.
b. A substance made up only of carbon and hydrogen.
c. A nuclear reaction in which two small nuclei join to form one large nucleus.
d. A black, oily liquid formed from the remains of ancient sea life.
e. Any plant or animal matter that is used as a source of energy.
f. Sulfur dioxide from burning fuels dissolved in rain water.
g. Energy obtained from moving water.
h. Energy obtained from heat within the earth's crust.
i. A device that changes sunlight directly into electricity.
j. A clear gas formed from the remains of ancient sea life.

REVIEW QUESTIONS

Choose the letter of the ending that best completes each of the following statements.

1. A source of energy that is not widely used now but may be used in the future is (a) nuclear energy (b) geothermal energy (c) solar energy (d) all of these.
2. Fusion reactions happen only in special places because (a) very high temperatures are needed (b) very low temperatures are needed (c) only a few atoms are available (d) only heavy atoms will work in this reaction.
3. One major source of energy at present is (a) biomass fuels (b) synfuels (c) natural gas (d) nuclear energy.

4. Coal, petroleum, and natural gas are all (a) liquid fuels (b) fossil fuels (c) future fuels (d) synthetic fuels.
5. Which of the following can be used to generate electricity? (a) hydropower (b) wind power (c) solar power (d) all of these.
6. At the present time, which fuel do we rely on for the greatest part of our energy supply? (a) coal (b) solar power (c) petroleum (d) hydropower.
7. Which of the following environmental problems are related to coal? (a) acid rain (b) strip mining (c) air pollution (d) all of these.
8. What percentage of the world's reserves of coal are located in the United States? (a) 8% (b) 28% (c) 58% (d) 78%.
9. Unconventional energy sources include (a) tar sands (b) oil shales (c) geopressured zones (d) all of these.
10. Heat from the earth's crust can be used to (a) heat homes (b) heat water (c) produce electricity (d) all of these.

REVIEW EXERCISES

Give complete but brief answers to each of the following exercises.

1. What are fossil fuels?
2. Explain three of the environmental problems related to making synfuels.
3. Name the three types of coal. Explain how each was formed.
4. Explain two direct uses of solar energy.
5. How does nuclear fusion differ from nuclear fission?
6. What are biomass fuels? Give two examples.
7. How does a breeder reactor differ from a standard nuclear reactor?
8. Name and describe the formation of two common hydrocarbon fuels.
9. Explain how a tidal power plant works. Why is tidal power not expected to play a major role in future energy sources?
10. Describe how hydropower and wind power may play an increasing role in future energy supplies.

UNIT 4: HOW IS OUR PLANET CHANGING?

CHAPTER 12 CHANGES HAPPENING NOW

1 EARTHQUAKES

In the mountains east of Naples, Italy, hundreds of towns and villages have clung to the steep slopes for centuries. Homes, churches, and other buildings are made mostly of stone and concrete. They seem to be part of the mountains themselves. One Sunday evening in November 1980, all of southern Italy was shaken by a violent earthquake. In less than a minute, thousands of people were killed by falling objects and collapsing walls. In some towns, over 80% of the buildings were destroyed. Hundreds of people were buried alive. Some were rescued, many were not. The victims were among the many millions who have died as a result of earthquakes. When you finish this section, you will be able to:

- Define the word *fault* and describe three types of motion along *faults*.

- Describe three kinds of earthquake waves and how they are recorded.

- Diagram the earth's interior and locate the core, mantle, lithosphere, and crust.

- Plot the location of several recent earthquakes.

Fault: A crack in the earth's surface along which there has been movement of one or both sides.

12-1. *When stresses build, the earth cracks and moves. This photo shows movement along the San Andreas Fault.*

For thousands of years, people have tried to explain why the earth's surface would not stay still. We now understand that the earth is a place of constant change. As the surface slowly moves, bends, twists, and turns, great pressures develop. Sooner or later, at the points where the pressures are greatest and the rocks weakest, something has to give. Solid rocks snap, slip, rise, or fall, and the earth shakes. These movements are called earthquakes.

The earth's surface is cracked and broken in many places. If one or both sides of such a break has moved, the crack is called a **fault.** Make fists with both your hands. Place your fists together with the knuckles of one hand between the knuckles of the other. Now try to move your fists sideways, in opposite directions, while pressing your fists together. Steadily increase the pressure and sideways movement until the knuckles slip.

You have just made a model of one type of earthquake motion. The place where your fingers came together represents the *fault*. Your fists are the blocks of land on either side of the fault. As you tried to move your fists to the side, pressure and friction between the two blocks kept them from sliding. As you increased the sideways force, you could probably feel the strain building up. Eventually the force overcame the friction and slid sideways. You could then feel the release of that energy as your fists and arms shook.

Earthquakes happen in the same general way. Stresses build within the rocks. Sooner or later the stresses overcome the resistance. The land slips along an old fault or creates a new one. The motion may be sideways, up or down, or a combination of both. During the earthquake in Italy, the land slipped several centimeters. During the San Francisco earthquake of 1906, the land slid sideways as much as three meters. In Alaska in 1964, some of the land moved upward as much as ten meters.

The point along the fault where the actual movement of the rock happens is called the **focus**. In some earthquakes, the *focus* is very close to the surface. At other times and places, the focus may be hundreds of kilometers beneath the surface. When the stored up energy is released, it creates energy waves in the surrounding rocks. Some of these waves are back and forth motions of the rock particles. The particles are pressed together, then released. This wave of energy travels away from the focus in all directions very quickly. This is called the *primary* or *P wave*.

The rocks are also shaken sideways. Imagine holding the end of a rope that you snap up and down. The wave moves away from you but the rope moves up and down. This motion starts a second, slower wave moving away from the focus in all directions. This type of wave is called the *secondary* or *S wave*. A third type of wave begins at the point on the surface directly over the focus. This point is called the **epicenter** (**ep**-uh-sent-ur). The third wave, called a *surface* or *L wave*, radiates away from the *epicenter* like ripples on a pond. The land surface actually moves in the same way water does as the wave passes. Most of the damage done by earthquakes is caused by these surface waves.

Primary and secondary waves are strong enough to travel right through the earth. Sensitive instruments called **seismographs** (**size**-muh-grafs) record the arrival and strength of these waves at *seismic* (**size**-mik) stations all around the world. *Seismographs* consist of a large mass that is suspended so that when the ground shakes, the mass remains still. A recording device firmly attached to solid rock moves with the shaking earth. A pen attached to the non-moving mass traces the movement of the recorder. Fig. 12-4 shows a seismograph recording of the arrival of these waves at the seismic station.

Because P and S waves travel at different speeds, it is

Focus: The point on a fault at which movement causes an earthquake.

Epicenter: The point on the earth's surface directly over the focus of an earthquake.

Seismograph: An instrument that records earthquake waves.

12-2. *Earthquakes send waves out in all directions from the focus.*

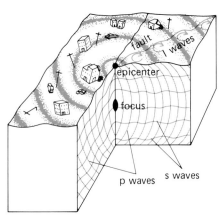

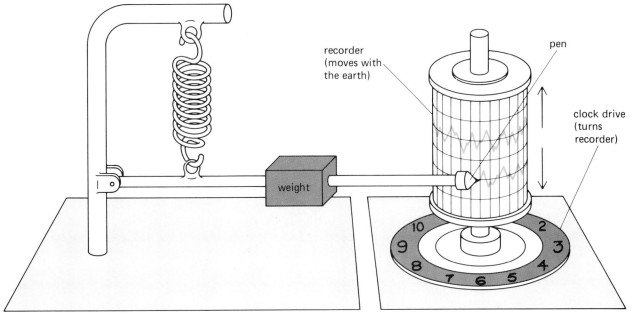

12-3. *This is a simple model of a seismograph. When earthquake waves arrive, the stationary mass traces the movement on the rotating drum.*

12-4. *This seismograph record shows the arrival of the primary, secondary, and surface waves from an earthquake.*

possible to determine how far from the seismograph the earthquake occurred. This is somewhat like the time between seeing lightning and hearing the thunder. The farther away from the source the greater the difference between the arrival times. Knowing the speed of the two waves makes it easy to calculate the distance. If three seismographs at different locations record the arrival of the waves, the precise location of the epicenter can be determined.

Determining their strength is an important part of studying earthquakes. One scale scientists use is based on the amount of damage done. However, a weak earthquake can do much damage if it occurs near a populated area. A strong earthquake in a remote area may cause little damage. A second scale, named the *Richter scale* after the scientist who developed it, is used to measure the intensity of the earthquake at its focus. This scale uses numbers beginning at 1 to express the energy released by an earthquake. Each number on the scale represents about ten times more energy than the next lower number. For example, the earthquake in Italy in 1980 registered 6.8 on the Richter scale. An earthquake at 8.8 would release 100 times more energy. So far the strongest quake recorded was a 9.5 on the Richter scale.

P and S waves can travel through the earth. They are among the most important clues we have as to what the

Chapter 12–1 Earthquakes

earth's interior is like. P waves can travel through solids or liquids. S waves can only travel through solids. Both types are bent and reflected as they pass through the earth. This information has helped us determine the earth's interior structure.

Until recently, the scientists' model of the interior of the earth was quite simple. The earth had three distinct areas, just like a hard boiled egg. There was a thin, brittle shell. Beneath that was a very thick zone that surrounded a two-layer center. As is the case with many scientific models, this one had to be changed.

Lithosphere: The strong, solid outer layer of the earth.

The present model is much more complex. Geologists still identify three main areas or shells, but each shell has layers. The outer shell is called the **lithosphere** (lith-uh-sfir). The name means "sphere of rocks." The *lithosphere* is made of strong, solid rocks. It varies in thickness from 100 to 200 km. The ocean floor and the continents make up a very thin layer of less dense rocks at the top of the lithosphere. This layer is called the **crust.** Beneath the oceans, the *crust* is between 5 and 10 km thick. Where the continents are, the crust is between 25 and 60 km thick. The rocks that make up the ocean crust and the continents are less dense than the rest of the lithosphere. They seem to ride on the lithosphere like rafts. Compared to the rest of the earth, the lithosphere is a thin shell.

Crust: The very thin top layer of the lithosphere.

Measurements have shown that the temperature of the earth rises as you go deeper. The source of much of this heat is thought to be the decay of radioactive elements. Below the lithosphere is a zone where the temperatures are so high the material there behaves like a thick hot fudge. This layer forms the top of the second shell. This second shell is called the **mantle.** It is about 2,900 km thick. The *mantle* makes up about 80% of the earth's volume and has almost two thirds of its mass. Even though the temperatures continue to increase with depth, the rest of the mantle behaves like a solid. The temperatures within the mantle may be as high as 3,000°C.

Mantle: A very thick layer of the earth reaching about 2,900 km beneath the lithosphere.

Core: The two layered center of the earth.

The very center of the earth is called the **core.** The *core's* diameter is about 6,940 km. It makes up about 20% of the earth's volume but has about one third of its mass. Only a mixture of the elements nickel and iron would be dense enough to have such a mass. Earthquake waves have shown that there are actually two layers to

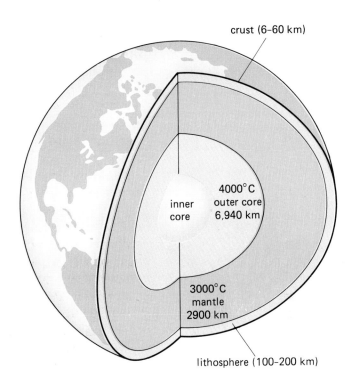

12-5. *The earth is composed of many layers. Earthquake waves have been used to determine their thickness and number.*

the core. The outer core is thought to be a liquid iron-nickel mass since S waves will not pass through it. The inner core is like a solid, hot ball. Temperatures in the core may reach 4,000°C.

Earthquakes are one of nature's most destructive acts. An earthquake which registered 6.6 on the Richter scale did over a billion dollars worth of damage and killed 62 people in the Los Angeles area in 1971. An earthquake in China in 1976 measured 8.2 and killed over 700,000 people. Because of this destructiveness, scientists around the world are looking for ways to predict quakes.

Most earthquakes happen in certain parts of the world. Many occur around the Pacific Ocean basin or down the middle of the Atlantic Ocean. Another common area for earthquakes to occur runs from the Mediterranean Sea through the Middle East to Southeast Asia.

The earth often signals approaching earthquakes. Sometimes a section of land begins to bulge in the months or even years before earthquakes occur. Certain gases are released into the ground water as the rocks are strained. These gases can be monitored in areas where earthquakes are expected. Many people have noticed strange behavior of animals shortly before earthquakes. In 1975, such signs caused Chinese scientists to evacuate several million people from their homes just hours before a major quake struck. Great damage was done, but since most people were away from buildings, many lives were saved. Yet there were no definite signs before the 1976 earthquake in China. Whether or not such advanced signals will prove reliable enough for prediction is something only time will tell. The day may come when such predictions are as common as weather forecasts.

SKILL BUILDING ACTIVITY

Materials
World map supplied by your teacher
Blue pencil
Lead pencil
Paper

PLOTTING AND INTERPRETING DATA

A. Obtain the materials listed in the margin.

B. Each year, thousands of earthquakes occur around the world. In this activity, you will use reference points to plot the epicenters of 25 earthquakes on a world map. The coordinates for locating these earthquakes are listed in Table 12-1.

C. Use the reference lines on the map and a blue pencil to mark and number the position of each earthquake. For example, the first earthquake is found by following line W across until it meets line 23. Mark and number this point.

D. Plot the locations of the remaining earthquakes.
  1. Do you see any pattern in these few scattered earthquakes?

E. Use a lead pencil to lightly connect earthquake 1 to 2, 3, 4, and 5 in sequence.

F. Connect earthquake 6 to 7, 8, 9, 10, 11, 12, and 16.

G. Connect 13 to 14, 15, 16, 17, 18, 19, 20, and 21.

H. Connect 22 to 23 and 24.

I. Connect 25 to 18.
  2. Which ocean has a series of earthquakes down its center?
  3. Which ocean seems to be split into three pieces by these lines?
  4. Which ocean seems to be surrounded by earthquakes?
  5. Which number represents the Italian earthquake of November 1980?
  6. Which represents the Los Angeles earthquake of 1971?
  7. If these lines represent faults, is the earth's crust a single, solid layer? If not, how would you describe it?
  8. How do these earthquakes match up with the earthquake areas discussed on page 295?

J. Save your map for section 2.

TABLE 12-1

Earthquake	Letter	Number	Earthquake	Letter	Number
1	W	23	14	G	17.5
2	S	22.5	15	A	14.5
3	P	22	16	B	11.5
4	J	24	17	D	7
5	B	25	18	G	6
6	D	20	19	O	4.5
7	J	20	20	N	3.5
8	S	17	21	Q	3.5
9	X	15	22	T	2
10	U	11	23	T	5
11	Q	9.5	24	R	7
12	I	12	25	D	3
13	M	18.5			

SUMMARY & QUESTIONS

The earth's surface is constantly shifting and moving. These movements cause strains in the rocks. Occasionally, the rocks break and move, causing earthquakes. Earthquake waves passing through the earth provide information about the earth's interior.

1. What is a fault? Explain how faults form and three ways in which they may move.
2. Name and describe each of the three types of earthquake waves.
3. Explain how a seismograph records earthquake waves.
4. Make a simple sketch of the major shells of the earth. Indicate the name and size of each shell.
5. Briefly describe what each of the major shells of the earth's interior is thought to be like. How do we know this information?
6. List some of the ways earthquakes may be predicted.
7. Explain how the Richter scale is used to measure earthquakes.

2 VOLCANOES

It was a picture postcard volcano. Snow-capped all year long, it stood almost 3,000 meters high. It seemed to watch over the green forests and blue lakes of the state of Washington that surrounded it. It had been quiet for over a hundred years. Then, in March 1980, it began to tremble. Small steam and ash eruptions continued through April and into May. Finally, at 8:32 A.M. on May 18, 1980, Mt. St. Helens went to pieces. Sixty-two people died. When you finish this section, you will be able to:

- Describe the structure of a volcano and the materials associated with an eruption.

- Explain how four kinds of volcanic structures are formed.

- Plot the locations of several active volcanoes on a world map.

Volcanoes can be dangerous. Over 200,000 people are known to have been killed by eruptions or their after-

Unit IV How Is Our Planet Changing?

effects in the past 500 years. One such aftereffect is a giant sea wave called a *tsunami* (soo-**nahm**-ee). In 1883, a volcanic island called Krakatoa (krak-uh-**toe**-uh) in southeast Asia exploded. The tsunami it created drowned 36,000 people in that area.

Volcanoes are also beneficial. Some scientists believe that our atmosphere and oceans were formed from gases that escaped from volcanoes. Volcanic soils are rich in minerals and make good farmlands. The minerals that volcanoes bring to the surface are important to industry. Recently, steam from volcanic areas has been used as an energy source. Volcanoes also give us important clues about the earth's interior.

As you can imagine, it is very difficult to study the interior of the earth. The deepest mines and oil wells only reach a few kilometers into the crust. A few drill holes have reached the bottom of the crust. But to study the materials that make up the lithosphere and the mantle, we have to use indirect methods such as earthquake waves or wait until that material comes to us. Fortunately, in some places it does.

In certain places around the world, hot molten material rises through the crustal rocks toward the surface. This red hot material is called **magma** (**mag**-muh). Very little is known about the formation of *magma*. Some of it seems to be material coming from pockets in the top, hot, molten layer of the mantle. In other places, a small part of the lithosphere may have remelted to form pockets of magma beneath the crust.

The composition of magma may be quite different

Magma: Molten rock beneath the earth's surface.

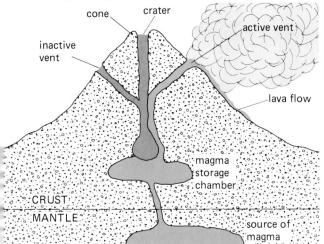

12-6. *(left) The source of the magma is deep within the earth.*

12-7. *(right) A typical volcano.*

Chapter 12–2 Volcanoes

299

12-8. *Shield volcanoes are broad and gently sloping. They usually erupt in lava flows.*

Lava: Magma that has flowed out on the surface.

Volcano: A structure that may result when magma reaches the earth's surface.

12-9. *Cinder cones are formed by magma with a lot of gas. The gas causes violent eruptions.*

from place to place. This gives us a clue as to where it was formed. However, magma is always very hot, usually about 1,000°C. This high temperature makes the magma less dense than the surrounding rocks. It rises toward the surface, melting a passage or squeezing through cracks. Usually a storage chamber of magma forms in the crustal rocks. Later, the magma may break through the crust and flow out onto the surface. As the magma flows out, some of the gases and steam that were in the magma escape. The red hot, molten rock that is left is called **lava.** The opening through which the magma reaches the surface is called the *vent*.

Magma may break through the crust underwater or on land. Sometimes the *lava* spreads over the land in sheets. Lava can also pile up to form a **volcano.** A *volcano* usually has a bowl-shaped *crater* around the vent. The mountain that builds up is called a *cone*. See Fig. 12-7 on page 299.

Volcanoes are not all alike. This is because of physical and chemical differences in the magma from which they were formed. If the magma was thick and heavy, the lava will flow slowly over the surface and pile up close to the vent. If the magma is thin and lighter in weight, the lava will flow quickly and spread out. Some lava flows can travel as fast as 50 to 60 km per hour.

The Hawaiian Islands are examples of volcanoes built up by flowing lava. The magma there does not contain much gas. As a result, the lava tends to ooze up and overflow the crater or break through the crater wall. Rivers of lava then flow out over the land. These flows may form a layer up to one or two meters thick. Over millions of years, a broad, gently sloping dome is built up. This is known as a *shield volcano*. The Hawaiian volcano Mauna Loa (**mou**-nuh **loe**-uh) is a shield that rises 10 km from the ocean floor and is 100 km in diameter.

A completely different type of volcano forms when the magma contains a lot of gas. The gases or steam from hot ground water cause pressure to build up. Material actually explodes out of the vent. This material is identified by its size. The smallest particles are called *volcanic dust* or *ash*. Slightly larger chunks, up to a few centimeters in size, are called *cinders*. Still larger chunks, some one meter or more in size, are called *bombs*. Most of the lighter material is carried farther away before it falls. As a result, cones made of cinders and bombs are often tall and narrow. They are called *cinder cones*. In 1963 and 1974, volcanoes near Iceland built these types of volcanic cones.

Most volcanoes, however, are a combination of lava flows and explosive eruptions. They are called *composite volcanoes*. See Fig. 12-7, page 299. The eruptions of lava make the cone broader and more gently sloped. The ash and cinder eruptions often build the cone higher. Mt. Hood in Oregon and Mt. St. Helens in Washington are composite volcanoes.

Often, volcanoes like this are quiet for many years. After a lava flow eruption, the cooling lava may seal the vent. This causes the gas and magma pressure to build up inside. Eventually, the pressure either blows out the seal or the side of the volcano in a violent ash and cinder eruption. This is what happened at Mt. St. Helens in 1980.

Mt. St. Helens had last erupted between 1831 and 1857. It was quiet until early 1980. Then swarms of small earthquakes were recorded. A very large bulge formed on the volcano's north side. These changes indicated that magma was pushing up under the cone. Small eruptions of steam and ash burst from the crater. Ash clouds rose as high as three kilometers into the sky.

Then, on the morning of May 18, a major earthquake occurred beneath the mountain. The shaking caused the bulge to break loose. It quickly slid down the north slope, bounced over ridges 350 meters high, and flowed 21 kilometers down a river valley. It buried the valley, lakes, homes, and people under 150 meters of debris.

When the bulge dropped away, it suddenly released the pressure on the gases and steam in the magma below. A sideways blast of ash, cinders, and steam roared as far as 28 km away from the volcano. The 300°C cloud de-

12-10. *(left) A landslide at Mt. St. Helens caused the bulge on the north side to break loose.*

12-11. *(middle) After the bulge broke off, gases began to escape from the opening.*

12-12. *(right) Ten seconds later, a major eruption blew out the entire side of the mountain.*

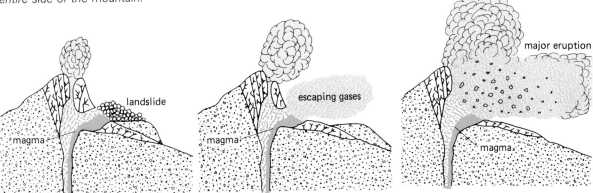

stroyed 550 km² of forest and almost everything in the way. Ash deposits from a few centimeters to several meters thick covered the land. Clouds of ash continued to rise from the crater as high as 18 km. When the cloud cleared, the top of Mt. St. Helens was gone. The cone was 400 m lower and had a new crater 750 m deep. Soon, thick lava began to plug the crater. The pressure was building again.

Mt. St. Helens and Mt. Hood are part of a chain of volcanoes that stretches from California into Canada. Many of these volcanoes have erupted in the past few thousand years. Since these volcanoes do not erupt frequently, they are said to be *dormant* or resting. Volcanoes that have not erupted for thousands of years are called *extinct*. However, there is always the chance that an extinct volcano could erupt again. There are over 400 *active* volcanoes on the earth today. They seem to be arranged in certain patterns around the world. Many are located in a belt around the edges of the Pacific Ocean. These two areas seem to be connected by another chain running through the Mediterranean Sea, the Middle East, and on to Southeast Asia. Most active volcanoes are in one of these areas.

Materials
Map from section 1
Red pencil
Pencil
Paper

COMBINING DATA

A. Obtain the materials listed in the margin.

B. There are several hundred active volcanoes in the world today. In this activity, you will plot the location of 25 of them. The coordinates for locating these volcanoes are listed in Table 12-2. Use the reference points and lines to locate these volcanoes on the map you used for the activity in section 1. Plot their position using a *red* pencil.

1. Is there any similarity in the pattern of volcanoes and earthquakes?
2. Which ocean has volcanoes that run north to south down its center?
3. Which volcanoes are part of the ring surrounding the Pacific Ocean?

C. Find the line you drew in section 1 connecting the earthquakes around the edge of the Pacific Ocean. Change this line so that it now connects earthquakes 6 through 16 with volcanoes 7 through 20.

4. What is the relationship between earthquakes and volcanoes and the mountain ranges?

D. Connect all the points down the center of the Atlantic Ocean in a single line.

TABLE 12-2

Volcano	Letter	Number
1	B	1.5
2	T	2
3	T	4
4	O	7.5
5	K	8
6	J	9
7	E	12.5
8	K	11
9	O	9.5
10	T	10.5
11	W	11.5
12	W	13
13	W	14.5
14	U	17
15	P	18
16	O	19
17	M	20
18	G	20.5
19	B	20.5
20	A	22.5
21	E	24.5
22	M	23
23	T	23
24	Y	23.5
25	Z	24.5

E. Draw a line from earthquake 24 to volcanoes 4, 5, 6, and 8.

F. Most earthquakes and volcanoes happen along the lines you have drawn on your map, but the pattern is not complete. To complete the pattern, connect earthquake 13 to volcanoes 15 and 17.

G. Connect volcano 16 to earthquake 3.

H. Connect earthquake 6 to earthquake 14.

I. Connect earthquake 5 to volcano 20.

J. Connect earthquake 23 to earthquake 25.

K. Connect earthquake 25 to volcano 1.

5. If earthquakes occur on faults and volcanoes break through the crust, is the earth's surface a stable, unchanging place?

6. Explain why some scientists compare the earth's surface to a cracked eggshell.

7. Which volcano represents Mt. St. Helens?

L. Keep this map for section 3.

SUMMARY & QUESTIONS

Volcanoes are among the most spectacular of nature's displays. Volcanic eruptions vary from quiet lava flows to violent explosions. The material that comes from volcanoes gives us some information about what the interior of the earth is like.

1. Define magma and lava. Describe how types of lava may differ from each other.
2. Name and describe four types of material that may explode out of a volcano.
3. Describe how three types of volcanoes form.
4. Why does some lava flow out of vents while other lava is blown out?

3 MOVING PLATES

Sherlock Holmes was one of the greatest of all detectives. He could take several bits of what seem like unrelated evidence and solve a mystery almost overnight. He would not only discover who did it, but also how and why it had happened. Being an earth scientist today is a lot like being a detective. Many pieces of what were once thought to be unrelated evidence have been shown to be part of a great puzzle. Many "earth detectives" are at work on one of the biggest puzzles in science today. Why is the surface of the earth so unstable? When you finish this section, you will be able to:

- Describe an early theory about the movements of the earth's crust.
- Explain what happens at *ridges*, *trenches*, and *fault zones*.
- Summarize the theory of *plate tectonics*.
- ▲ Describe and diagram plate actions at several typical boundaries.

Even a quick look at a globe or map will show how South America and Africa seem to fit together. This fact was noted almost as soon as good maps were available. At the time, nobody really believed that the two had once been together. It was just one of those unusual observations. As scientists and explorers learned more about the earth's surface, other puzzling questions arose. Why are the remains of ancient sea animals found at the tops of mountains? Why are some rocks and fossils in South America so much like rocks and fossils in Africa? Why are coal deposits, which must have formed in warm areas, now found in Antarctica? How can the pattern of earthquake and volcano belts be explained?

Several scientists offered explanations for one or the other of these questions. However, it wasn't until 1912 that a single idea that could explain them all was proposed. In that year, Alfred Wegener, a German scientist, proposed the idea of *continental drift*.

Wegener said that about 200 million years ago, all the continents were joined together in one large mass. He called this supercontinent Pangaea (pan-**jee**-uh). He also said that some force caused this supercontinent to break up. Its pieces then drifted to where they are today. If such a supercontinent had existed, it would explain why the same rocks and fossils can now be found both in South America and Africa. It would also explain why coal deposits are found in the Antarctic and evidence of past ice sheets is found in areas that are hot today. According to Wegener, as North and South America drifted westward, their front edges bent and folded into the Rockies and the Andes Mountains. The Himalayas formed as India moved northward to collide with Asia.

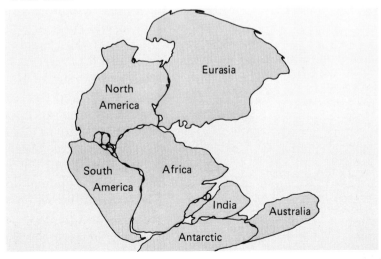

12-13. *Wegener proposed that 200 million years ago all the continents were together in a supercontinent he called Pangaea.*

12-14. *The Mid-Atlantic Ridge is only a portion of the entire mid-ocean ridge system.*

Rift: A long, deep valley-like crack in the earth's surface.

Mid-ocean ridge: A chain of underwater mountains found near the center of the oceans.

12-15. *Molten rock becomes magnetized in one direction then hardens. It is later split and moved away from the ridge on both sides.*

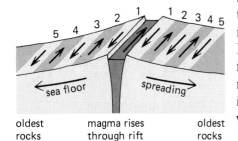

However, Wegener could not explain what force was causing the movement. Except for a few supporters, his ideas were soon ignored.

After World War II, scientists increased their study of the world's oceans. They discovered some evidence that aroused interest in Wegener's ideas again. New equipment for studying the ocean bottom had been developed during the war. Now efforts were begun to map this unexplored area. Early in the 1950's, a massive chain of undersea mountains was discovered in the Atlantic Ocean. It was called the Mid-Atlantic Ridge. Closer examination showed that the ridge had a narrow, steep walled valley that ran right down its length. Such a long, deep valley is called a **rift**. It was soon discovered that the ridge continued around Africa into the Indian Ocean and then into the Pacific. A branch from the Indian Ocean runs up into the Red Sea and the *Rift Valley* of East Africa. This entire chain of mountains, the biggest on earth, is now called the **mid-ocean ridge.** It seems to circle the earth like the seam on a baseball. This ridge system is still being studied very carefully. Other clues have been discovered here. The ridge is the site of many active volcanoes. It is also the location of thousands of earthquakes each year. Most of these earthquakes are small and their focus is fairly close to the surface. The rocks of the ocean floors are youngest at the ridge and get older as you go away from the ridge toward the continents. Instruments have shown that a lot of heat is rising from the mantle toward the surface at the ridges. Near the continents the heat flow is much less.

All this evidence was combined into a new idea called *sea-floor spreading*. It states that the rift valleys in the *mid-ocean ridge* are places where the sea floors are splitting apart. Hot magma from the top layer of the mantle rises to fill in the cracks. This material cools and hardens to become part of the ocean floor. Later this material splits apart and the process continues. The rate of splitting is different at different ridges. In the North Atlantic, it is about 2 to 3 cm per year. At one point in the South Pacific, the ridge is separating about 18 cm per year. This idea can account for the volcanoes, earthquakes, young rocks, and high heat flow at the ridges. More proof of this idea was found when rock samples from the ocean floor were tested in labs.

The rocks on the ocean floor were formed as magma

cooled. When the magma was still molten, the iron minerals in it were affected by the earth's magnetic field. These minerals became magnetized in the direction of the north pole. When the magma hardened, they were locked into that position. When rocks of different ages were compared, it was found that some had iron minerals magnetized in the opposite direction. It appears that the earth's magnetic field has flip-flopped, north to south, many times. Rocks on either side of the ridge can be divided into parallel stripes based on the direction of their magnetism. Moreover the stripes on opposite sides can be matched up by age. Since each stripe formed at the ridge, then split, matching their ages and magnetism changes proved the sea floor was spreading.

If the sea floor was spreading, why wasn't the earth getting bigger? The idea of sea floor spreading also said that parts of the floor were plunging back into the mantle where there were deep ocean *trenches*. Here they are remelted in the hot mantle. It was found that earthquakes occur deeper under the edges of the continents than under the edges of the oceans. This indicated that sections of the sea floor were sinking. It also suggested why rocks from the ocean floor were not as old as some continental rocks.

Finally, in the late 1960's, all of these ideas were included in an overall theory now known as **plate tectonics** (tek-**ton**-iks). This theory seems to explain many related problems so well that many scientists consider it a major step in understanding our planet. These ideas are still

Plate tectonics: The theory that the earth's surface is made up of large moving plates.

12-16. *The Juan De Fuca plate forms at a ridge, moves eastward and is pushed down under North America. Melted rock works its way up to the surface under Mt. St. Helens and other volcanoes.*

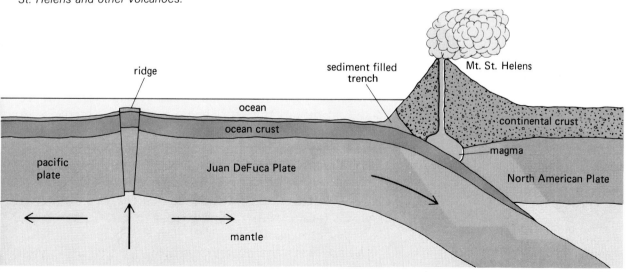

Chapter 12-3 Moving Plates

Plate: A section on the earth's lithosphere and crust.

12-17. *(top) Ridge formation.*

12-18. *(middle) Trench formation.*

12-19. *(bottom) Fault zone.*

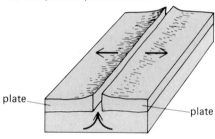

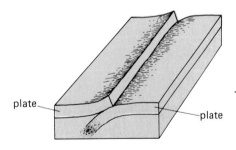

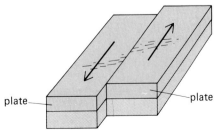

theories, however, because scientists have not yet solved the problem of what forces cause the movement. In addition, much more supporting data has to be gathered.

The theory of *plate tectonics* views the earth's crust as being much like a cracked eggshell. The lithosphere is broken into about eight large and several smaller sections called **plates.** The continents and ocean crust ride on the plates like rafts frozen in chunks of ice. The edges of the *plates* meet at one of three types of boundaries. One type of boundary is a *ridge*. Here spreading takes place. Magma from the mantle rises into the rift valley between the plates. New material is added to the edge of each plate. The Mid-Atlantic Ridge is a good example of this type of boundary. The rift in the Red Sea and East Africa is thought to be forming a new ocean.

A second boundary is a *trench*. Trenches are found where one plate is being forced downward by collision with another plate. For example, just off the west coast of Oregon and Washington is the small Juan De Fuca (**hwahn** day **foo**-kuh) plate. It seems to be moving downward at an angle under the North American plate to the east. As it moves downward, earthquakes occur at deeper and deeper points under the North American plate. The less dense rocks of the continent slide up and over the more dense ocean crust. The continental edges are bent and folded into mountains. Pockets of magma form along the boundary between the plate bottoms. This magma rises through the continental rocks to form a chain of volcanoes. One of these volcanoes is Mt. St. Helens.

A third type of boundary occurs where two plates move sideways in opposite directions. This is called a *fault zone*. An example of this is the San Andreas Fault in California. The front edge of the North American continent has slid up and over part of the Pacific plate. The Pacific plate is moving northwestward along this fault. As a result, all the land west of the fault is being carried to the northwest. This movement is the cause of the earthquakes along this fault.

There are still problems with the plate tectonics theory. The biggest problem is that the force causing all the motion is not fully understood. Many scientists believe that differences in heat and density in the mantle cause material there to flow, thus moving parts of the crust. Other causes have also been suggested. Scientists are still seeking the answer.

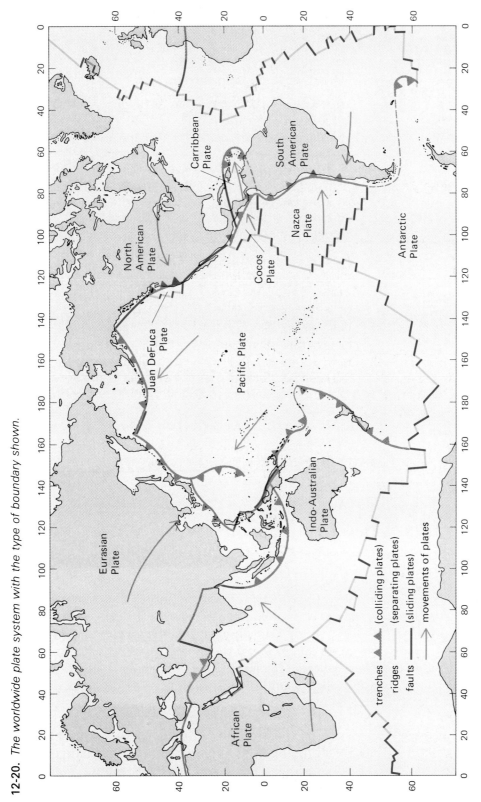

12-20. The worldwide plate system with the type of boundary shown.

Chapter 12–3 Moving Plates

SKILL BUILDING ACTIVITY

Materials
Map from sections 1 & 2
Pencil
Paper

COMPARING DATA

A. Obtain the materials listed in the margin.

B. Figure 12-20 on page 309 shows the major plate boundaries as they are known today.

1. Which plate do you live on?
2. In which general direction is this plate thought to be moving?
3. Which plate(s) is it moving away from?
4. Which plate(s) is it colliding with?
5. Explain one current idea about what forces cause this movement.

C. Compare this map with the map you made in sections 1 and 2.

6. How do the earthquake and volcano belts compare with the plate boundaries?
7. Which of the three types of plate boundaries is found along the Mid-Atlantic Ridge?
8. Explain what is happening here. Use diagrams.
9. Which type of boundary is found along the west coast of South America?
10. Explain what is happening here. Use diagrams.
11. What type of boundary is found along the west coast of Canada?
12. Explain what is happening here. Use diagrams.
13. Why are there volcanoes on the Mid-Atlantic Ridge?
14. Why are there earthquakes in Italy?

SUMMARY & QUESTIONS

In the past twenty-five years, many ideas on how the earth's crust behaves have been drastically changed. A new theory has been proposed that seems to explain many of the major questions scientists have about earthquakes, volcanoes, and drifting continents. Like all theories, this one too will have to stand the test of time.

1. Explain Wegener's theory of continental drift.
2. Using diagrams, explain what is happening at ocean ridges, trenches, and fault zones.
3. Briefly summarize the theory of plate tectonics.

4 BUILDING FORCES

The rocks of the earth's crust bend and fold along the edges of the plates. There are other movements taking place also. Some happen far from the plate edges. The Scandinavian countries are one example. Sweden and Finland have been steadily rising for over 10,000 years. In some places, the land has risen over 200 meters. What causes such movements? Will they continue? When you finish this section, you will be able to:

● Describe how the theory of *isostasy* explains some crustal movements.

● Compare the formation of *folded, faulted,* and *domed mountains*.

● Name one *plateau* and describe how it formed.

▲ Give evidence that the land in a certain area is rising.

Have you ever tried to keep a block of wood under water? It is a difficult job. As you push the wood downward, you can feel the force of the water pushing it up

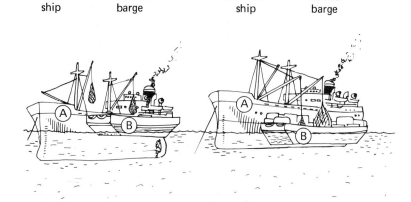

12-21. *The earth's crust floating on the mantle behaves like ships floating on water.*

Isostasy: The state of balance between different parts of the lightweight crust as it floats on the heavy mantle.

Joint: A crack in rocks with no movement along either side of the crack.

again. When wood is put into water, it doesn't sink completely. It sinks until a balance is reached between it and the water. The reason for this state of balance is the difference in their densities.

Thousands of years ago, Scandinavia was covered by a thick ice sheet. The mass of the ice forced the crust deeper into the denser mantle. Then the ice melted. The mantle has been slowly pushing the land upward since. This motion will continue until a state of balance between the crust and mantle is reached again. This state of balance is called **isostasy** (ie-**soss**-tuh-see).

Observing floating ships shows us how the crust behaves as it floats on the heavy mantle rock. See Fig. 12-21. When cargo is transferred from ship A to ship B, ship A rises in the water and ship B sinks. The crust, floating on the mantle, behaves in a similar way. For example, rock and soil move from one area on the crust to another when rock falls from the mountains to the land below. Because of *isostasy*, the area that loses the material rises. The area that gains the material sinks.

The rising and falling of parts of the crust take place very slowly and put a strain on the crustal rock over a long period of time. How do you suppose the rocks respond to this strain? Sometimes they crack, like a piece of glass. If there is no movement along such a crack, it is called a **joint**. A crack in which there is movement along either side is called a *fault*. Small *joints* can be seen in many rocks. Large systems of joints may cover many square kilometers.

Sometimes large blocks of rock may be tilted, like a row of books that fell over. The result is a set of tilted blocks separated by faults. Mountains made in this way

12-22. *The Grand Teton Mountains.*

Fault-block mountains: Mountains formed when large blocks of rock are tilted over.

Thrust fault: A slanting fault in which one slab of rock is pushed up over the other.

Folding: The bending of rocks under steady pressure without breaking.

12-23. *(left) The Appalachian Mountains.*

12-24. *(right) Folds, faults, and thrust faults are found in the crust.*

are called **fault-block mountains.** The Grand Tetons are *fault-block mountains.*

Some forces squeeze rock together. The faults caused by these forces are slanted. One slab of rock slides up over the next. This type of fault is a **thrust fault.** *Thrust faulting* causes the area of the crust to be decreased. The Appalachian (app–ah–**lay**–shin) Mountains contain many large thrust faults.

Forces that squeeze rocks together do not always cause faults. Some rocks may yield to strain by **folding.** Slow, steady pressure can make some rocks bend without breaking. *Folds* in rocks may be small wrinkles. They also can be large enough to form the foundations of mountains. The Appalachian Mountains, for example, contain the remains of many great folds. During their long history the Appalachians have been worn down. Today we see long parallel ridges and valleys formed by the tilted rock layers of ancient folds. See Fig. 12-23.

Scattered all over the earth are very large regions of land that are elevated above the level of the surrounding

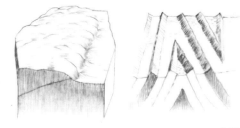

Chapter 12–4 Building Forces

Plateau: A large region of elevated land.

12-25. *The Columbia Plateau.*

Dome mountains: Mountains produced when forces below the surface lift a part of the crust.

12-26. *(left) Death Valley.*

12-27. *(right) The Black Hills.*

crust. A region of land that is elevated is called a **plateau** (pla-**toe**). The Grand Canyon is found on the Colorado Plateau. *Plateaus* can be caused by the forces that cause faults and folds. In this case, large parts of the crust are bent upward. Plateaus can also be formed when lava pours out and covers a large part of the land surface. The Columbia Plateau of Washington State is an example of a lava plateau.

While large areas of land are being uplifted into plateaus, some small areas are sinking. Death Valley in California is an example of such a place. The large block that makes up the floor of Death Valley is slowly tilting as one end sinks. See Fig. 12-26.

Some of the forces that cause strain in the crust come from movement of the crustal plates. At the mid-ocean ridges, hot mantle material rises. When the mantle material cools, it forms new sea floor and pushes the crustal plates. The plates bend and twist as they are driven away from the mid-ocean ridges.

Magma may also push up and lift the crust at places within the interior of the plates. Sometimes such forces from below can raise a part of the crust to make **dome mountains**. An example of *dome mountains* are the Black Hills of South Dakota. These mountains are like a blister in the midst of the Great Plains that surround them.

ACTIVITY

Materials
Pencil
Paper

Forces acting on the earth cause the crust to rise and sink. The motion helps us to understand how mountains, valleys, and plateaus are formed. However, the movement of the land is a very slow process. If it takes millions of years for mountains to be formed, how can we be sure that the crust is moving? Is there any evidence of such a motion? This activity shows you some evidence that the motion of the crust forms land features.

A. Obtain the materials listed in the margin.

A railroad travels across the mountains through the Cajon Pass, 80 kilometers (km) east of Los Angeles. The pass is really a valley. It is 1.2 km high. It cuts through the mountains into the high desert region.

Survey teams have studied a 22-mile stretch of track in Cajon Pass. The elevation of 26 spots on the eastern side of the pass were measured. According to measurements

12-28. *The Cajon Pass.*

Chapter 12-4 Building Forces

made since 1906, it is estimated that the eastern side of the pass is rising at a rate of 0.14 m per century.

1. About how many meters is the ground on the eastern side of Cajon Pass rising every year?

2. How many meters would it rise in 70 years?

On the western side of Cajon Pass the elevation of 25 spots was measured. The survey teams found that the western side of the pass also is rising. The elevation of the western side increases at a rate of 0.78 m per century.

3. Which side of Cajon Pass is rising at a more rapid rate?

4. What is the difference, in meters per century, between the two rates?

If this pattern of rising continues, the region around Cajon Pass will form a higher and higher mountain.

5. If the pattern continues at the same rate, which side will be steeper?

B. Look carefully at Fig. 12-28.

6. What evidence is there in Fig. 12-28 that shows this may be happening?

Mount Whitney is located just over 160 km due north of Cajon Pass. Mount Whitney is 3,120 m higher than Cajon Pass.

7. The west side of Cajon Pass is rising at a rate of 0.78 m per century. How many centuries will it take for the west side of Cajon Pass to rise 3,120 m?

SUMMARY & QUESTIONS

The solid land is not as quiet as it seems. Huge forces are at work to strain the rock. Sometimes the rock cracks and shifts its position. Other rocks bend without breaking. Whole regions are lifted, producing large, flat highlands. Other parts of the crust are slowly sinking. These changes in the earth's surface are a result of natural processes.

1. Name two processes that strain the crust.
2. Compare the formation of folded, faulted, and domed mountains.
3. Define *plateau*. In what two ways can plateaus be formed?
4. Cite evidence to support the idea that the land is rising at certain places.

5 WEARING AWAY THE LAND

New buildings are constructed, but as they grow older their surfaces wear away. The same is true for the features that make up the shape of the land. When you finish this section, you will be able to:

● Define *weathering* and *erosion*.

● Identify four forces that erode the land.

● Describe how one land feature has been formed.

▲ Relate the appearance of mountains to their age.

Weathering: The processes that break rock into small particles.

Erosion: All the processes that cause rock to be carried away.

No matter where you look at the land you see a battleground. On one side of the constant battle are the forces beneath the surface. These powerful forces cause the crust to be faulted, folded, tilted, and lifted. On the other side of the battle are the natural processes of **weathering** and **erosion** (ee-row-shun). The term *weathering* refers to all the processes that break rock into small particles. Everything that happens to cause rock to be carried away is called *erosion*.

Weathering and erosion, along with folding and faulting, shape the mountains, hills, plateaus, and plains that cover the earth. If there was no weathering and erosion, the earth would look very different. It might appear as rough and forbidding as the moon. On the other hand, weathering and erosion alone could flatten the continents within 25 million years. These two forces have been acting together to shape the land for a very long period of time.

Many of the forces that lift and twist the crust are hidden beneath the surface. The process of erosion, however, can be seen. Some erosion takes place when *gravity* causes loose rock to move downhill. The most effective tool of erosion is *running water*. Streams of

12-29. *The Mississippi River.*

running water wear away and move rock. The Mississippi River moves about 600 million tons of rock particles each year. This *load* is rock that has been worn away from the land by streams that feed into the river. The 600 million tons of rock particles are dumped into the Gulf of Mexico.

At some time in history, different parts of the earth were covered by **glaciers** (glay–shur). These great moving masses of ice have left their marks on almost all parts of the earth. A *glacier* is formed when snow turns into ice and begins to move slowly downhill because of its great weight. As it moves, the glacier works like a giant bulldozer, scraping, grinding, and plowing rock.

Another tool of erosion is the wind. The wind moves small particles of rock. Sand dunes at the shore or in a desert are produced by wind action.

Glacier: A large, slowly moving mass of ice on the land.

12-30. *(left) A glacier scrapes and plows rock.*

12-31. *(right) Sand dunes are formed by the wind.*

Unit IV   How Is Our Planet Changing?

12-32. (left) What forces have shaped this land?

12-33. (right) The Presidential Mountain Range in New England.

Along the coastline, the sea shapes the land. Waves pound on the rocks along the shore. Continual pounding reduces the rock to small pieces. The small pieces roll or are carried down to the sea floor. The sea acts as a dumping ground for material eroded from the land. The eroded material might later be lifted above the water by faulting or folding action.

Look at Fig. 12-32. Notice the steep slopes and narrow valleys through which the stream flows. Can you tell what forces have shaped this land? It appears that this rock has not been affected very much by erosion. Why is this so? The rock might be very resistant to erosion. The mountains may be rising rapidly. This might be a new stream. Can you think of any other reasons?

Look at Fig. 12-33. Notice the slopes and valleys. How do they differ from those in Fig. 12-32? Erosion has had a great effect on this mountain range. The wide valleys and gentle slopes of these mountains are evidence of the activity of erosion over a long period of time.

Peneplain: An almost completely flat surface produced by erosion.

Monadnock: An isolated hill on a peneplain.

12-34. *A monadnock in North Carolina.*

In some areas, erosion over a very long time has produced a surface that is almost completely flat. Look at Fig. 12-34. This is called a **peneplain** (**pen**-eh-plain). Although it is almost flat, a *peneplain* may have isolated hills. The hills are made of resistant rock which has not been worn like the surrounding land. These resistant knobs of rock are called **monadnocks** (mow-**nad**-nock). *Monadnocks* give a clue to the rock under the peneplain. This rock is often folded and faulted.

The process of erosion levels off the surface of the crust. First it wears away the features of the land. Then the agents of erosion deposit materials in the low places. At any particular time each part of the earth's land surfaces are at some stage in the battle between erosion and the building forces at work on the crust.

Chapter 12–5 Wearing Away the Land

ACTIVITY

Materials
Pencil
Paper

At any time a mountain is in some stage of development. It may be rising due to some activity of the crust, while at the same time being worn down by weathering and erosion processes. In this activity you will identify pictures showing some changes that occur in the mountains. You also will identify evidence of mountain building, as well as the depositing of eroded material as the land is worn down.

A. Obtain the materials listed in the margin.

A common feature of cuts made through mountains for highways and railroads is the visible layers of several kinds of rock. At the same time, natural cuts are made by erosion processes.

B. Look at Fig. 12-35. Write the answers to the following questions.

1. Which photograph shows layers of rock exposed by a cut not made by natural forces?

2. Which photograph shows layers of rock exposed by a natural cut made by a river?

The most common way that layers of earth are formed is by the wearing away of existing land and the depositing of this material into bodies of water.

C. Look again at Fig. 12-35 to answer these questions.

3. Which photograph shows how earth has been moved by machine to build a mound?

4. Which photograph

A

B

C

D
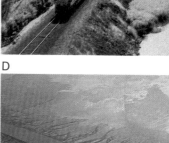

12-35. *Which photographs show natural features?*

Unit IV How Is Our Planet Changing?

shows a river moving material on the earth's surface?

Much of the land that filled the Grand Canyon in the past has been transported by the Colorado River to its *delta* in the Gulf of California. This is enough material to make a mountain at least 2,000 m high and some 70 km long. One reason the material transported by the Colorado River does not show this piling-up effect is that the lowlands are sinking as more layers of earth are deposited. In this manner thousands of meters of layered earth can be deposited over a very long period of time. As the lower layers sink to deeper levels, they are pressed by tremendous forces into rock layers.

D. Look at Fig. 12-36 to answer the following questions.

5. Which two photos show evidence of tremendous forces from inside the earth?

Layers of rock formed by wearing down the land features become mountains when lifted up to great heights by the forces at work inside the earth. In time these rugged youthful mountains show the effect of wear and become mature mountains. Thus the never ending cycle continues.

E. Look at Fig. 12-36.

6. Which photograph shows relatively young mountains?

7. Which photograph shows mature mountains?

A

B

C

D

12-36. *Look at these natural scenes.*

Chapter 12–5 Wearing Away the Land

Can you look at a building and tell something about its age? A building is always a battle between repair and decay. The earth's land surfaces also are the result of the work of two opposing forces. Movements of the crust lift the land to make features such as mountains. The agents of weathering and erosion work to level the land. The shape of the land at any time and place is a stage in the battle between the forces that lift the land and the forces that wear it away.

1. Define *weathering* and *erosion*.
2. Name tools of weathering and erosion that are continually shaping the land.
3. Describe and give evidence to show how a peneplain is formed.
4. Describe the difference in the appearances of youthful mountains and mature mountains.

VOCABULARY REVIEW

Match the description in Column II with the term it describes in Column I.

Column I
1. plate
2. crust
3. mantle
4. core
5. fault
6. glacier
7. peneplain
8. fault-block mountains
9. folding
10. isostasy

Column II
a. Solid outside layer of the earth.
b. The center part of the earth.
c. The result of movement of rock along either side of a crack.
d. A secton of the earth's crust.
e. Thick layer of earth reaching to 2,900 km below the surface.
f. A state of balance existing as the crust floats on the mantle.
g. The bending of rocks without breaking.
h. Produced when large blocks of rock are tilted.
i. A large, slowly moving mass of ice on the land.
j. An almost completely flat surface produced by erosion.

REVIEW QUESTIONS

Choose the letter that best completes each of the following statements.

1. A system of mountains extending around the earth under water is called the (a) under-water chain (b) mid-ocean ridge (c) sea trench (d) continental drift.
2. Which two of the following are evidence in support of plate tectonics? (a) mid-ocean ridges (b) volcanoes (c) age of rock on the sea floor (d) earthquakes.
3. The three layers of the earth are (a) focus, mantle, core (b) mantle, core, crust (c) focus, core, crust (d) focus, mantle, crust.
4. A place where the rock on either side of a crack has moved is called (a) a displacement (b) a fault (c) a crust (d) a breakup.
5. Rocks along a fault may have moved (a) upward (b) downward (c) sideways (d) any of these motions is possible.
6. Large earthquakes usually occur (a) along faults (b) along river beds (c) in the middle of the ocean (d) in desert areas.
7. The three kinds of earthquake waves are (a) primary, secondary, and ocean (b) primary, ocean, and surface (c) ocean, secondary, and surface (d) primary, secondary, and surface.
8. Which of the following features was caused by falling land? (a) Death Valley (b) Grand Canyon (c) Cajon Pass (d) Colorado Plateau.
9. The process that breaks rock into small particles is called (a) erosion (b) weathering (c) glaciation (d) gravity.
10. An almost completely flat surface produced by erosion is called a (a) peneplain (b) plateau (c) monadnock (d) valley.

REVIEW EXERCISES

Give complete but brief answers to each of the following exercises.

1. What happens at the mid-ocean ridge and how is this related to crustal plates?
2. Explain why scientists think the continents may have drifted apart.
3. Describe the earth's interior in terms of sections and temperature.
4. Define *fault*.
5. What relationship is there between faults and earthquakes?
6. How do the Colorado Plateau and Death Valley differ in the land movements that caused them?
7. Describe some of the features of Cajon Pass and how scientists think these features were formed.
8. Name four forces that shape the land.
9. Describe and tell how a monadnock is formed.
10. Describe some characteristics of mature mountains.

SPECIAL FEATURE: MINERALS

(above) · **The gold rush brought people to the West.**

Gold was discovered in California in 1848 and again in Alaska in 1898. "Gold fever" swept across the nation. Men and women from all parts of the country left their jobs, homes, and families to seek their fortune on the wild, unsettled frontier. Some spent their lives searching for such treasures. Most failed. But many, more important discoveries were made. Large amounts of minerals were found in the rocks. Minerals that were important to the country's industries.

Minerals are nonliving natural substances found in the earth. A mineral always has the same properties no matter where it is found. There are about 2,500 known minerals. Minerals are compounds made from natural elements, and many are similar. Similar minerals can be placed in groups. The figures on these pages show examples of mineral groups and some of the properties of minerals.

If you were a prospector, how would you know if you had found gold? Would the color indicate it was gold? Could you tell it was gold by the way it reflected light? These properties would be useful, but they also might be misleading. Many prospectors found what they thought was gold, but what turned out to be the mineral iron pyrite. This mistake was so common that iron pyrite became known as "fool's gold." The best way to identify a mineral is to use a combination of the following properties.

(right) **Minerals provide the materials for industry (clockwise, from left, limonite, bauxite, zincite, hematite, sulfur, and magnetite).**

(top) **Native elements are sometimes found as pure elements.**

Color can be a useful property. However, many minerals, such as quartz, have a variety of colors, and many different minerals may have the same color. See Fig. 4.

Streak refers to the color of the powdered mineral. Scratching the mineral on an unglazed tile usually leaves a powder. It may not be the same color as the mineral sample. But the streak color is the same for all types of a certain mineral. See Fig. 5.

(above) **Some minerals, such as these examples of quartz, may have a variety of color.**

(right) **Streak color helps to identify a mineral.**

325

(above) **Minerals may have a metallic, glassy, waxy, pearly, or dull luster.**

(below) **Many minerals have beautiful crystal shapes.**

Luster refers to the way a mineral reflects light. Minerals may have a metallic, glassy, waxy, or pearly luster. Some minerals do not reflect light well. They have a dull luster. See Fig. 6.

The *crystal shape* of a mineral is caused by the way its atoms are arranged. Most mineral crystals are small. Gems are large, beautiful, and rare crystals. See Fig. 7.

Hardness is a measure of a mineral's resistance to being scratched. If mineral A can scratch mineral B, A is harder than B. Diamond is the hardest mineral; talc is the softest. See Fig. 8.

How a mineral breaks is an important identifying property. An uneven break is called a *fracture*. A smooth break is called *cleavage*. Cleavage may be in one, two, or three directions, and it always produces a surface that reflects light. See Fig. 9.

(left) **Minerals shown in order of hardness:** top row, left to right, talc (the softest mineral), gypsum, calcite, fluorite; bottom row, left to right, albite, feldspar, rock crystal quartz, topaz, corundum, diamond (the hardest mineral).

(below left) **Some minerals fracture. Others may split in one, two, or three directions.**

(below right) **Loadstone, a magnetic mineral, attracts iron filings.**

By using the identifying properties outlined above, and also some special properties, such as magnetism, geologists can identify a sample of mineral.

CHAPTER 13 CHANGES IN THE ROCKS

1 MAGMA AND IGNEOUS ROCKS

Rock collecting can be an interesting hobby. Rock collectors soon discover that their hobby is not a simple one. It is very difficult to sort rocks because there are so many different kinds. Sometimes it is hard to tell one kind from another! Once the collector knows what the rock samples are, how can the rocks be grouped? A stamp collector can organize stamps according to the country the stamps come from. Organizing rocks is not as easy. In order to organize a rock collection, the collector must know about the classification system of rocks. You will learn about the grouping of rock in this section. When you finish this section, you will be able to:

- Name the three major groups of rock and briefly describe how each is formed.
- Define *igneous rock*.
- Classify igneous rock into two major types.
- Demonstrate that the rate of cooling affects the size of rock crystals.

Rock: A piece of the earth that is usually made up of two or more minerals mixed together.

A rock is a piece of the earth. *Rocks* are almost always made up of two or more minerals mixed together and come in all sizes, shapes, and colors. Some rocks are denser than others, some are harder than others. There seems to be no limit to the variety of rocks found in the earth's crust. Careful study shows that there are some important differences among rocks. These differences result from the way the rocks were formed.

Geologists classify rock into three large groups: *igneous* (**ig–nee**–us) *rock*, *sedimentary* (said–eh–**men**–tar–ee) *rock*, and *metamorphic* (met–ah–**mor**–fick) *rock*. Igneous rock is formed when molten material cools.

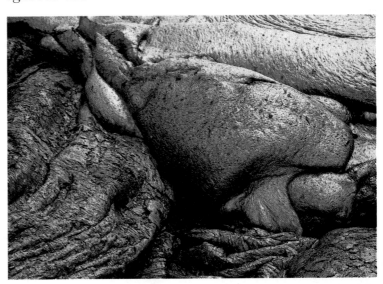

13-1. *Igneous rock forms as lava hardens.*

Chapter 13–1 Magma and Igneous Rock

Most sedimentary rock is formed when small pieces of rock become joined together. Metamorphic rock is rock that has been changed by the action of great heat or pressure. In this section you will learn about igneous rock.

Igneous rock: Rock formed when molten material cools.

Rock formed from molten material is called **igneous rock.** The word *igneous* means "coming from fire." The lava that pours out of a volcanic eruption begins to cool and harden as it comes in contact with the air. See Fig. 13-1. Hardened lava is a form of *igneous rock*. The appearance of the rock depends largely on how fast the lava cools. When a molten material cools slowly, large crystals form. Rock formed by slow cooling of molten material has large crystal grains.

Materials
Candle
Test tube, 13 × 100 mm
Test tube holder
Small plastic spoon
Index cards, 2
Stearic acid
Pencil
Paper

In this activity you will see how the rate at which molten materials cool affects crystal size. Because lava is extremely hot in the molten state you will use a substitute. Stearic (**steer**–ech) acid melts at about 69.6°C and forms crystals upon cooling. It is an excellent substitute for lava.

A. Obtain the materials listed in the margin.

B. Fill the test tube about one-third full of PDB. This material can easily be broken up into small pieces by hand.

C. Attach a candle to one of the index cards so the candle stands on its own. A few drops of melted candle wax dropped onto the card will hold the candle when the wax hardens.

D. Attach the test tube holder to the test tube near the open end of the test tube. The test tube holder keeps your fingers away from the candle flame.

E. Heat the test tube until the PDB is completely melted. Hold the test tube about 1 cm above the flame as shown in Fig. 13-2. **CAUTION:** POINT THE OPEN END OF THE TEST TUBE AWAY FROM YOURSELF AND OTHERS IN THE ROOM. The test tube will blacken if held too close to the flame. If the test tube blackens, stop heating and wipe it clean with a paper towel. Then continue heating but hold the test tube a little higher above the flame.

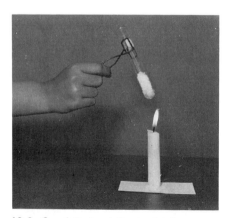

13-2. *Carefully heat the test tube.*

13-3. *Lava may be blown high into the air.*

F. When completely molten, pour about half of the PDB onto an index card. Pour the remaining molten material into the small plastic spoon. Lay the spoon on the table so that the molten material does not spill. The best way to do this is to prop up the handle of the spoon by laying it across the test tube or test tube holder.

1. Which sample is cooling the fastest: the one on the index card or the one in the spoon?

2. Which sample best represents lava which has poured out onto the earth's surface?

G. Look closely at the sample on the card while the other sample continues to cool.

3. Are crystals present?

4. Describe the shape of the crystals.

H. Locate one of the longest crystals on the card. Measure its length using a metric ruler.

5. What is its length in millimeters?

I. After the PDB in the spoon is completely hardened, look closely at its crystals.

6. Describe the shape of the crystals.

7. In general, are the crystals larger than those on the index card?

8. What is the length of the largest crystal you can find on the spoon?

9. How does the rate at which PDB cools affect its crystal size?

10. Based on your observations in this activity, how do you think the rate at which lava cools affects crystal size in igneous rock?

13-4. *Obsidian is formed from magma that cooled quickly. Obsidian is sometimes called volcanic glass.*

Volcanic lava usually contains gases. If there is a small amount of gas, the lava pours out quietly. If there is a large amount of gas, the lava comes out as an explosion. Molten lava may be blown high into the air. See Fig. 13-3. Tiny pieces of hardened lava fall back to the ground as volcanic ash. As volcanic ash piles up on the ground, the pieces are squeezed together to form rock called *tuff* (**tough**). *Pumice* (**pum**-iss) is formed when gas bubbles are trapped in hardened lava. Pumice looks like a hard sponge and can float on water. Some magma cools so quickly that there is no time for crystals to form. As a result the hardened lava looks like dark-colored glass. It is called *obsidian* (ob-**sid**-ee-an).

Not all lava comes to the surface of the crust through volcanoes. It can come to the surface through many fractures. Then the lava spreads out in great sheets. Such an outpouring may cover the existing landscape to form a *lava plateau*. A huge lava plateau covers parts of Washington, Oregon, and Idaho. See Fig. 13-5.

Chapter 13–1 Magma and Igneous Rock

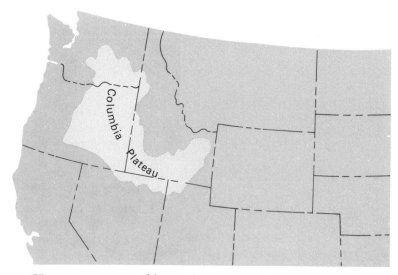

13-5. *The Columbia Plateau is a lava plateau.*

Huge amounts of lava also reach the surface beneath the sea along the mid-ocean ridges. The cooling of this lava creates new sea floor. This slowly pushes apart the great crustal plates. Rock formed when lava comes to the surface is called *extrusive* (ex–**true**–siv) *igneous rock*. *Extrusive* means "pushed out." What about the huge amount of magma that never flows out onto the surface? This magma can also cool into rock. Igneous rock formed beneath the surface of the earth is called *intrusive* (in-**true**-siv) *igneous rock*. *Intrusive* means "forced in." See Fig. 13-6. Lava in the throat of an inactive volcano has hardened. Erosion has worn away the mountain around the body of hardened lava, exposing the intrusive igneous rock.

13-6. *Shiprock in New Mexico.*

13-7. (left) A dike is formed by intrusive igneous rock.

13-8. (right) The Sierra Nevada batholith.

Batholith: A large body of igneous rock formed underground.

13-9. Granite is often found in batholiths.

Dikes and *sills* are bodies of igneous rock formed underground. They form when magma is forced into cracks in the existing rock and hardens. If erosion wears away the surrounding rock, dikes and sills are exposed at the surface. Dikes cut across the layers of existing rock. Sills run parallel with the existing layers.

There are some very large bodies of hardened magma within the crust called **batholiths** (**bath**-oh-lith). *Batholiths* may stretch for hundreds of miles under the surface. Batholiths usually form the cores of mountain ranges. The upward movement of magma in the crust seems to be one of the steps in mountain bulding. The Sierra Nevadas are the result of a great batholith. See Fig. 13-8. A part of the batholith may be exposed as the mountains erode away.

In general, intrusive igneous rock cools very slowly. Magma in batholiths probably takes many thousands of years to cool. As a result, the rock formed this way usually has large crystals. Granite, a kind of rock with large mineral crystals, is an example of the kind of rock found in batholiths. Fig. 13-9 shows the large grains in granite rock.

Chapter 13–1 Magma and Igneous Rock

SUMMARY & QUESTIONS

Every piece of rock has a story to tell. The origin of a rock is an important part in its history. One of the best ways to sort the different kinds of rock is to classify them according to their origins. Igneous rock is formed directly from molten material. The cooling of the molten material can take place on the surface or beneath the ground. The size of the mineral crystals shows how fast the rock has cooled.

1. List the three groups of rock. Describe how each group is formed.
2. What does the term *igneous* mean?
3. What are the two major types of igneous rock and how is each type formed?
4. How does the rate at which rock cools affect the size of its crystals?

2 SEDIMENTS

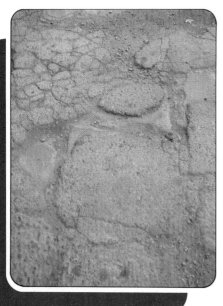

This concrete pavement was once a complete and unbroken surface. Over a period of time, it has been worn as shown in the photograph. The concrete has been eroded bit by bit. It is likely that all rock began as igneous rock. The forces of erosion have been at work on exposed rock for billions of years. Like the pavement, the original igneous rock has been worn away. In this section you will see how the erosion of existing rock can cause a new kind of rock to be formed. When you finish this section, you will be able to:

● Define *sediment*.

● Describe three ways sediments are formed.

● Name one rock formed by each of the three kinds of sediments.

▲ Interpret information from sedimentary layers.

Unit IV   How Is Our Planet Changing?

There is evidence that about four billion years ago most of the earth's surface was covered with molten rock. The first rock of the crust was igneous rock formed as the molten material cooled. The igneous rock formed a thick layer covering the surface of the young earth. Since its formation, the igneous rock on the continents has been exposed to the agents of erosion.

The chief agent of erosion is running water. Water flowing over the surface can pick up small pieces of rock. Also, small amounts of many of the minerals in rock dissolve in the water. The small rock pieces and dissolved mineral carried by running water is called a *load*. Every stream from the smallest trickle to the biggest river carries a load. For example, near its mouth, the Mississippi River moves a load of about 15 tons each second.

The amount of material carried by a stream depends upon how fast the water is moving. Doubling the speed of the water in a stream increases the size of the load it can carry by four times! When a stream flows into a larger body of water such as a river, lake, or ocean, it slows down. When it slows down the stream can carry less material. Part of the load falls to the bottom. This material that settles out of water is called **sediment** (**said**–eh–ment). *Sediment* is deposited where streams of running water slow down.

Sediment: Any substance that settles out of water.

13-10. *This flooding stream is carrying a large amount of material.*

13-11. *Layers of sediment.*

The rock pieces in sediments are not all the same size. The smallest rock fragments make up what is commonly called *mud*. Mud is the most abundant of all sediments. It is a mixture of clay and silt particles. Larger pieces of broken rock make up *sand*. One fourth of all sediments is sand. Coarse pieces of *gravel* are also found in sediments. Gravel sediments are abundant in some places, but usually make up only a small portion of all sediments.

All sediments are deposited in layers when they settle out of a body of water. Generally, the larger rock particles are the first to fall to the bottom. Gravel settles out before sand, sand before silt and clay. Because the different particles settle out at different rates, layers are formed.

Most sediments accumulate very slowly. Many thousands of years may be needed to build a layer of sediment that is only a few centimeters thick. Each layer will grow in thickness until conditions change. For example, a layer of mud might settle out in quiet water that is not moving. When the water begins to move, the layer of mud stops growing or a new deposit is started.

Layers of sediment usually become covered over by other layers. Particles are squeezed together. Sometimes, minerals fill the space between the particles. When this happens the sediments are cemented together. The rock made when a layer of sediment becomes solid is called a **sedimentary rock**.

Shale (**shail**) is the most common of all *sedimentary rocks*. See Fig. 13-12. Shale is formed from clay particles. When grains of sand become cemented together, *sandstone* is formed. See Fig. 13-13. Gravel and large round particles become cemented together to make *conglomerates* (kon-**glom**-err-ret). See Fig. 13-14.

Sedimentary rock: A kind of rock made when a layer of sediment becomes solid.

13-12. (left) Shale.

13-13. (center) Sandstone.

13-14. (right) Conglomerate.

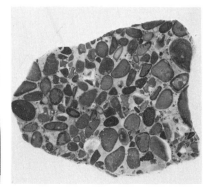

13-15. *Gypsum.*

13-16. *Limestone.*

Not all sedimentary rocks are made from rock pieces. There are two other types of sediment. Dissolved minerals form *chemical sediments*. A body of water may contain a large amount of dissolved minerals. When the water evaporates, the dissolved mineral matter remains as a deposit. For example, if the water supply to a lake is cut off, the water in the lake will eventually disappear. As the water evaporates, the lake becomes a salt lake. When the lake is completely dry, a deposit of minerals remains. *Gypsum* (jip–some) and *rock salt* are sedimentary rocks formed in this way.

The third type of sediments are called *organic* (ore–**gan**–ick) *sediments*. The word *organic* means "life." Organic sediments are made from the remains of living things. Clams, oysters, snails, coral, and some types of fish are among the animals that live in salt water. Each of these has a hard shell or a skeleton. When these animals die, their soft body parts decay. Their hard parts, however, fall to the bottom of the sea. Layers of such sediment become a sedimentary rock called *limestone*. Limestone may also contain mud and sand. See Fig. 13-16.

Sedimentary rocks are an important tool in the study of the earth's history. They are a record of the conditions in the earth's past. *Coal* is a sedimentary rock that comes from the remains of plants that lived in swamps. The coal found in Antarctica shows that the climate of that land was different in the past.

Sedimentary rocks tell of seas once spread over the present continents. They tell of the advance and retreat of glaciers. They tell of the shifting of sands on ancient deserts. Sedimentary rocks also contain the record of past life. The fossil remains of many plants and animals are preserved in the sedimentary layers.

One of the first things a student in geology learns to do is to study drawings that represent a profile of a measured section of land. Figure 13-17 on page 338 shows such a drawing. The layers of the sedimentary rock shown vary in thickness and composition.

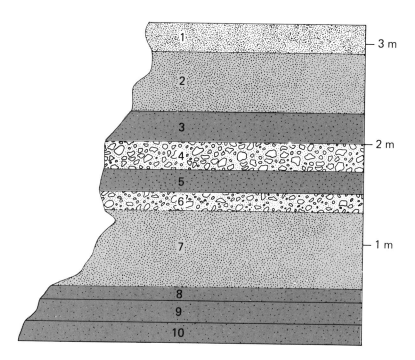

13-17. *Profile of a section of land. Layer 1: gypsum; 2, 7: shale; 3, 5, 8, 9, 10: sandstone; 4, 6: conglomerate.*

Materials
Pencil
Paper

A. Obtain the materials listed in the margin.

B. Look at Fig. 13-17. Answer the following questions.

 1. What kind of rock is found at the 1 m level?
 2. What kind of rock is found just above the 2 m level?

When students in geology study a section of rock layers, they may also take samples of the rock from each layer back to the laboratory for analysis. One property that is usually checked is the grain size. Figure 13-17 shows where these samples were taken. The numbers in Fig. 13-17, starting with *1* in the top layer of gypsum, show the 10 places samples were taken. Figure 13-18 on page 339 gives the grain sizes of the samples in order of the finest grains, found in the shale samples, to the coarsest grain, found in the conglomerate.

C. Study Figs. 13-17 and 13-18. Answer the following questions.

 3. Why is the shale found at the 1 m level slightly finer grained than the shale located near the top at the 2.5 m level?
 4. Do all the sandstone layers have the same grain size?
 5. Which kind of rock layer is a chemical sediment?
 6. Which rock layer was originally deposited as a very fine silt?

Much more can be found out about the history of a section of land. Remember that larger grains of sediment require faster running water to carry them. Figure 13-18 graphs the grain size of each sample alongside the layer in which the sample was found.

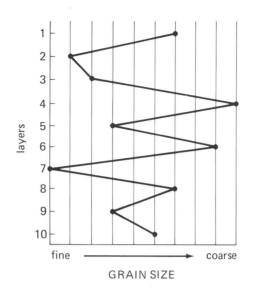
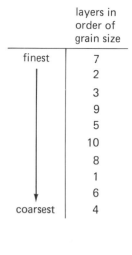

13-18. *Grain size of the samples.*

D. Study Fig. 13-18. Answer the following questions.

7. What kind of rock was formed by sediments transported by the fastest moving water?

8. How many times during the history of this section did the water show such a swift current?

9. As the grain size of the sedimentary rock became finer, what must have happened to the speed of the current of water that delivered the sediment?

10. Which kind of rock layer was laid down when the water was very calm? Such a condition would be present in a large inland shallow sea.

11. How many times did the water become this calm?

12. How can you account for the layer of gypsum being just above the last shale layer deposited?

SUMMARY & QUESTIONS

The earth's surface is no longer covered with a layer of igneous rock. There are processes at work on the surface that reduce igneous rock to sediments. Sediments are moved from one place and deposited in another. Layers of sediments can become changed into many different kinds of sedimentary rocks. These rocks contain a record of much of the earth's history.

1. Define *sediment*.
2. Briefly describe three ways sediments are made.
3. Name one rock formed by each kind of sediment named in Question 2.
4. State two things you can tell from sedimentary layers.

3 CHANGED ROCKS

The lump of material shown in the photograph is the remains of a car. Applying pressure to that car has certainly changed it! Rock in the crust can be changed in a similar way. In this section you will learn about the ways in which rock is changed. When you finish this section, you will be able to:

● Define *metamorphic*.

● Name the two forces that produce metamorphic rock and explain briefly how each may occur.

● Describe the *rock cycle*.

▲ Demonstrate an effect of pressure on sedimentary rock.

ook carefully at the remains of one of the junk cars shown in the photograph above. Is there anything in the lump of twisted metal that tells you what kind of car it was? Now look at Fig. 13-19. This rock is like that car; it

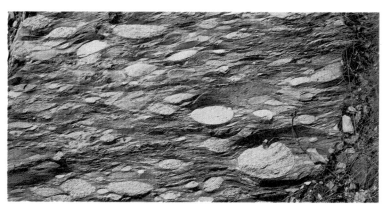

13-19. *This rock has been squeezed by great pressure.*

Metamorphic rock: Rock that has changed by the action of heat or pressure without melting.

has been squeezed by great pressure. The rock was once a sedimentary rock. The sediments have been flattened and pressed tightly together. The original sediments can still be seen. The rock is a **metamorphic rock.** *Metamorphic* means "changed in form." Great pressure has changed a sedimentary rock into this *metamorphic rock.*

Both igneous and sedimentary rock can become metamorphic rock. Many igneous rocks and all sedimentary rocks are formed on or near the earth's surface. How, then, are they put under pressure great enough to change them? There are several ways this can be done. For example, rocks on the surface may be covered by layers of sediment. The weight of these layers might provide the needed pressure. Movements of the crust also twist, flatten, and stretch the mineral grains that make up the rock.

Materials
Clay, 2.5 cm cube
Bread
Eyedropper
Waxed paper
Pencil
Paper

In this activity you will use clay to simulate one way rocks are changed in the earth.

A. Obtain the materials listed in the margin.

B. Form about 20 small spheres from the piece of clay. Each sphere is to represent a piece of sediment.

C. Put the spheres together loosely in a pile. Call this the original form of the clay.

D. Press the pile of clay spheres with the palm of your hand, but not hard enough to remove all the space between the spheres.

E. Take the pile apart carefully and inspect the spheres.
  1. How would you describe the shape of the spheres now?
  2. What caused the present shape (heat or pressure)?
  3. Compare the space between particles in this flattened model with the space in the original form.
  4. Where on the earth could this type of thing happen to rocks?

F. Look carefully at Fig. 13-19 on page 340.
  5. What evidence do you find that the rock in Fig. 13-19 has been under great pressure?

6. What is the name given to a rock that has undergone change such as this?

Scientists often work with models and try to discover things in an indirect way. For instance, to test the affect of great pressure on the amount of water a rock can hold, a scientist might do the following experiment.

G. Place a 1-cm square of bread on a piece of waxed paper.

H. Fill an eyedropper with water. Carefully release one drop at a time onto the bread.

I. Count the number of drops of water that the bread can soak up before water starts coming from the lower edges onto the waxed paper.

7. How many drops does it hold?

J. Flatten a second 1-cm square of bread until you have pushed most of the air from it. This reduces the space between particles in the bread.

K. Count the number of drops this piece of bread holds just as you did with the first piece.

8. How many drops does it hold?

The space between particles in rocks are also changed when under great pressure.

9. How would great pressure affect the amount of water a rock can hold?

Heat is another agent that acts to change rocks. The interior of the earth is hot. Rock buried under several kilometers of earth can be affected by heat as well as pressure. A rock under high pressure can be heated to a high temperature without melting. When this happens, the rock can change. The minerals in that rock may grow into large crystals. New minerals may form as atoms rearrange. Sometimes gases trapped in the rock are released. New gases or liquids from surrounding rocks may mix with the rock material. Rocks changed in this fashion are often hard to trace to their original form.

Metamorphic rock is often found near mountains. If batholiths are formed during mountain building, hot magma comes in contact with rock material. The heat from this magma seems to change the surrounding rock. Folding and faulting often occur along with the mountain building. These processes twist and squeeze existing rock material. Metamorphic rock formed at the roots of mountains is exposed only after millions of years of erosion.

13-20. *(left)* Quartzite.

13-21. *(right)* Slate.

13-22. *(left)* Marble.

13-23. *(right)* Gneiss.

The figures above show common metamorphic rocks. *Quartzite* (**kwort**–zite) is the metamorphic rock made from sandstone. Sand grains in quartzite are fused together more firmly than they are in sandstone. Quartzite is harder than sandstone.

Slate is a metamorphic rock made from clay or shale. Slate can be found as black, brown, red, green, or blue rock.

Marble is another metamorphic rock that can vary in color. Marble is made from limestone. The color of marble depends on the minerals mixed in with the limestone.

Gneiss (**nice**) is one metamorphic rock with a streaked or banded appearance. The changes that produce gneiss are so great that it is hard to determine the original rock material. Without the bands, most gneiss looks like granite. Probably many different rocks can form into gneiss. In the same way, many fine-grained rocks can be formed into the metamorphic rock called *schist* (**shist**).

The formation of metamorphic rock is one step in an endless process that has been going on for billions of

Chapter 13—4 Changed Rocks

Rock cycle: The endless process by which rocks are formed, destroyed, and formed again in the earth's crust.

years. The process is called the **rock cycle.** This process consists of the mixing and reusing of the earth's rock material. For example, hot magma exists beneath the earth's surface. When the magma is forced to the surface, it is called lava. Lava cools and hardens, forming extrusive igneous rock. Agents of erosion act on this rock. The magma in the crust may also harden below the surface, forming dikes or batholiths. Such intrusive igneous bodies are brought to the surface by erosion or crustal movement. Agents of erosion act on these bodies as soon as they reach the surface.

Erosion results in sediments. Sedimentary rock is formed from deposits of sediments. This rock becomes buried by other layers. Sedimentary rock may remain buried, or may be brought to the surface, where it is eroded.

Any rock that is buried below the surface can be changed by heat and pressure into metamorphic rock. The new metamorphic rock eventually is brought to the surface. Erosion then acts upon this form of rock.

Scientists believe that a large portion of the *rock cycle* is taking place at the edges of the crustal plates. Magma is coming to the surface along the mid-ocean ridges and near trench areas. In the trenches, old rock is being forced down into the mantle. In the mantle it is melted and mixed with other rock material. This material may then make its way back to the crust as new lava or magma. Such recycling has been going on since the earth was formed. Figure 13-24 represents the rock cycle.

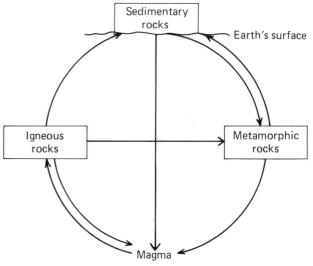

13-24. *The rock cycle.*

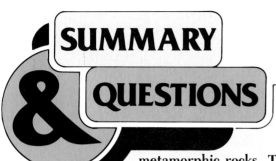

SUMMARY & QUESTIONS

Metamorphic rock is formed by the action of heat and pressure on existing rock. Slate, gneiss, marble, and schist are some common metamorphic rocks. The formation of metamorphic rock is one step in the rock cycle. The rock cycle is the name given to the process that recycles the rock material of the earth.

1. What does *metamorphic* mean?
2. What two conditions produce metamorphic rock?
3. Diagram the rock cycle.
4. What effect does pressure have on the air space in sedimentary rock?

VOCABULARY REVIEW

Match the description in Column II with the term it describes in Column I.

Column I
1. rock cycle
2. fracture
3. rock
4. sedimentary rocks
5. igneous rocks
6. cleavage
7. metamorphic rocks
8. mineral
9. batholith
10. sediment

Column II
a. A mineral that breaks unevenly has this.
b. A piece of the earth made up of two or more minerals.
c. Rocks that have changed due to heat or pressure without melting.
d. The continuous process by which rocks form, change, are destroyed, and are formed again.
e. A mineral that breaks with smooth, flat surfaces has this property.
f. Rocks that form when a layer of sediment is squeezed and becomes solid.
g. A very large mass of igneous rock that formed underground.
h. Any material that settles out of water.
i. Rocks that form when hot molten material cools and becomes solid.
j. A substance found in the earth that always has the same properties.

Chapter 13—Review

REVIEW QUESTIONS

Choose the word or phrase that best completes each sentence.

1. A rock that forms from hot magma deep underground is called (a) sedimentary (b) igneous (c) metamorphic (d) extrusive.
2. Rocks are changed by (a) heat and cold (b) pressure and cold (c) heat and pressure (d) water and ice.
3. Which of the following are properties of minerals? (a) color and streak (b) luster and hardness (c) cleavage or fracture (d) all of these.
4. Which of the following is not a metamorphic rock? (a) marble (b) limestone (c) gneiss (d) slate.
5. Sedimentary rocks may form from (a) rock fragments (b) chemicals (c) organic remains (d) all of these.
6. An igneous rock that cools slowly will probably have (a) large crystals (b) small crystals (c) no crystals (d) the same size crystals as a rock that cooled quickly.
7. Which of the following would be classified as an organic sedimentary rock? (a) marble (b) basalt (c) sandstone (d) coal.
8. Metamorphic rocks may form from (a) igneous rocks (b) sedimentary rocks (c) metamorphic rocks (d) all of these may be changed to metamorphic forms.
9. A sedimentary rock is formed when (a) molten material cools (b) material settles from water (c) an igneous rock is changed (d) a rock is brought under great pressure.
10. Which of the following is an example of an intrusive igneous rock? (a) granite (b) obsidian (c) pumice (d) tuff.

REVIEW EXERCISES

Give complete but brief answers to each of the following exercises.

1. Explain the differences between rocks and minerals.
2. Briefly describe how each of the three major groups of rocks are formed.
3. Compare and contrast dikes, sills, batholiths, and lava plateaus.
4. Explain the effect cooling rate has on crystal size in igneous rocks.
5. List three groups of sediments and describe how each forms.
6. Explain two factors that determine what sediment load a stream can carry.
7. Name the original rocks from which the metamorphic rocks quartzite, slate, gneiss, and marble were formed.
8. Compare and contrast the mineral properties of cleavage and fracture.
9. What physical changes are caused in rock by heat and pressure?
10. Diagram the rock cycle.

CHAPTER 14 CHANGES THROUGH TIME

1 FOSSILS

In a small park in Los Angeles, California, is an unusual museum exhibit. On the shore of a small pond, stand models of an elephant-like animal and its small calf. A few meters from shore, the animal's mate is trapped in the sticky tar that underlies the pond. Soon, exhausted by its efforts to get free, it will collapse and die. The body will sink into the tar. This great animal will join the millions of others that have become trapped in the La Brea (lah **bray**-uh) tar pits. When you finish this section, you will be able to:

14-1. *These are bones found in the La Brea tar pits.*

Extinct: An organism that no longer exists anywhere on the earth.

Fossil: Any indication of life that existed long ago.

- List and explain four types of *fossils*.
- Describe several ways *fossils* can be formed.
- Describe how *fossils* can be used to learn about the earth's history.
- Make models of one type of *fossil*.

The elephant-like animals described on page 347 are called mammoths (**mam**-uths) by scientists. These animals lived in North America about 40,000 years ago. They have been **extinct** for thousands of years. The exhibit reconstructs a scene that may have happened many times in the past. Sometimes, other animals tried to eat those trapped in the tar. Often, they became trapped themselves. So, over the centuries, thousands of bodies sank into the sticky tar. Oil in the tar seeped into the bones and preserved them.

Bones like these, or any other evidence of ancient life, are called **fossils**. *Fossils* may include bones, teeth, shells, leaves, branches, footprints, pollen grains, and sometimes even the skin or fur of an animal. Thousands of fossils have been recovered from the tar of the La Brea tar pits. These include the bones of many animals that are no longer found anywhere in the world. These *extinct*

14-2. *From the fossil evidence found by scientists, an artist drew this picture of the animals found in the tar pits.*

Unit IV How Is Our Planet Changing?

14-3. *This is a reconstructed skeleton from bones found in the tar pits.*

Petrified: Living matter has been replaced by minerals and "turned to stone."

14-4. *Some fossils are the actual remains of the organism. This ant was preserved in amber.*

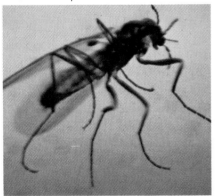

animals include saber-toothed cats, early types of camels and horses, mammoths, and giant ground sloths.

Being trapped and preserved in tar is one of the more unusual ways living things can become fossils. Scientists are often able to locate all the bones of an animal's body in the tar. Then they are able to reconstruct the skeleton and determine what the animal looked like. Hundreds of complete skeletons have been reconstructed from the bones in the tar pits.

Most fossils of animals and plants are not this complete. Animal flesh may be eaten. Flesh, leaves, and wood will rot quickly if exposed to air and bacteria. Occasionally, organisms are buried rapidly or fall into water that slows decay. Then there is a chance that part of the organism will be preserved.

Most fossils are parts of ocean organisms that were buried by sediments soon after death. Usually, hard parts such as bones, teeth, shells, or wood have the best chance of becoming fossils. Rapid burial and the presence of hard parts are important factors in the formation of fossils. Sometimes, conditions are such that even soft bodies such as worms or jellyfish are preserved.

There are several major types of fossils. First, there are those that are the *original remains* of the organism. These sometimes include the whole organism. Complete insects have been found trapped in the sticky sap of pine trees. The bodies of woolly mammoths have been found frozen in mud or soil in Siberia and Alaska. These animals have been preserved right down to the hair on their bodies and with food still in their stomachs. Pollen grains from flowers are often found as fossils. The bones in the tar pits also belong in this group of fossils.

A second group of fossils include those in which the original material has been *replaced by minerals*. Even deeply buried bone and shell material will decay over long periods of time. Sometimes, the molecules of the organic matter are slowly replaced by minerals dissolved in the groundwater. In some cases, the replacement is so exact that details such as growth rings of trees are preserved. The original is, in a sense, "turned to stone." **Petrified** wood, dinosaur bones, and most marine fossils are found in this form.

Shells are often buried in soft mud or fine sand. The mud or sand is packed tightly into every groove and around every bump on the shell. Slowly the mud or sand

Chapter 14–1 Fossils

14-5. *Casts are formed when impressions are filled in.*

is hardened into shale or sandstone. The shell decays but the impression of its surface remains in the stone. This fossil is called a *mold*. The hollow mold may later be filled in by minerals or sediment, forming a *cast* of the original shell. These molds and casts are very common types of fossils. They are often exact copies of the surface of the original organism.

In a few, very unusual instances, volcanic ash or even lava has buried animals, trees, and even people. As the ash hardens, a mold of the bodies may form. The bodies decay but the molds remain. Scientists found such molds at Pompeii, Italy. This city was buried by volcanic ash in A.D. 79. They poured plaster into the molds and formed casts of the people's bodies. The molds are fossils. The plaster casts are not considered fossils. Natural casts would be.

A fourth kind of fossil includes all the *indirect evidence* that life was once present. Dinosaur footprints in mud are a good example. Tracks, trails, burrows of worms, ants, and rodents all indicate where these animals once lived. Thus, these are all fossils.

Many scientists believe fossils tell us much about ancient life and environments. For example, in Nebraska in 1979, scientists uncovered hundreds of fossil skeletons in a layer of volcanic ash. They were able to identify the skeletons as belonging to horses, camels, rhinos, and tiny saber-toothed deer. There were also wading birds and water turtles. Some birds had mice and lizards in their stomachs. A few rhinos still had grass seeds in their throats. Lab work indicated that the ash was 10 million

14-6. *This cast of a human was made from molds found at Pompeii, Italy.*

years old. How this dating was done will be explained in the next lesson.

What can scientists learn from a situation such as this? First, rhinos and camels are not now found in North America. Their bones have not been found in rocks here that are less than 10,000 years old. Therefore, they must have become extinct in North America long ago. Horses also became extinct here. The horses ridden by the American Indians were descendants of horses brought from Europe by the Spanish. Therefore, a group of such fossils in the same layer indicates that all these animals lived in North America at the same time long ago.

Many scientists feel that fossils give us a record of the kinds of organisms that lived in certain places at certain times in the past. Secondly, these scientists believe that arranging fossils by their ages provides a record of the order in which living things appeared on the earth. This indicates that simple forms of life appeared first while more complex forms appeared later. Many types of fossil organisms no longer exist anywhere on the earth. Thus, the record may also show the extinction of organisms including dinosaurs and mammoths. That record, however, is incomplete. Not all organisms became fossils and not all fossils have been found.

Thirdly, the types of fossil organisms tell us what that area was like at the time the organism lived. Rhinos, camels, and horses are grazing animals. Areas where their fossils are found now must have been grassy plains at the time they lived. The grass seeds in the rhino's mouth and the type of teeth they had support this idea.

14-7. *The fossils of these animals were found in volcanic ash 10 million years old.*

The wading birds, water turtles, and microscopic water organisms found here indicate that the animals died in or near a water hole. If fossil clams, oysters, and other mollusks were found in an area, it would indicate that a shallow sea was once present. The types of plant fossils found in an area can indicate the climate in that location long ago.

Lastly, many scientists believe that comparing fossils of similar organisms from rocks of different ages can tell us how life has changed. For example, the fossil horses found in this deposit in Nebraska have three toes. Horses of today have only one toe. Some scientists would say that this is evidence that organisms changed over long periods of time. In fact, many horse fossils have been found covering about 50 million years of time. These fossils appear to show a gradual increase in size, the change from three toes to one toe, and several other changes.

Finding such numbers of complete fossil skeletons in one location is very unusual. These animals in Nebraska all died and were buried by the volcanic ash in a short period of time. The bones in the California tar pits accumulated over thousands of years. It is possible that some of the remains of the thousands of animals and plants killed by the eruption of Mt. St. Helens may become fossilized. Some future scientist may discover these fossils and use them to interpret what the area and organisms were like in 1980.

Materials
Modeling clay
Half pint milk carton
Scissors
Plaster of Paris
Vegetable oil
Cotton swab
Styrofoam cup
Object to be "fossilized"

A. Obtain the materials listed in the margin.

B. Cut the top and bottom off of a half pint milk container so that only the four sides remain.

C. Flatten the clay so that it is bigger than the bottom of the milk carton.

D. Brush a thin layer of oil onto the surface of the object you are "fossilizing." Press the object into the clay until a good, deep impression is made. Carefully remove the object.

E. Brush a layer of oil onto the clay.

F. Position the sides of the milk carton around the impression. Press the carton into the clay slightly so nothing can leak out of the sides or bottom.

G. Stir enough plaster of Paris into 1/2 cup of water to make a thick creamy liquid.

H. Pour the plaster into the

container to form a layer about 2 cm thick. Allow the plaster to dry about 30 minutes before removing the carton and clay.

1. Which of the fossil types is represented by the impression in the clay?

2. Explain how a similar process might happen in nature.

3. Which type is represented by the shape of the hardened plaster?

4. Explain how this process might happen in nature.

5. Suppose you found the footprints of a dinosaur in stone. What steps could you follow to make a mold and a cast of this footprint?

SUMMARY & QUESTIONS

Fossils provide evidence from which we can learn many things about life in the past. The remains of living things can be preserved in many ways. Sometimes a fossil is the original organism. Other times only a hint of its existence, such as a footprint, is preserved.

1. Each of the pictures below represents one of the major types of fossils. Name each type and describe how the fossil formed.

A   B   C   D

2. Explain three ways fossils can be used to learn about living things and their environments in the past.

2 READING EARTH'S DIARY

Many people like to keep a diary for many reasons. It provides a record of the events that happen over a period of time. Many famous people keep diaries so that they can later write books about their experiences. Teenagers also like to record their experiences and feelings. Reading over these notes later helps them to understand themselves better. At least two diaries of teenagers have become famous. Anne Frank recorded the events of her short life during World War II. Laura Ingalls Wilder's diary became the basis of a series of books and a television series. The earth also keeps a diary. With a little experience, you too can learn to read it. When you finish this section, you will be able to:

- List and explain several ideas that are helpful in reading and interpreting the earth's diary.

- Define what is meant by *relative age* and *absolute age* of rocks.

- Explain how radioactivity can be used to determine the age of rocks.

- ▲ Determine the *relative age* of several layers of rocks and groups of fossils.

14-8. *Volcanoes and earthquakes acted on the crust in the past as well as the present. This is uniformitarianism.*

Uniformitarianism: The idea that the processes that are at work on the earth today have acted throughout the earth's history.

Law of superposition: The idea that in undisturbed layers of rock, the oldest layers are at the bottom and youngest layers at the top.

14-9. *Erosion has always been at work. This glacier will erode the rock beneath it.*

Year after year, the surface of the earth is constantly changing. Forces such as volcanoes, earthquakes, and colliding plates build up, twist, and bend the surface. Erosion by wind, water, and ice slowly carves it and wears it down. These processes have been at work on the earth's surface since it was formed. Yet it wasn't until the eighteenth century that this was realized. At that time a Scottish geologist, James Hutton, put into words what now seems obvious. The processes that are at work on the earth today also acted on the earth in the past. This is called **uniformitarianism** (*yoo*-nuh-*for*-muh-**ter**-ee-uh-*niz*-um). In other words, earthquakes, volcanoes, running water, wind, and ice have effected the earth's surface throughout the earth's history. If we understand how these processes are effecting the surface now, we can understand how they effected the surface millions of years ago. *Uniformitarianism* is an important tool in reading earth's diary.

Imagine that in your attic you find a carton that contains family papers. There are about twenty books in which are recorded important family events during the past 160 years. As you lift the carton, it breaks. The diaries tumble out onto the floor. You notice that the top edge of several diaries have been burned so that the dates of each entry are destroyed. As you pick up one old diary, the binding breaks. Pages fall out in bunches and singly. Several bunches fall together, landing in a pile on the floor. Other bunches scatter away from the pile. Single pages drift several feet away.

How are you going to put this book together again? How will you decide which order the pages belong in? Which side of a single page is the front? Which is the back? Are any pages missing? This problem is not unlike trying to read earth's diary. The "pages" are scattered in many places. Some have been "lost." In places, part of the record is out of order. The pieces have to be put together again.

One clue to the order of the pages in earth's diary is the **law of superposition** (**soo**-pur-puh-**zish**-un). This applies to layers of sedimentary rocks. These layers form one at a time, on top of each other, in oceans, lakes, and rivers. They may also form from wind-blown sand. In each case, the layer on the bottom was formed first.

Chapter 14–2 Reading Earth's Diary

14-10. *The law of superposition says the layer on the bottom is the oldest.*

Therefore, it is the oldest. Layers on top are younger. When you look at layers in a cliff, the *law of superposition* tells you that as you go down the wall the layers get older and older.

You could use this idea with the bunches of pages from your family history. Chances are the pile has most of these bunches in order, one on top of the other. *However*, is the sequence in order with the oldest pages on the bottom, or did it turn over so that the oldest pages are on top? Surprisingly, even rock layers can be turned over after they form. The forces of plate tectonics are great enough to bend and fold layers. Uniformitarianism helps scientists solve this problem.

If you pour a bucket of gravel into water, the biggest particles settle first and the smallest settle last. The same thing happens each time we try this. Since this is what happens now, uniformitarianism tells us it also happened this way in the past. So, by examining the particles in a layer, you can tell if they are right side up or not.

These and other clues help geologists determine the **relative age** of the rocks. *Relative age* does not refer to actual years. It is simply an "older" or "younger" scale. For example, if some of the pages from the family history tell about your grandmother's school days and others about her own children, you can determine which are older and which are younger pages. The relative ages of rock layers, faults, lava flows, and magma bodies can be determined this way.

Look at Fig. 14-12. Assume the layers have not been turned over. Which layer is the oldest? Which is the youngest? When did the faulting happen? When did the magma push into the rocks? The law of superposition helps answer these questions.

How do you know when pages are missing? With the

Relative age: a time scale in which objects or events are older than or younger than others. Does not give exact age.

14-11. *Can you determine the relative ages of these rocks?*

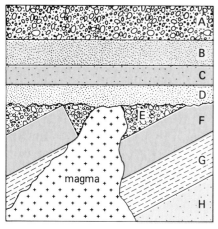

Unit IV How Is Our Planet Changing?

14-12. *Fossil correlation allows distant rock layers to be correctly matched.*

Unconformity: A boundary in rocks that indicates a gap in the rock record. Usually due to erosion of some rocks.

Fossil correlation: Determining the relative ages of rocks by matching the fossils they contain.

Absolute age: A term referring to age in terms of approximate years as determined by radioactive dating.

14-13. *a) tree dies b) one half-life = 5,800 years c) two half-lives = 11,600 years d) three half-lives = 17,400 years*

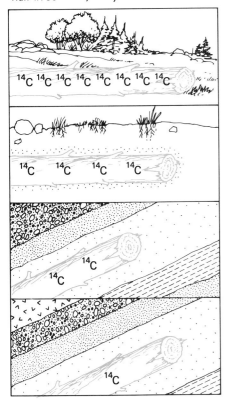

family history you would know if the story did not make sense from one page to another. With rocks it is more difficult to determine when pages are missing. Sometimes the top layers in a sequence are eroded away. Later, new layers are formed on top of this erosion surface. Thus, "pages are missing" from earth's diary. If you know what to look for, you can find clues to this break in the record. These breaks are called **unconformities** (un-kun-**for**-mut-ees). Sometimes the breaks are obvious. Fig. 14-12 shows tilted layers beneath flat ones. The lower layers were tilted, then eroded before the top ones were formed. Many *unconformities* are not this obvious.

The law of superposition will not help with pages of the family history that are not in the main pile. You would have to look for clues within the pages as to their age. This is also true with rocks. How would you determine where layers found far away fit into the history? Are they older, younger, or the same age as those in Fig. 14-12? Uniformitarianism makes this problem even harder. For example, sandstones formed 10 million years ago were made the same way as sandstones that are 100 million years old. How do you determine the relative ages of these rocks if they look the same?

In cases like this, geologists can use fossils if there are any. Some fossil organisms only existed during a small part of the earth's history. Thus, any rocks that hold these fossils must be of the same age. Sometimes the order in which a series of fossils appears in the layers is a useful matching tool. In other cases, certain groups of fossils only appear together in rocks of certain ages. Fossils are the most important clues for matching widely separated layers and determining relative ages of rocks. This is called **fossil correlation** (kor-uh-**lay**-shun).

All of these clues have been used to determine the relative age of rocks and fossils from around the world. But relative age gives no good indication of age in years. For this, we need to determine the **absolute age** of the rocks.

Chapter 14–2 Reading Earth's Diary

Half-life: The length of time it takes one-half of the radioactive elements in a rock or fossil to change into another element.

Several methods of determining *absolute age* have been tried but the best seems to be by using radioactive elements. The length of time it takes one-half of the radioactive elements in a rock or a fossil to change into another element is called its **half-life.** These rates are measurable. By determining how much radioactive material is left in a fossil, geologists can tell approximately when the organism died. These elements provide us with a way to determine the absolute ages of rocks and fossils. However, because many things can affect the amounts of these elements in the rocks, these dates are given as approximate ages.

Materials
Master sheet of cliff diagrams
Scissors
Glue or tape
Paper
Pencil

The figures below represent rock layers and fossils found in four different cliff faces. Some layers contain identical fossils and, thus, may be parts of the same layer. Your task will be to match the layers between cliffs and determine the relative ages of rocks and fossils.

A. Obtain the materials listed in the margin.

B. Cut each cliff model into separate cards representing separate layers. *Do not cut the figure that is on this page.*

C. Arrange the cards in order in a column beneath the cliff number. *These cards must stay in the same column in proper order throughout the Activity.*

1. Explain the law of superposition.

2. Do any of the bottom cards (layers) in each column match?

3. Which layer seems to match layer J?

4. Since layer E is beneath layer D, what does superposition tell you about the relative ages of layer J and layers E and O?

D. Move cards (layers) E and O down your desk toward you until they form a separate

14-14.

	Cliff I		Cliff II		Cliff III		Cliff IV
A		F		K		P	
B		G		L		Q	
C		H		M		R	
D		I		N		S	
E		J		O		T	

horizontal layer. *Keep them in their columns.*

E. Move layers D and J downward so they form a second layer above layers E and O.

F. Look at layers C, I, N, and T. Determine by matching fossils which letters will form the next layer.

5. What is the process of matching by fossils called?

G. Continue moving cards downward until all the layers have been matched properly. Have your teacher check them before gluing them in their matching layers on your answer sheet. Number these matched layers from oldest to youngest.

6. Which layer contains the oldest fossils?

7. Which fossils does the youngest layer contain?

8. What might the gap between layers B and C be called? Explain how it may have formed.

9. Identify any other gaps that might fit this description.

H. Assume that the volcanic ash in layers H, N, and S has been dated at 325 million years.

10. What can you say about the age of fossil 1?

SUMMARY & QUESTIONS

The rocks of the earth's crust contain a record of the earth's history. This history can be read much like a diary if we can determine the proper order of the "pages" that form it. Geologists have discovered several clues to help interpret and date this history.

1. Explain the concept of uniformitarianism. How does this help us interpret the rock record?
2. How does the law of superposition allow us to determine the relative age of rock layers?
3. What is the difference between relative and absolute ages of rocks?
4. What is meant by the half-life of a radioactive material?
5. A sample of rock is found to contain one eighth of the amount of radioactive material expected. If the radioactive material has a half-life of 4,000 years, how old is the sample?
6. How can fossils be used to determine the relative ages of rocks?

3 THE EARTH THROUGH TIME

How old is the earth? There is a lot of disagreement between scientists over this question. However, most scientists would agree that the earth is several billion years old. If the earth has such a long history, how can we organize the rock and fossil evidence so that we can understand it? When you finish this section, you will be able to:

● Explain the structure of the geologic time scale.

● Name four major *eras* of earth history and list them in proper order.

● Briefly describe the major changes in the earth's appearance and life forms that apparently occurred during these *eras*.

▲ Construct and label a model of the geologic time scale.

What does a billion of something look like? A million of anything is hard enough to picture. If this page were solid letters it would hold 2,475 letters. If we started with page 1, the millionth letter would be the forty-fifth letter in the second line on page 405. Imagine that each letter represents one year. All the years since about

490 B.C. would be on the first page. The history of the United States since before the Revolutionary War would fit on the top four lines.

What about a billion years? A billion is a thousand million. We would need over 670 copies of this book to hold one billion letters. Most scientists agree that the earth is at least 4.5 billion years old. It would take over 3,000 of these books to hold that many letters.

Geologists have divided the earth's history into four main units called **eras.** Like the units in this book, the *eras* vary in length depending on the amount of material they cover. The oldest era covers the time from the earth's formation about 4.5 billion years ago up to 600 million years ago. This is called the *Precambrian*

Era: A major division of the geologic time scale, marked by major changes in living things or the earth's surface.

14-15. *The geologic time scale. This scale is based on fossil evidence and radioactive dating.*

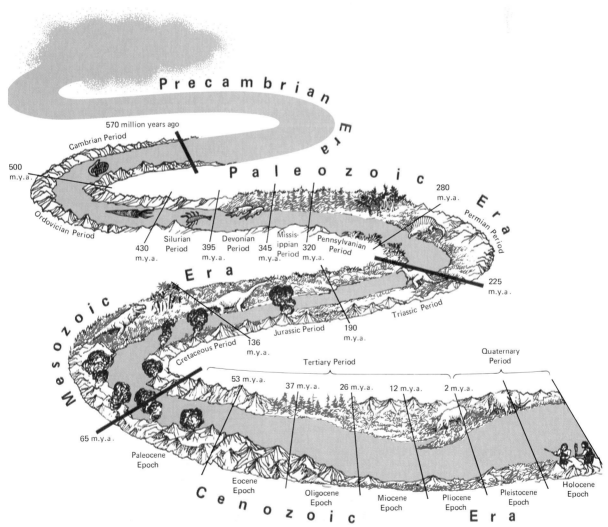

Chapter 14-3 The Earth Through Time

14-16. *The Grand Canyon holds a record of millions of years of earth history.*

14-17. *Precambrian era rocks make up the "nuclei" of all the continents.*

(pree-**kam**-bree-un) era. The Precambrian era lasted almost 4 billion years.

The last 600 million years have been divided into three eras. The *Paleozoic* (pay-lee-uh-**zoe**-ik) era began about 600 million years ago and lasted about 345 million years. The name means "ancient life." The next era is called the *Mesozoic* (mez-uh-**zoe**-ik) era. It means "middle life." This era began about 225 million years ago and lasted about 160 million years. The last era began about 65 million years ago and is called the *Cenozoic* (sen-uh-**zoe**-ik) era. This is the era we are living in today.

As you can see, the Precambrian is much longer than the other eras combined. One reason is that erosion and mountain building in recent eras have destroyed much of that part of the record. Therefore, there is little evidence on which to divide the Precambrian into smaller units. Remember, these time units are only the geologist's way of organizing earth history in order to understand it.

A more complete record still exists for the more recent eras. Thus, it is possible to break them into smaller time units. Just as the units of this text are divided into chapters, eras of earth history are divided into *periods*. The record of the Cenozoic era, being the most recent, is the most complete. The periods in this era are divided into smaller units called *epochs*. This is similar to the chapters of this book being divided into sections. Fig. 14-15 on page 361 shows a more complete picture of these time periods.

Throughout the rock record there is evidence of major changes in the earth's crust or in the forms of life. These changes were chosen as boundaries between the time periods. The geologic time scale shown in Fig. 14-15 was established before the twentieth century began. However, it wasn't until the process of radioactive dating was developed that the approximate dates were added.

We will probably never understand more than a general outline of the earth's history. Too much of the record has been destroyed or is still missing. Scientists do not always agree as to the meaning of new evidence. However, piece by piece, the general outline accepted by most scientists is filled in. New ideas such as plate tectonics modify our understanding of that history. The summary of earth history given here is only a general look at the theories supported by many scientists.

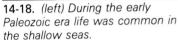
14-18. *(left) During the early Paleozoic era life was common in the shallow seas.*

14-19. *(right) Amphibians and insects were found in the swampy forests of the Late Paleozoic era.*

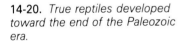
14-20. *True reptiles developed toward the end of the Paleozoic era.*

Most scientists believe the earth and the rest of the solar system formed about 4.5 billion years ago. There are no records of this time in earth rocks. However, samples from the moon have been dated at about this time. The oldest rocks known on the earth are found in western Greenland. These are about 3.9 billion years old.

Precambrian era rocks seem to form the nuclei of each of the continents. Volcanoes and mountain building caused the continents to grow along their edges. The greatest part of the earth's surface was covered by oceans during the Precambrian era. It was in these oceans that life appears to have begun. We do not know exactly how or when life began. We may never know. There are some fossils of simple life forms that have been dated as old as 3.2 billion years. The early fossils resemble bacteria and very simple plantlike organisms.

About 600 million years ago, many complex organisms lived in the oceans. Many had hard parts that were easily fossilized. Early geologists chose the appearance of these fossils as the boundary between the Precambrian and Paleozoic eras. At the time, they had no evidence of earlier life although many felt it must have existed. We have since found the fossils of much older and simpler life forms in Precambrian rocks.

The early Paleozoic era was marked by widespread shallow seas covering the continents. In these shallow seas, many new forms of life developed. Sponges, jellyfish, clams, starfish, and worms are all found as fossils in rocks of this age. So too are the first true *fish*. Also during the Paleozoic era, simple plants began to grow on land. Insects soon appear in the fossil record of this time.

During the late Paleozoic era, the shallow seas flooded central North America several times. Thick layers of sediments were formed here. These rocks are rich with

Chapter 14-3 The Earth Through Time

14-21. *The Mesozoic era is often called the Age of the Reptiles. Dinosaurs were the most important reptile during this time.*

fossils. The evidence indicates that fish continued to develop. One group that appeared at this time was sharks. Another group contained lungfish able to obtain oxygen directly from air. These fish were able to travel short distances over land on their fins. This ability seems to have led to the development of land animals.

The land plants of the late Paleozoic era lived in great swampy forests of scale trees and giant ferns. The remains of these plants were buried and later became the great coal deposits we mine today for fuel.

Evidence suggests that the first *amphibians* (am-fib-ee-unz) lived in these swampy forests. Adult amphibians breathe air through lungs but lay their eggs in water. This dependence kept them close to the swamps. Soon, the earliest evidence of *reptiles* seems to appear in the fossil record. Reptiles are able to live completely out of water all the time.

14-22. *Birds developed during the Mesozoic era.*

While these life forms were developing, the continents were moving together. The ocean floors between them were squeezed and folded into mountains. The supercontinent we call *Pangaea* was forming. The Appalachian mountains were apparently formed by the coming together of Pangaea. The shallow seas were drained back into the deeper oceans. Thousands of types of shallow water organisms could not adapt to the changes and became extinct. These events mark the boundary between the Paleozoic and Mesozoic eras.

The Mesozoic era is often called the Age of Reptiles. Evidence indicates that during its 160-million-year length, new forms of life spread across the supercontinent. Reptiles became the dominant type of life. Turtles, crocodiles, lizards, and dinosaurs developed. More new types of life appeared including *birds* and *mammals*. True seed plants spread over the land. Among them were many of the trees we know today.

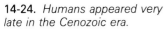

14-23. *The Cenozoic era is called the Age of the Mammals. Many strange kinds developed and became extinct.*

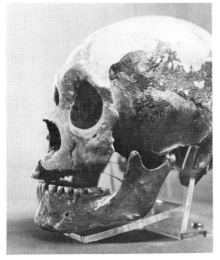

14-24. *Humans appeared very late in the Cenozoic era.*

The most important crustal event of the Mesozoic era is thought to be the breakup of Pangaea. Scientists believe that slowly, over millions of years, North America pulled away from South America and Africa. The Rockies were formed as North America pushed westward. Then Australia and Antarctica pulled away from the rest of Pangaea. Later, South America and Africa separated.

These movements of the crust probably caused changes in sea level. Almost half the land was at one time or another under shallow seas. This produced the wet and warm climate so favorable to the reptiles. Geologists say this era ended with the rapid extinction of the dinosaurs and many other forms of life. Exactly what may have caused this rapid extinction is not known. There may have been extensive climate changes caused by changes in sea level. Some scientists favor theories based on changes in solar radiation.

The last 65 million years form the Cenozoic era. This era is known as the Age of Mammals. The mammals became the dominant form of life on the continents. Late in the era evidence of a group of mammals called humans appears.

Life in the Cenozoic era was heavily influenced by the climate. World-wide cooling has been the rule rather than the exception. The continuing motions of the crustal plates may be the cause of these changes. There is evidence that at least four times in the past 2 million years massive ice sheets have advanced and retreated across the northern continents. Those animals and plants that could adjust to the changing climate survived. Many forms of life disappeared. Many scientists feel the appearance of humans may have been directly related to these extinctions.

Chapter 14-3 The Earth Through Time

Materials
3 index cards
Scissors
3 large rubber bands
2 copies of text
Pencil
Paper

In this activity, you will construct a model of the geologic time scale. For this model, one numbered text page will represent one million years. There are 600 numbered pages in this text.

A. Obtain the materials listed in the margin. Work with a partner so there is one text to read and one to use in the activity.

1. How many years will this book represent?

2. At this scale, can the book contain all of earth's history?

B. Cut three index cards into 4 strips each 3 cm wide. Label the end of one strip for each event in Table 1.

C. Place these strips, like book marks, at the page number that matches the age given. For example, the ice ages began 2 million years ago. The marker would be between pages 2 and 3.

3. How long did each of the geologic eras last?

4. Which eras can you fit into this book?

D. Place a rubber band around the pages that would represent the Cenozoic, Mesozoic, and Paleozoic eras.

5. How many of these books would be needed to represent all of earth's history?

TABLE 1

1. Ice Ages begin: 2 million years ago (MYA)	10. Fish appear: 500 MYA
2. Earliest humanlike fossils: 3.5 MYA	11. Fossils with hard parts: 600 MYA
3. Dinosaurs disappear: 65 MYA	12. First known animals: 1.2 billion years ago (BYA)
4. Birds appear: 160 MYA	13. First known plants: 3.2 BYA
5. Mammals appear: 180 MYA	14. Oldest earth rocks: 3.9 BYA
6. Dinosaurs appear: 225 MYA	15. Age of moon rocks: 4.3 BYA
7. Reptiles appear: 305 MYA	
8. Amphibians appear: 400 MYA	
9. Land plants appear: 440 MYA	

E. A row of books has been placed at the front of the room. These books, plus yours, represent a model of the 4.5 billion years of earth history.

F. Place your marked book at the front of the row and answer these questions.

6. According to the scale discussed in the section, most of earth history belongs to which era?

7. On this scale, how many years passed between the forming of the earth and the appearance of life?

8. How many years passed between the appearance of early plants and early animals?

9. How long between the earliest animals and humans?

10. How long did the Age of the Dinosaurs last?

11. How long ago did dinosaurs become extinct?

12. Could early humans have possibly seen living dinosaurs? Explain.

SUMMARY & QUESTIONS

The most widely accepted estimate of the age of the earth is about 4.5 billion years. Throughout this time, the earth has been constantly changing. Fossils of simple life forms appear in very old rocks. Fossils of complex forms are only found in much younger rocks. The earliest human-like fossils are in rocks only a few million years old.

1. Arrange these terms in proper sequence: period, era, epoch, time. What evidence did geologists use to establish the beginning and end of these units of time?
2. List the four eras of earth history in order, oldest to youngest. Include the length of each era in years.
3. What major crustal event occurred during the Mesozoic era?
4. Which era shows evidence of the appearance of land plants?
5. During which era are we living today?

VOCABULARY REVIEW

Match the description in Column II with the term it describes in Column I.

Column I
1. era
2. half-life
3. fossil
4. extinct
5. unconformity
6. relative age

Column II
a. Any indication of life that existed long ago.
b. A major division of the geologic time scale.
c. An erosional boundary that represents a gap in the rock record.
d. An organism that no longer exists anywhere on earth.
e. The time it takes for one half of any amount of a radioactive element to become stable.
f. An older/younger type of time scale with no reference to actual years.

Chapter 14–Review

REVIEW QUESTIONS

Choose the letter of the response that best completes each of the following statements.

1. An insect trapped in sap from a tree is an example of which type of fossil? (a) original remains (b) replacement (c) mold and cast (d) indirect evidence.
2. Which of these eras has the least complete fossil record? (a) Cenozoic (b) Mesozoic (c) Paleozoic (d) Precambrian.
3. Which of the following is an example of fossilization by replacement of minerals? (a) petrified wood (b) footprints (c) stone tools (d) coal.
4. Which of these organisms is now extinct? (a) wooly mammoth (b) saber-toothed cat (c) three-toed horse (d) all of these are extinct.
5. According to uniformitarianism, which of these forces were at work on the earth long ago. (a) volcanoes (b) earthquakes (c) erosion (d) all of these.
6. Determining the relative ages of distant rocks can be done by using (a) superposition (b) uniformitarianism (c) fossil correlation (d) all of these.
7. Which of these is the correct order? (a) era, period, epoch (b) era, chapter, period (c) era, epoch, period (d) era, epoch, chapter.
8. In which era do true fish first appear in the rock record? (a) Precambrian (b) Paleozoic (c) Mesozoic (d) Cenozoic.

REVIEW EXERCISES

Give complete but brief answers to each of the following.

1. Briefly explain uniformitarianism.
2. How can the law of superposition be used to determine the relative ages of rocks?
3. How do the relative and absolute time charts differ?
4. List four types of information scientists can gain by studying fossils.
5. How can radioactive decay be used to establish the age of a fossil?
6. Explain four ways organisms can become fossilized.
7. Why is the rock record of Precambrian time so poor?
8. How are the boundaries between the time periods established?
9. During which era did each of these forms of life appear? Amphibians, land plants, reptiles, life in the sea, birds, mammals.
10. Explain how unconformities form.

CHAPTER 15 CHANGES IN THE ATMOSPHERE

1 WEATHER FACTORS: HEAT AND PRESSURE

The photo above shows one type of weather balloon. Early one June morning, a fifteen-story-high balloon rose into the skies west of St. Louis, Missouri. Beneath the balloon swung a platform crammed with 2,000 kg of scientific instruments and four scientists. The scientists were to study the polluted air above it. Throughout the day, the crew measured air currents and pollution levels. They also collected air samples to be more closely examined in the laboratory. The balloon landed the next morning over 300 km east of the city. When you finish this section, you will be able to:

- Describe the structure and composition of the earth's *atmosphere*.

- Explain how the sun affects air temperature and pressure.

- Diagram the major wind belts of the earth.

- Demonstrate the effect of the earth's shape on the distribution of solar energy.

Atmosphere: The layer of gases that surrounds the earth.

If the scientists had taken their air samples in a place free from pollution, their tests would have had the following results. About 79% of each sample would have been the gas *nitrogen*. Nitrogen is the most abundant gas in the earth's **atmosphere.** The second most common gas is *oxygen* that would have made up about 20% of the sample. The remaining 1% would have contained mostly argon, carbon dioxide, and water vapor.

The samples taken over St. Louis showed the effects of human activity on the atmosphere. Waste chemicals from power plants, automobiles, and industry continued to increase as the day wore on. Chemical changes were occurring in these materials due to the action of sunlight. That night, a gentle wind blew the balloon and pollution eastward. This cleared the air over St. Louis, but it simply passed the pollution to people living to the east.

Studies conducted by balloons, planes, and rockets have helped us learn a lot about our atmosphere. For example, everyone knows that the higher you go in a balloon, the lower the air temperature is. At first, this may seem strange since you are getting closer to the sun that is the source of the earth's heat. However, scientists have learned that solar energy warms the air very little as it passes through it. Instead, the energy warms the surface of the earth, which in turn warms the air near it.

For years, scientists felt that as you went higher, the temperature would continue to drop steadily until it reached the deep cold of space. Surprisingly, this was found not to be the case. There are several levels at which the temperature changes reverse direction. Scientists use these temperature changes to divide the atmosphere into layers.

15-1. *Pollutants affect the composition of the atmosphere and may be destroying the ozone layer.*

Troposphere: The most dense layer of the atmosphere. It lies closest to the earth.

Stratosphere: The atmospheric layer above the troposphere.

The layer closest to the earth is called the **troposphere**. This is the layer in which we live. Because of the pull of the earth's gravity, about 90% of the gases that make up the atmosphere are found in this layer. It is in the troposphere that most of the changes we call weather occur. Temperatures in the troposphere average about 15°C near the surface and decrease to about −55°C as you move upward.

At a height of about 11 km, the temperature starts to increase. This marks the beginning of the second layer, called the **stratosphere**. The boundary is not sharp. The temperature gradually starts to increase with altitude to a high of about 10°C. Air in the *stratosphere* is thinner than in the troposphere. It contains very little moisture and dust. As a result, there is practically no weather there.

Around 50 km above earth's surface, the stratosphere contains a high percentage of a gas called *ozone*. Ozone is a form of oxygen that absorbs most of the harmful ultraviolet rays from the sun. Absorbing this energy is what causes the higher temperatures in this layer. Many scientists are concerned that pollution in the atmosphere may be breaking down the protective layer of ozone. If this happens, strong ultraviolet rays could be harmful to life on earth. It could also change the earth's climate.

The stratosphere also contains a broad, fast flowing "river" of air circulating around the world. This is called the *jet stream*. The jet stream plays an important role in

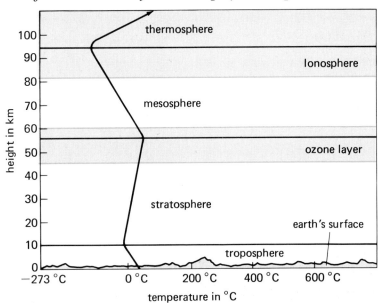

15-2. *The layers of the atmosphere are determined by temperature changes.*

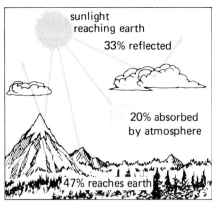

15-3. Only 47% of the solar rays that reach the earth actually reach the surface.

Mesosphere: The atmospheric layer above the stratosphere.

Thermosphere: The atmospheric layer above the mesosphere.

Ionosphere: A region of the atmosphere containing many electrically charged particles.

15-4. The equator receives direct rays of sunlight, but the poles receive slanting rays.

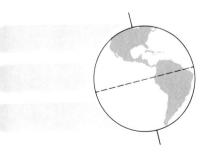

controlling the weather in the troposphere below. Jet planes often fly in the stratosphere to avoid bad weather or to be "carried" by the jet stream.

Above the ozone layer of the stratosphere, temperatures begin to drop once more. This is the beginning of the **mesosphere.** At the top of the *mesosphere*, about 80 km above the earth, temperatures reach −75°C. These temperatures start to rise again in the top layer of the atmosphere, the **thermosphere.** There is no exact end to the *thermosphere*. The gases just continue to spread out until you are in space. This is thought to occur somewhere around 600 km high. Even though the gases are so thinly spread here, the temperature of the thermosphere may reach 2,000°C. This is due to the solar radiation these gases absorb.

When solar energy is absorbed by air molecules, the atoms either gain or lose electrons. Many of the gas molecules between 80 and 400 km have become electrically charged particles called *ions*. This part of the mesosphere and the thermosphere is called the **ionosphere.** The *ionosphere* is important in communications. It can reflect many types of radio waves, allowing them to be "bounced" around the world.

About 20% of the solar radiation that reaches the earth is absorbed by the gases in the atmosphere. Another 33% is reflected back into space by clouds, snow, and ice. Most of the 47% that reaches the surface of the earth is visible light. This radiation is absorbed by the land and oceans. However, it is not evenly distributed over the earth's surface.

Several factors affect the spread of this radiation over the earth. The first is the earth's shape. Since the earth is a sphere, the sun's rays strike different places at different angles. Near the equator the sun passes almost directly overhead. Here these direct rays warm the surface the most. North and south of the equator, the surface of the sphere curves away from the sun. Thus, the rays strike at a lower angle and spread out over greater distances. As a result, these locations receive less solar energy than the equator. The poles receive the least energy.

A second factor is that land and water do not absorb energy at the same rate. Land warms up faster than water does. At night or in the winter, the land loses its heat and cools down faster than the water. Other factors include the tilt of the earth's axis, its day and night periods, and its path around the sun. These factors are explained in Chapter 16.

This unequal spreading of radiation causes unequal heating of the earth's surface. Since the atmosphere is heated by the surface beneath, it too is heated unequally. Air near the equator is heated more than air near the poles. Heated air expands. This means that warm air at the equator is less dense than cold air at the poles. The density of the air determines the force with which it presses down on the earth's surface. This force is measured as *air pressure*. Cold air presses down on the earth with a greater pressure than does warm air. Cold air is said to have a *high pressure*. Warm air is said to have a *low pressure*.

These density differences are also the reason why air moves. In general, air flows from a region of high pressure toward one of low pressure. Thus, cold, dense air from the poles would flow toward the equator along the surface of the earth. At the equator, it would be warmed and expand. The warm air would then be pushed upward by more cold air moving in behind it. The warm air would flow back toward the poles at the top of the troposphere. Here it would cool, become denser, sink, and flow back toward the equator. Fig. 15-4 shows this general circulation. This pattern is affected by many factors. In the Northern Hemisphere, the path of any object traveling north or south above the earth appears to curve to its right. Actually, the object travels in a straight line, the earth simply moves beneath it. Of course, this is more visible when the object is traveling long distances. This

15-5. *Cold, dense polar air flows toward the equator. Warm, less dense air rises and flows toward the poles.*

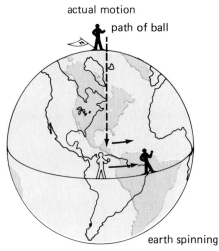
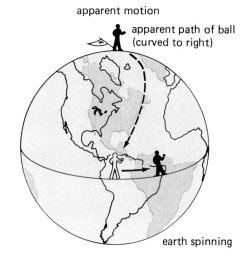

15-6. *The Coriolis effect is a result of the earth's rotation.*

Coriolis effect: The apparent curving of the path of moving objects as a result of the earth's rotation.

is called the **Coriolis** (kore-ee-**oe**-lus) **effect.** In the Southern Hemisphere, the effect is an apparent curve to the left.

The *Coriolis effect* breaks the general circulation into a series of smaller wind belts. In each hemisphere, rising air near the equator produces a calm region called the *doldrums* (**dole**-drums). Here there are only weak surface winds.

In the upper parts of the troposphere, the warm air rising from the doldrums turns north and south. As this air moves toward the poles, it is cooled. About one fourth of the way to the poles, most of this air sinks back to the surface. In this area of sinking air, there is also little surface wind. At the surface, the sinking air divides. Some of it flows toward the equator. The Coriolis effect turns this air so that it blows from the east. This creates the warm and steady belt of *trade winds*.

The rest of the sinking air moves along the surface toward the poles. The Coriolis effect and other forces turn this wind so it blows from the west. These wind belts are called the *westerlies*. It is very hard to explain why this belt exists. We will just say that it does. Most of the United States is beneath this belt so most of our weather moves from west to east.

The westerlies blow northeastward until they meet with very cold air moving along the surface toward the equator. This is the cold sinking air from the poles. It too has been turned by the Coriolis effect until it seems to come from the east. This belt of winds is called the *polar easterlies*. The movements of the major wind belts are a major factor in controlling the weather.

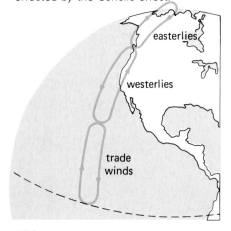

15-7. *The earth's wind belts are effected by the Coriolis effect.*

Materials
Flashlight
Cardboard tube, 15 cm long
Masking or electrical tape
Paper
Pencil

A. Obtain the materials listed in the margin.

B. Tape the cardboard or paper tube over the end of the flashlight so that it throws a circular beam of light.

C. Place an "x" about 1 cm in from the bottom of your paper. With the light on, hold the flashlight so that the end of the tube is about 25 cm above the x. Match the lower edge of the circle of light with the x. Trace the border of the circle of light on your paper.

D. Keep the end of the tube the same distance from the x. Also keep the lower edge of the light beam on the x throughout the remaining steps.

E. Tilt the flashlight toward you so that the beam strikes the paper at an angle of about 30° from the overhead position. Trace the edge of the lighted area.

F. Repeat the procedure at angles of 60° and 90° from overhead.

1. At which position is the light energy concentrated in the smallest area?
2. At which position is the light energy spread over the largest area?
3. Which position is similar to the way land at the equator receives energy from the sun?
4. Which is similar to land at the poles?
5. What causes these differences in how energy spreads out over the earth?
6. Explain the effect of this unequal spreading of energy on the earth's atmosphere.

The atmosphere is made up mostly of nitrogen and oxygen with small amounts of other gases. It can be divided into several layers based on temperature differences. Unequal heating of the earth's surface as well as the earth's spinning has caused the formation of several major belts of wind in the atmosphere.

1. Name five gases found in the atmosphere. List the percent of air by volume each occupies.
2. Explain how the atmosphere is divided into layers.
3. What effect does the rotation of the earth have on the wind? What is this effect called?
4. How does the unequal heating of the earth due to its shape affect the atmosphere?

Chapter 15–1 Weather Factors: Heat and Pressure

2 WEATHER FACTORS: MOISTURE

Each day, about a trillion metric tons of water move from the earth's surface into the atmosphere. Each day, about the same amount returns to the earth as some form of precipitation. Yet many parts of the world, including parts of the United States, face serious water shortages. How does this moisture move from the surface to the atmosphere? What causes it to return to the surface? Why are there so many different types of precipitation? When you finish this section, you will be able to:

- Define *evaporation, condensation, humidity,* and *dew point,* and explain how they are related.
- Name the conditions necessary for clouds to form.
- Describe the major types of clouds and precipitation.
- Determine the *dew point* in your classroom on a given day.

Evaporation: Changing water from a liquid to a gas.

Humidity: The amount of water vapor in the air.

Relative humidity: The amount of water vapor in the air compared with the amount of water vapor the air could hold at that temperature.

Water is an important weather factor. As it is heated by energy from the sun, water **evaporates.** It *evaporates* from lakes, streams, rivers, and plants, but mostly from oceans. The amount of this water vapor in the air is called the **humidity.** The amount of moisture the air can hold is determined by the air temperature. Warm air can hold more moisture than cold air.

Air does not always have as much moisture as it could hold. The weather forecaster uses the term **relative humidity** to compare the amount of moisture the air does have with the amount it could hold. For example, if a certain volume of air could hold 40 g of water but only has 20 g, the *relative humidity* would be 50 percent. If the same mass of air has 30 g of moisture, the relative humidity would be 75 percent. Only 10 g of water would mean the relative humidity was a low 25 percent. A relative humidity of 60 percent is most comfortable at normal room temperature.

If the relative humidity reaches 100 percent, the air is said to be *saturated*. It is holding as much moisture as it can at that temperature. Sometimes a mass of air may become saturated if its temperature drops. The amount of moisture stays the same, but cold air holds less moisture than warm air. As the temperature drops the relative humidity increases. The temperature at which the relative humidity reaches 100 percent is called the **dew point**. If the temperature continues to drop, the water vapor begins to change back into a liquid or solid form of water. This process is called **condensation**.

Dew point: The temperature at which the air becomes saturated with water vapor.

Condensation: The process by which a gas turns into a liquid.

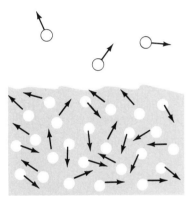

15-8. *(above) Fast-moving molecules move out of the liquid.*

15-9. *(right) Condensation.*

A. Obtain the materials listed in the margin on page 378.

B. Fill a shiny can about half full with water. Keep the outside surface clean. Do not breathe on it.

C. Measure the temperature of the air in your room.

Chapter 15–2 Weather Factors: Moisture

Materials
Shiny can
Thermometer
Ice
Water
Stirring rod
Pencil
Paper

15-10.

Meteorology: The branch of science that studies the atmosphere.

1. What is the room temperature?

D. Measure the temperature of the water in the can.
2. What is the temperature of the water?
3. Is dew forming on the can?
4. Is the temperature of the water above or below the dew point?

E. If dew has not formed on the can, add a piece of ice to the water. Stir with a stirring rod. Watch for dew to form on the can. Do not breathe on the can. As soon as dew forms, remove the ice and then measure the temperature of the water.
5. What is the temperature at which dew appears?

F. Stir the water until the dew disappears. Read the temperature of the water.
6. What is the temperature at which the dew disappears?

G. If your answers to Questions 5 and 6 differ by more than 2 or 3 degrees, repeat Steps D and E. If they are within 2 or 3 degrees of each other, find their average. This is the dew point.
7. What is the dew point?

Air may be cooled as it rises to flow over mountains. It may be pushed upward and cooled by denser air moving in under it. If the air is cooled to the *dew point* and if there are solid particles such as dust or salt crystals in the air, visible water droplets will form. The solid particles give the moisture a surface on which to *condense*. Visible droplets form clouds.

Clouds come in many shapes and sizes. A system of names, based on the appearance of clouds, is used in **meteorology** (me-tee-ore-**ahl**-oh-gee). Three major classifications of clouds are *stratus* (**stra**-tus) *clouds* (spread out), *cumulus* (**kue**-mew-lus) *clouds* (piled up), and *cirrus* (**sear**-us) *clouds* (curled). Stratus clouds are flat broad layers. They form a gray blanket over the entire sky. Cumulus clouds are individual clouds with rounded tops and flat bottoms. Their tops often resemble heads of cauliflower. Cirrus clouds are found high in the atmosphere. Cirrus clouds are made up of ice crys-

15-11. *Stratus clouds.*

tals. They resemble feathers or curls of hair and have a delicate appearance.

Some cloud formations are combinations of these classes. For example, a layer of stratus clouds may have a very bumpy surface, like many cumulus clouds put together. These clouds are called *stratocumulus* (stra-toe-**kue**-mew-lus).

Sometimes the dew point is reached right at ground level. This may happen as warm air flows over cold ground or snow. The cloud that forms is a *fog*.

Moisture may condense on cold objects such as metals,

15-12. *(left) Cumulus clouds.*

15-13. *(right) Cirrus clouds.*

Chapter 15–2 Weather Factors: Moisture

15-14. *(left)* Fog.

15-15. *(right)* Dew.

rocks, and plants. This moisture is called *dew*. If the dew point is below 0°C, *frost* will form. Frost is solid ice formed directly on the object. It is not frozen dew.

Cloud droplets are very small. Even slight movement of the air keeps them floating. Sometimes, however, the drops are drawn together by electric charges, and these new, heavier drops fall as *rain* or *drizzle*. In colder regions this precipitation begins as ice crystals high in the clouds. The ice crystals grow into *snowflakes* and begin to fall. If the air is cold enough, the flakes will stay snow all the way to the ground. However, if they pass through a warm layer, they will melt and become rain. Rain may pass through another cold layer and freeze. It is then called *sleet*. In violent storms, raindrops may be tossed high enough in the cloud to freeze. As they fall again, a new layer of water forms. If they are tossed up and down several times, *hail* forms. Hail pellets are made of layers of ice.

15-16. *(left)* Frost.

15-17. *(right)* A hailstone.

SUMMARY & QUESTIONS

The big storm of February 6 and 7, 1978, drew its moisture from the warm waters of the Gulf of Mexico. As it moved northward along the Atlantic coast, vast amounts of this moisture condensed because of cooler temperatures. This moisture fell as snow, sleet, and rain all along the eastern seaboard. In many places the amounts set new records and caused many problems.

1. Describe the three major types of clouds.
2. How do humidity and relative humidity differ?
3. List several ways that air may be cooled to the dew point.
4. Name the conditions necessary for clouds to form.

3 WEATHER CHANGES

On January 31, 1977, weather forecasters discovered an unusual weather situation. They had just put together information from satellite pictures and ground weather stations. As they studied these maps and charts, they found that everyone of the original 48 states had snow on the ground at the same time! This was the first time such a thing had happened since records had been kept.

People in Miami had been shocked to see snow falling. Why is it that we expect certain types of weather in certain parts of the country? What factors determine the general weather patterns of the United States? When you finish this section, you will be able to:

● Describe the major types of *air masses* that affect North America.

● Explain the formation of weather *fronts* and the type of weather they bring.

● Describe *cyclones* and *anticyclones*.

▲ Use information on a weather map to make a forecast.

Why is snow in Miami so unusual? The air over the southeastern United States has plenty of moisture. However, it is seldom cold enough to form snow. Air over the northern states is known for its cold winter temperatures. It appears that there are different masses of air with different properties that affect the weather.

Scientists have found that large **air masses** tend to take on the properties of regions where they are located. These regions are called *source areas*. There are four types of source areas that affect air over North America.

In general, *air masses* that form over land are dry,

Air mass: A large mass of air that has taken on the temperature and humidity of a part of the earth's surface.

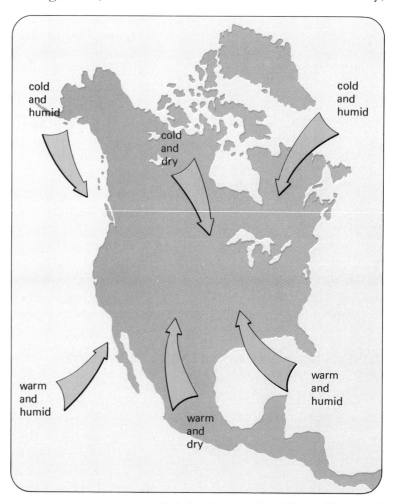

15-18. *The source areas shown here affect weather in North America.*

382 Unit IV How Is Our Planet Changing?

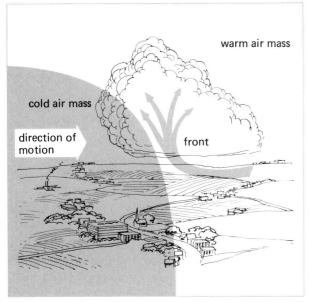
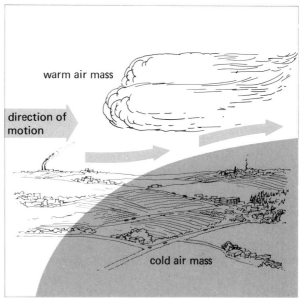

15-19. *(left) A cold front.*

15-20. *(right) A warm front.*

Front: The boundary separating two air masses.

those that form over oceans are humid. Secondly, air masses that form in northern regions are cold and those that form in regions close to the equator are warm. Figure 15-18 shows the source areas of the major air masses that affect North America.

As air masses move, they meet each other. The air in them does not mix easily. This is mainly because of the differences in temperature between the two bodies of air. Cold air has a greater density than warm air. Mixing air with different densities is like trying to mix oil with water. Because they don't mix easily, the boundary between the two air masses tends to be sharp and distinct. Boundaries separating air masses are called **fronts**. It is usually possible to locate *fronts* by measuring the differences in temperature, pressure, and wind direction on the opposite sides of a front.

Most fronts form when one air mass moves into a region occupied by another air mass. If a cold air mass invades warmer air, a *cold front* is created. The colder air is denser and pushes under the warmer air like a wedge. The lifting of the warm air usually causes cumulus-type clouds to form along the cold front. The clouds along a cold front usually bring heavy rain. Cold fronts move quickly because the dense, cold air pushes easily into the warmer air. The stormy weather connected with a cold front usually passes quickly. The air mass behind a cold front brings cool, dry weather.

Chapter 15–3 Weather Changes

When a warm air mass pushes into an area of cold air, a *warm front* is formed. Because it is less dense, the invading warm air rides up over the colder air. As the warm air is lifted, cirrus clouds are formed high in the sky. As time passes, the cloud cover becomes lower and thicker, until a solid sheet of stratus clouds covers the sky. A long period of gentle rain may occur, followed by slow clearing and rising temperatures. The weather changes that follow a warm front are not as noticeable as the changes following a cold front.

The boundary between the cold, northern air masses and the warm, southern air masses is called the **polar front**. Often the smooth flow of the winds past each other at the *polar front* is disturbed. A disturbance, or interference, in the flow can be caused by many things. For example, interference can be caused by mountains or by the differences in temperature between land and sea surfaces. Any interference can produce a small wave or bulge in the polar front. At the bulge, the warm air pushes into the cold air, making a warm front. The cold air moves around into the warm air, forming a cold front. The faster moving cold front soon catches up with the warm front. The warm moist air is squeezed upward leaving only cold air near the ground.

Lifting of the warm, moist air almost always causes a storm center with dark clouds and heavy precipitation. A storm center formed in this way is called a **cyclone** (si-klone). The air in a *cyclone* tends to have an upward movement. This results in an area of low pressure. Because of this low pressure area, a cyclone is also called a *low*. Winds tend to blow in toward the area of low pressure in a cyclone. In the Northern Hemisphere

Polar front: The boundary where cold polar air meets warm air.

Cyclone: An area of low pressure with winds circling into the center.

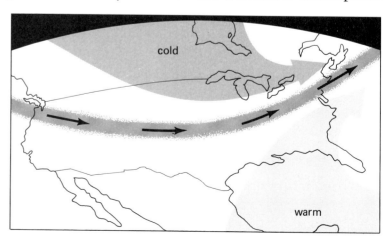

15-21. *A polar front.*

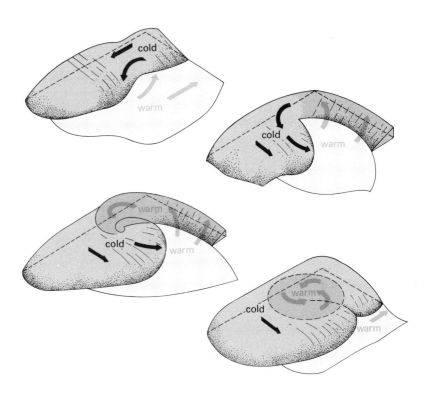

15-22. *The formation of a cyclone.*

15-23. *A cyclone.*

Anticyclone: A large mass of air spinning out of a high pressure area.

these winds are turned to the right by the Coriolis effect. A cyclone in the Northern Hemisphere becomes a giant swirl, slowly twisting in a counterclockwise direction. Lows in the Southern Hemisphere are giant swirls slowly twisting in a clockwise direction.

Cyclones can develop in a few hours or they may take many days to form. Once formed, a cyclone may remain in one place for many hours or it may move at a speed of up to 40 km/hr. A cyclone may grow until it is more than 1,000 km in diameter, or it may remain very small. A large area of low pressure often brings rain or snow and unsettled weather for days as it moves across the land.

A mass of cool, dry air may also develop a spinning motion. The greater density and higher pressure of cool air causes an outward flow of air. Winds created by this air movement are given a clockwise twist by the Coriolis effect. The result is a large, slowly spinning body of air called an **anticyclone** (an-tee-si-klone). In the Northern Hemisphere *anticyclones* turn in a clockwise direction. Their motion is counterclockwise in the Southern Hemisphere. Because they are formed by cool air that is usually not very moist, anticyclones often bring fair, dry weather. Anticyclones are also called *highs*.

Chapter 15–3 Weather Changes

ACTIVITY

Materials
Ruler, 30-cm
Pencil
Paper

Nearly all newspapers have a section that gives weather information and a weather forecast. Often there is a map that shows the weather for a very large area. In this activity you will interpret such a map.

A. Obtain the materials listed in the margin.

B. Look at the map on page 387. Use the legend to answer the following questions.

  1. In what part of the United States are there cloudy, overcast skies with rain?
  2. Is there a warm front and a cold front involved in this weather condition?

As you may recall, a cold front is the leading edge of a mass of cold air and a warm front is the leading edge of a mass of warm air. The most violent kinds of weather result when two such masses come in contact with each other.

C. Look again at the map.
  3. Is the cold front in the midwestern states the only cold front shown on the map?
  4. What kind of weather is found to the east and west of the weather front found in the northwestern U.S.?

High and low pressure areas generally sweep across the United States from the Northwest to the Southeast. There usually are several high pressure areas and low pressure areas scattered over the continental United States at any one time. Clear, cool weather is found around high pressure areas.

  5. How many high pressure areas are shown on the map?
  6. How many low pressure areas are shown on the map?
  7. Which kind of pressure area is causing all the weather fronts?

The entire weather system found over the midwestern states is moving eastward. The warm front is moving in a northerly direction, and the cold front is moving in a southeastern direction. The weather conditions along the two fronts are moving with the fronts. It is possible to predict the weather if the speed of the fronts is known.

D. Suppose the low pressure area in the Midwest is moving eastward at 16 km/hr. Use this information and the map to answer the following questions.

  8. How far would the low pressure area move in 24 hours?
  9. Point A is Toledo, Ohio. What change in weather would you predict for Toledo in 24 hours?
  10. Point B is Nashville, Tennessee. What kind of weather would you predict for Nashville in 24 hours?
  11. Can a long-range forecast be made from this information?

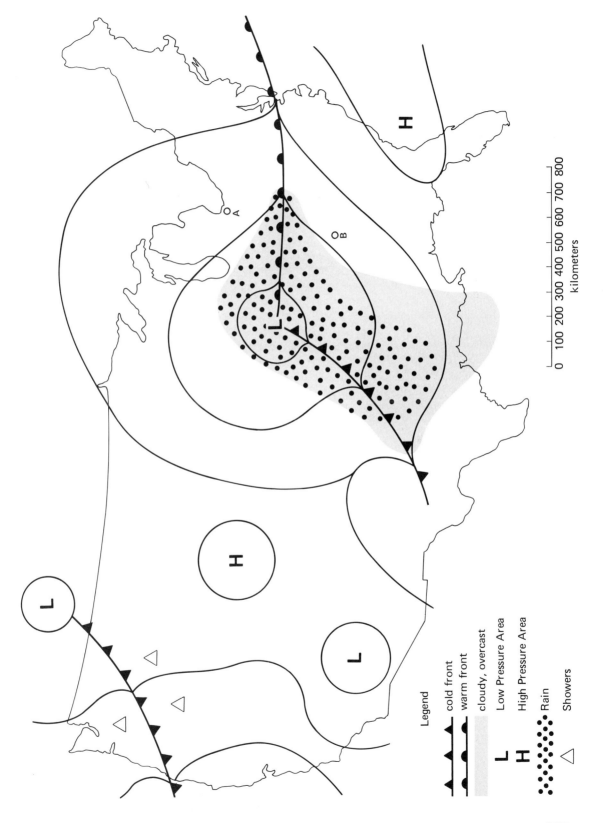

Chapter 15–3 Weather Changes 387

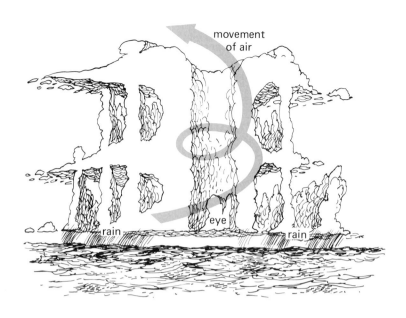

15-24. *A hurricane is an intense cyclone.*

15-25. *A hurricane.*

Hurricane: A small but intense cyclonic storm formed over warm parts of the sea.

Hurricanes (**her**-rih-kane) are cyclonic storms formed over warm ocean water near the equator. *Hurricanes* are usually smaller, but more intense, than cyclones formed near the polar front. Hurricanes that affect North America are produced in the late summer and early fall. It seems that during this time of year, conditions in the atmosphere cause rapid upward motion of very warm, moist air. As the air rises, its moisture condenses. More moist air is drawn into the developing storm until a great column of rapidly rising air is created. The rotation of the earth twists the rising column. When fully developed, hurricanes in the Northern Hemisphere have bands of clouds spinning in a counterclockwise direction around a center of calm air. The center of a hurricane is called the *eye*. Wind circling the eye may reach speeds greater than 100 km/hr. As hurricanes come near land, winds fling destructive waves onto the shore. Heavy rain causes flooding. Hurricanes die out as they move over land or cooler water and their supply of warm, moist air is cut off.

The most destructive of all weather disturbances are *tornadoes* (tore-**nay**-doe). Tornadoes are small, but extremely violent, storms that form along cold fronts. Scientists believe that tornadoes are formed when, for unknown reasons, fast moving cold air rides up over warm, moist air. The warm air rises rapidly, creating an area of very low pressure. Air rushes in from the sides of the low pressure area, giving the rising column of air a twisting motion. The result is a funnel-shaped cloud. Tornadoes vary in size. Some are small, with diameters of tens of meters. The diameter of others can be hundreds of meters. Tornadoes usually last only a few minutes. Their winds, sometimes as fast as 800 km/hr, can destroy almost everything in their path. Within the funnel, the pressure is so low that buildings may explode like a bursting balloon, as the air trapped inside pushes out.

15-26. *A tornado.*

Within the atmosphere, there are large bodies of air with uniform properties. These air masses are produced when air remains over one area for a long period of time. Boundaries between different air masses can be identified because different air masses do not mix easily. These boundaries, or fronts, are associated with certain kinds of weather. Once in a while, unusual motions of air masses can bring unexpected weather conditions such as snow in Miami.

1. Define air mass and front.
2. Describe the different types of air masses that affect North America.
3. Compare a cyclone to an anticyclone.
4. How are hurricanes and tornadoes different?

VOCABULARY REVIEW

Match the description in Column II with the term it describes in Column I.

Column I
1. meteorology
2. air mass
3. front
4. dew point
5. polar front
6. cyclone
7. hurricane
8. anticyclone
9. relative humidity
10. Coriolis effect

Column II
a. An area of low pressure with winds circling into the center.
b. A large body of air that has taken on the temperature and humidity of part of the earth's surface.
c. The boundary separating two neighboring air masses.
d. The branch of science that studies the atmosphere.
e. A small but intense cyclonic storm formed over warm parts of the sea.
f. The boundary where cold polar air meets warm air.
g. A large mass of air spinning out of a high pressure area.
h. The amount of water vapor in the air compared to the amount the air could hold at that temperature.
i. The temperature at which air becomes saturated with water vapor.
j. Curve in the path of a moving object due to the earth's motion.

REVIEW QUESTIONS

Choose the letter of the response that best completes each of the following statements.

1. Which of the following best describes air in which clouds might form? (a) dry and warm with particles (b) moist and cool with particles (c) moist and cool without particles (d) dry and cool without particles.
2. Which type of cloud resembles a flat layer? (a) cirrus (b) stratus (c) cumulus (d) cumulonimbus.
3. Where two air masses meet (a) a storm always occurs (b) it always rains (c) it never rains (d) a front always forms.

4. The gas found in air in the greatest amount is (a) oxygen (b) nitrogen (c) argon (d) carbon dioxide.
5. A cyclone usually develops where (a) a polar air mass meets a cold air mass (b) a polar air mass meets a warm air mass (c) two warm air masses meet (d) a tropical air mass meets a warm air mass.
6. Wind belts on the earth are the result of air moving and (a) ultraviolet rays (b) the Coriolis effect (c) the greenhouse effect (d) absorption.
7. The wind belts just north and south of the equator are called the (a) doldrums (b) westerlies (c) polar easterlies (d) trade winds.
8. Water vapor becomes part of the atmosphere because of (a) evaporation (b) condensation (c) humidity (d) dew point.
9. If the amount of water vapor in the air remains constant while the temperature decreases, the relative humidity will (a) decrease (b) remain the same (c) increase (d) not be affected.

REVIEW EXERCISES

Give complete but brief answers to each of the following exercises.

1. What is needed for clouds to form?
2. Name and describe the appearance of the three classes of clouds.
3. Give two things that can happen to cloud droplets in order for precipitation to take place.
4. How are air masses and fronts related?
5. Describe four kinds of air masses that affect North America.
6. List five gases found in the atmosphere.
7. What is the Coriolis effect?
8. Name the major wind belts of the earth.
9. Explain how humidity, relative humidity, and dew point are related.
10. If one cubic meter of air contains 1.5 g of water vapor, but it could hold 2.0 g, what is the relative humidity?

CAREERS:
WEATHER WATCH

All over the world weather observers watch the sky and their weather instruments carefully. A good system of observation and weather forecasting is vital to society. Farmers depend on accurate forecasts when they plan their work each day. They also consult long-term forecasts before selecting crops and planning schedules for planting and harvesting. Industry relies on the long-term forecast too. Based on such weather forecasts, the factories and plants order the fuels they will need in the months ahead.

(left) **Weather observers check their instruments carefully at the same time each day.**

(lower left) **New methods of weather control are checked in the field. This technician is working on a study of cloud seeding.**

(right) **Information sent in by weather observers is recorded onto maps by workers at a weather bureau.**

CHAPTER 16 OUR CHANGING FRONTIERS

1 THE OCEANS

Have you ever heard of *oceanauts*? Jacques Yves Cousteau uses the term to describe a new breed of scientists. They are exploring ways to live and work beneath the ocean surface. More than 70 percent of the earth's surface is covered by ocean. Oceanauts are working to learn more about this frontier. When you finish this section, you will be able to:

● Describe the composition of sea water.

- List and describe the features of the sea floor.
- Explain how differences in temperatures and *salinity* of sea water produce ocean *currents*.
- Describe tides.
- ▲ Demonstrate a density current.

16-1. *Why does ocean water taste salty?*

Salinity: The number of grams of dissolved salts in 1,000 g of sea water.

Young children often ask, "Why is ocean water salty?" The oceans are a large settling basin for all the materials washed into them by rivers and streams. This material contains pebbles, sand, silt, clay, and dissolved minerals from the breakdown of the rocks. Most of the large particles settle close to the edges of the ocean. The fine material is carried further out to sea. The dissolved materials make sea water taste salty. When the ocean water evaporates, the dissolved materials are left behind. Each time the water falls on the land it washes a new load of material into the sea. Over the centuries the saltiness of ocean water has increased.

Out of every 1,000 g of ocean water, 35 g is dissolved salts. This equals 3.5 percent. This figure is called the **salinity** of the water. The most common of these materials are chloride and sodium. You may recognize these two substances as the ones that make up table salt. However, there are many other dissolved materials present. See Fig. 16-2. The constant motion of the oceans keeps the salts pretty well mixed through the water. However, the *salinity* may differ from place to place because of different rates of evaporation.

Oceanographers are interested in the shape of the

16-2. *The salts in sea water.*

Unit IV How Is Our Planet Changing?

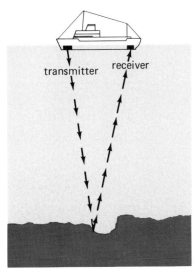

16-3. (above) Echo sounding has provided a profile of the sea floor.

16-4. (right) A sea floor profile made by an echo sounder.

ocean bottom. Exploring the bottom is very difficult because of the depth of the ocean. One method used to explore the ocean floor is *echo sounding*. The echo sounder sends out sound waves that bounce off the ocean bottom. The time it takes the echo to reach the bottom and return to the ship is measured. Dividing this measurement by two gives the time it took the sound to reach the bottom. From this information, the depth of the water can be determined. Using this data, profiles of the bottom can be drawn.

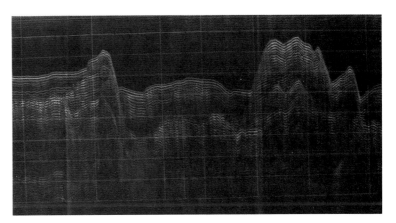

Imagine that we could take a trip across the ocean bottom from North America to Europe. As we go away from the shoreline, we travel over a gently sloping **continental shelf**. These shelves are found off most coastlines but vary in their width.

In some places off the Atlantic coast of North America, the *continental shelf* is up to 200 km wide. On the Pacific coast there is only a narrow shelf. Continental shelves are built up by the sediments washed into the oceans from the land.

At the edge of the shelf the ocean bottom slopes downward more steeply. This part of the bottom is called the **continental slope**. In many places the slopes and shelves are cut by huge *submarine canyons*. These canyons have probably been formed by underwater landslides of sediment built up on the shelves. At the base of the slopes is a deposit built up by the sediments that have come to rest here. These deposits along the bottom of the slope are called **continental rises**. The *continental slopes* represent the true edges of the continents. Along the Atlantic basin they may represent the cracks at which an ancient supercontinent broke apart. Along some continental

Continental shelf: The flattened top part of the continental slope formed by sediments from the continent.

Continental slope: The sloping part of the sea floor that marks the boundary between the floor and the continents.

Continental rise: An apron of sediments deposited along the base of the continental slope.

Chapter 16–1 The Oceans

16-5. *Continental shelf, slope, and rise can be found off the shores of most continents.*

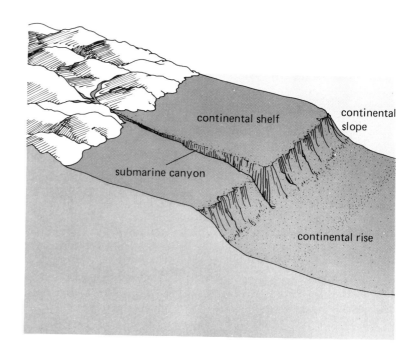

Trench: A deep valley on the ocean floor. Found along the edges of the ocean basin.

slopes, the bottom continues to drop into an ocean **trench**. This is the case off the Pacific coast of South America. These *trenches* are the deepest parts of the ocean basins.

16-6. *The North Atlantic Ocean floor.*

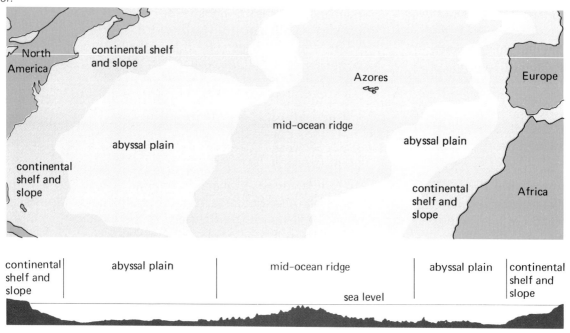

Unit IV   How Is Our Planet Changing?

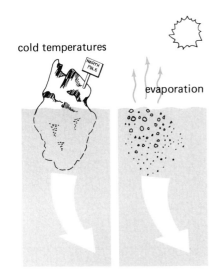

16-7. *How does the density of sea water increase?*

Abyssal plain: A wide deposit of sediment that covers the deepest part of the ocean basin.

Mid-ocean ridge: A chain of volcanic mountains that rise from the floor of the ocean.

Current: Water moving in a particular direction.

You already know that differences in temperature cause density currents (winds) in air. The same is true in water.

Turbidity current: A current of sediment-laden water.

Chapter 16–1 The Oceans

There is no trench off the Atlantic coast. Instead, the bottom levels off into broad **abyssal plains**. These plains are formed from deposits of sediment. Much of the plain is covered with a sediment called red clay. This is made from the smallest particles carried by streams and dust blown by wind or from volcanoes. It is not usually red or made of clay. In other places the sediment is called ooze. Ooze is formed by the very slow piling up of the remains of tiny living things. In all, the *abyssal plains* form over 60 percent of the ocean floor.

At this point the bottom begins to rise toward the **mid-ocean ridge**. The *mid-ocean ridge* is a continuous chain of mountains rising 2 to 4 km above the ocean floor. The chain runs down the Atlantic Ocean, around Africa into the Indian Ocean, and then into the Pacific Ocean. In some places peaks reach the surface to form islands. The Azores are islands on the mid-ocean ridge.

According to the theory of plate tectonics, the mid-ocean ridges are where new crust is formed. Hot liquid magma from below the crust pushed up near the center of the ridges. As the magma hardens, new sea floor is created. The newly made crust cools and shrinks as it moves away from both sides of the ridges. This causes the sea floor to slope away from both sides of the mid-ocean ridge.

After crossing the ridge, our trip continues as we descend to another abyssal plain. Eventually, we reach the continental slope and shelf that form the western edge of Europe.

Scientists are also interested in the movement of water in the oceans. There are several possible reasons for the movements of ocean water within the ocean basins. Some **currents** are caused by density differences. Others are caused by the wind or by the earth's rotation. Cold, dense water in the Arctic and Antarctic oceans sinks, forming *currents* that flow toward the equator along the ocean bottom. Warmer water currents move along the surface from the equator toward the poles.

As warm water evaporates, the salts stay behind. This increases the salinity of the water. As salinity increases, density increases. The saltier surface water sinks below the less salty water under it.

Water carrying a lot of sediment is more dense than the water around it. Masses of this water rush down the continental slopes onto the deep ocean floor. These **turbidity currents** may be started by storms or earth-

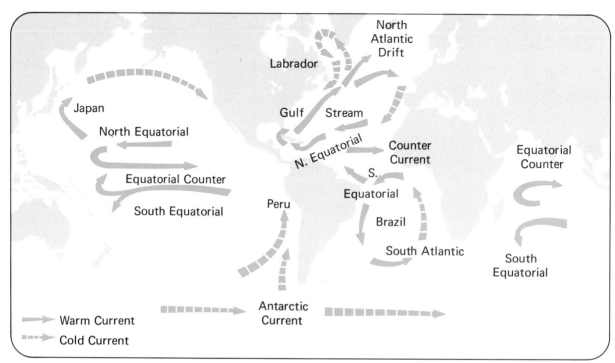

16-8. *The principle ocean currents of the world.*

quakes. They may be the main force cutting submarine canyons.

Surface currents are generally caused by the wind. These currents are influenced by the Coriolis effect. Thus in the Northern Hemisphere the currents are turned to their right. In the Southern Hemisphere they turn to the left. Land masses act as barriers to currents and cause them to turn. Figure 16-8 shows the major ocean currents of the world.

Materials
Small glass containers, 2
Food coloring
Table salt (sodium chloride)
Water
Liter container
Teaspoon
Pencil
Paper

Ocean water and tap water have different densities. In this activity, you will see what happens when water of different densities is mixed.

A. Obtain the materials listed in the margin.

B. Prepare a salt-water solution by dissolving 5 teaspoons (tsp) of table salt in 1 L of water.

C. Fill half of one of the small containers with the salt water.

D. Add 3 or 4 drops of food coloring to the salt water and mix it well to color the water.

398   Unit IV   How Is Our Planet Changing?

16-9.

E. Fill half of the second small container with tap water. Allow this container of water to sit undisturbed for several minutes. Do not move this container.

F. Carefully pour one third of the colored water down the side of the container of tap water.

G. Let the mixture set a minute. If the colored water mixes into the tap water, throw the mixture away. Repeat Steps E and F until the two solutions do not mix together.

H. Look at the salt water and tap water in the container to answer the following questions.
  1. Which is denser, tap water or salt water?

I. Without lifting it from the table, slowly tilt the container holding the colored water and tap water to one side. Hold it this way until the colored water layer settles in a new position. Then quickly, but *gently*, place the container flat on the table. Watch the colored layer for a few minutes.
  2. What happened to the salt-water layer when the container was laid flat?

The two layers would stay separated for a long time if no currents were present in either of them. When you first poured the salt water into the tap water, you probably saw the effect of currents caused by the sinking of the salt water. These currents are formed because of the greater density of the salt water, which causes it to settle at the bottom.

The downward flow of the higher density salt water in the lower density tap water is called a density current. Such density currents are common in lakes and in the ocean.

J. Repeat Step I. This time, carefully watch the border between the salt water and the tap water.
  3. Did you see a wave travel across the top of the salt-water layer?

K. Repeat Step I again. Watch carefully for the wave on top of the salt water.
  4. Would you say it travels faster or slower than waves you have seen on the surface of the water?
  5. Does the wave bounce off the sides of the container?

L. Repeat Step I as often as time permits.
  6. What effect do the density currents have on the mixing of salt water and tap water?

Although you may not be familiar with currents in large bodies of water, you are most likely familiar with waves and **tides**. Waves and *tides* can be observed on

Tide: A regular rise and fall of water level in the sea.

Chapter 16–1 The Oceans

large lakes and ocean shores. Sea level, in most places, rises and falls regularly. This motion is called a tide. A tide is a very long, regular wave. The crest of the wave brings *high tide*, the trough of the wave brings *low tide*. High and low tides are about six hours apart. Tides are the result of the gravitational attractions between the earth, the moon, and the sun, and the shape of the ocean basins.

16-10. *Tides in the Bay of Fundy.*

SUMMARY & QUESTIONS

The oceans of the world represent one of the final frontiers for exploration. The work of dedicated scientists is slowly opening this frontier.

1. Explain how you would answer the question, "Why is the ocean salty?"
2. Describe several of the major features of the ocean basins.
3. Explain how temperature and salinity cause ocean currents.
4. Name two factors that result in tides.

2 THE EARTH IN SPACE

On a wide, gently rolling plain in southern England there is an ancient monument.

Huge stone pillars, some more than 4 m high, stand in two rough circles.

Some are connected to the next pillar by a large stone laid across their tops. Within the circles are several other stones. At the very center is a large single stone called the Altar Stone. Surrounding the rings are several circles of holes and a few scattered stones. Does the arrangement have some meaning? This is the mystery of Stonehenge. When you finish this section, you will be able to:

● Describe the position of the earth in the *universe*.

● Describe the way the earth moves around the sun.

● Explain the causes of the seasons.

▲ Use a shadow stick to measure time.

The people who built Stonehenge may have been the first to observe the motions of the sky. They may have

Astronomers: A scientist who studies the stars, planets, and other heavenly bodies.

Galaxy: A huge system of stars.

Universe: All the galaxies and the space between them.

wondered where the earth is located in space and how large space is. **Astronomers** have been asking these questions for thousands of years. We now know that the earth and at least eight other planets circle one star. This star is the sun. *Astronomers* believe that our sun is one of many stars in a **galaxy**. They also believe that there are many *galaxies* in the **universe**. The galaxy called M31 may be the closest neighbor to our galaxy. The existence and motion of heavenly bodies in space pose questions for astronomers to answer now and in the future.

The sun appears to move across the sky each day from east to west. The daily movement of the sun probably

16-11. *The M31 Galaxy, often called Andromeda.*

interested the people who built Stonehenge. If you were to stand at the Altar Stone on the first day of summer and look to the northeast, you would see the sun rise directly over a stone called the Heel Stone. On the first day of summer, the sun follows its highest path across the sky. The day also has the greatest number of daylight hours

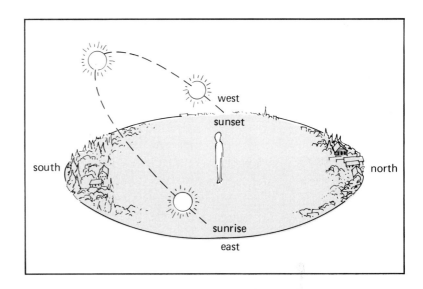

16-12. *The sun appears to move across the sky each day.*

Summer solstice: The day on which the sun traces its highest path across the sky.

Winter solstice: The day on which the sun traces its lowest path across the sky.

Vernal equinox: The first day of spring. Daylight and darkness are of equal length.

Autumnal equinox: The first day of fall. Daylight and darkness are of equal length.

Axis: The imaginary line through the center of the earth about which the earth rotates.

of any day in the year. This day is called the **summer solstice**. It occurs on June 21 or 22 each year. To the people who built Stonehenge it may have been an important day of celebration.

For the next six months, the sun travels a lower and lower path across the sky. The daylight hours are fewer and fewer. On December 21 or 22, the sun travels its lowest path across the sky. This is the **winter solstice**. After the *winter solstice* the sun's path moves higher in the sky each day until the next *summer solstice*. The days get longer between the winter and summer solstices.

Twice a year, on March 21 or 22 and again on September 22 or 23, the days and nights are of equal length. These days are called the **vernal equinox**, or first day of spring, and the **autumnal equinox**, or the first day of fall.

For centuries these daily and yearly motions of the sun were considered to be proof that the sun traveled around the earth. Then in the 1500's, Copernicus, a Polish scientist, proposed that the earth and the other planets actually moved around the sun.

Why then does the sun appear to move across the sky? The best explanation is that the earth is rotating like a wheel on its **axis**. The earth's *axis* is an imaginary line running through the center of the earth. We call the tips of the axis the North Pole and South Pole. Rotation on this axis causes each place to pass through a period in the sunlight and a period in the darkness. We call one complete turn on the axis a day.

Why then does the path of the sun across the sky

Chapter 16–2   The Earth in Space

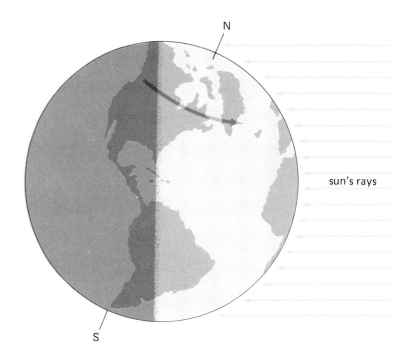

16-13. *The rotation of the earth produces day and night.*

Orbit: The curved path of one object around another object. The earth is in orbit around the sun.

change position during the year? The earth's **orbit** around the sun is a slightly flattened circle called an *ellipse*. If the earth's axis were at a right angle to the plane of its *orbit*, sunlight would fall from pole to pole on one side of the earth. Since the earth turns, every place would spend half of each day in the sunlight and half in the dark. Day and night would be of equal length

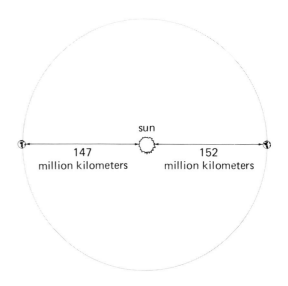

16-14. *The earth's orbit is a slightly flattened circle called an ellipse.*

404

Unit IV How Is Our Planet Changing?

all the time. Observations show that the axis is tipped at an angle of about 23½°. As a result, the North Pole is tilted toward the sun during part of the earth's orbit. It is tilted away from the sun during another part of the orbit. See Fig. 16-15. In the Northern Hemisphere, the North Pole is tilted most toward the sun at the summer solstice. The sun, at this time, follows its most northerly path across the sky. For everyone within the area of the Arctic Circle, the sun does not set at all at the summer solstice.

In the Northern Hemisphere, the North Pole is tilted most away from the sun at the winter solstice. On this date the sun follows its most southerly path across the

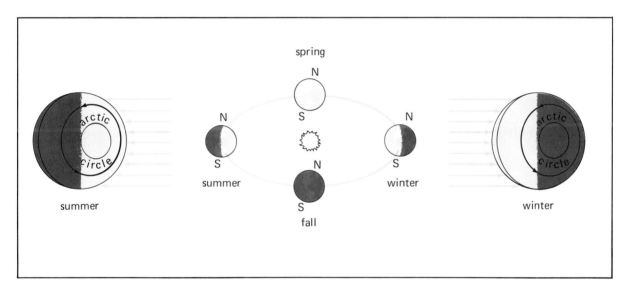

16-15. *Locate the equinoxes and the solstices in this figure.*

sky, as seen from the Northern Hemisphere. It is the shortest day and longest night of the year. Within the Arctic Circle, the sun does not rise at all.

Halfway between the summer solstice and the winter solstice the poles are tilted neither toward nor away from the sun. The days and nights are of equal length all over the earth at this time. These times are called *equinoxes*.

The seasons of the year are a result of the earth's movements in its orbit around the sun. In June, the Northern Hemisphere is tilted toward the sun. The Northern Hemisphere receives more solar energy at this time than at any other time of the year. The large amount of solar energy at this time creates the warm summer sea-

Chapter 16–2 The Earth in Space

son for the Northern Hemisphere. As the earth moves on through its orbit, summer fades into autumn. Six months later, at the winter solstice, the Southern Hemisphere is receiving the greater amount of solar energy. The Southern Hemisphere is in its warm summer season. The Northern Hemisphere is in its cool winter season. As the earth moves toward the vernal equinox, the northern half of the earth gradually begins to receive more solar energy. At this time, the northern half of the earth begins to get warmer. In June the summer solstice occurs again. The earth has completed one full revolution around the sun. It takes the earth one full year to complete its revolution. It is now in the same place in its orbit as when we began to trace the seasons. The four seasons have passed. The seasons result from the different angles at which the sun's rays reach the earth's surface.

One thing the people of Stonehenge probably did not keep track of was the time of day. The ancient Egyptians did, however. They used the shadows cast by a stick to divide up the day. In this activity you will also use a shadow stick.

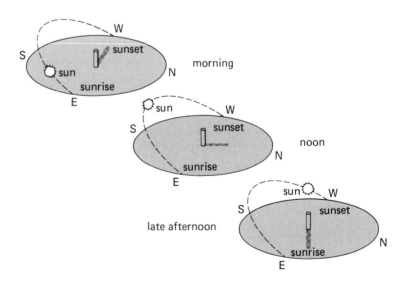

16-16. *Shadows change their positions during the day.*

Materials
Pencil
Paper

A. Obtain the materials listed in the margin.

The three diagrams in Fig. 16-16 show how the shadow of a stick changes with the position of the sun in the morning, at noon, and in the afternoon.

1. At what time of day is the shadow shortest?

2. When the sun is in the east, in what direction would the shadow point?

B. Look at the diagrams of the shadow stick and its shadow in Fig. 16-17. All the diagrams in Fig. 16-17 are for the same location and represent what you would see while facing *east*. The diagrams are not in the order the shadows would occur during the day. Look at Fig. 16-17 and answer the following questions. REMEMBER you are facing *east*.

3. Which shadow is the one cast in early morning?

4. Which shadow is the one most likely cast at noon?

5. Which shadow is the one most likely cast in the late afternoon?

6. List the letter of each diagram in the order in which you think the shadows would have occurred during a day.

7. Describe how a shadow stick could be used to tell time.

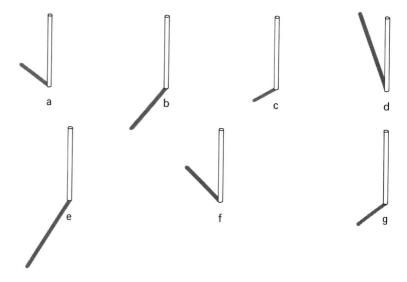

16-17. *Put these shadows in the order you would see them during one day.*

The shadow stick could also be made into a sundial and used to tell time. When the sun was highest in the sky, the shadow position would be marked twelve noon. A numbered scale on each side of this mark would show daylight hours before and after noon. However, this shadow stick sundial would not agree with clocks. What would happen if everyone set their clocks for noon when the sun was highest in the sky? People a short distance away from each other would set their clocks for noon at a slightly different time. What other problems would occur if we used shadow sticks to tell time?

SUMMARY & QUESTIONS

Early peoples kept track of the apparent motions of the sun and stars even though they did not understand them. These observations may be explained by the theory that the earth is moving in an orbit around the sun.

1. Why does the sun appear to move across the sky each day?
2. How would conditions on earth be changed if the earth were not tilted?
3. Describe the locations of the earth in the universe.
4. Explain the terms *solstice* and *equinox*.

3 THE MOON

The piece of rock in the photograph is from the moon. The study of rock samples brought back from the moon has added to the understanding of the way the planets were formed. Information about the origin and past history of the moon is helpful in understanding the earth's past. For this reason, the study of the moon is an important part of the study of the earth. When you finish this section, you will be able to:

- Describe several hypotheses on the formation of the moon.

- Compare and contrast several conditions on the moon with those on the earth.

- Identify several features of the moon's surface and explain their formation.

▲ Observe and record the phases of the moon.

If it were not already the earth's only natural satellite, the moon could qualify as a small planet. In fact, some astronomers feel that at one time it was just that. They feel that the earth's gravity captured the moon and pulled it into its present orbit about the earth.

There are several hypotheses about how the earth came to have a moon that are considered more likely. They all suggest that the earth and the moon were formed from the same cloud of dust and gas. One idea suggests that the two were formed at the same time. The others state that the moon pulled away from the earth at some point in its early history. Which of these hypotheses, if any, is correct we do not yet know. However, studies of moon rocks show that the isotopes of elements in them are very similar to those in earth rocks. This seems to indicate that the two formed from the same mass of dust and gas.

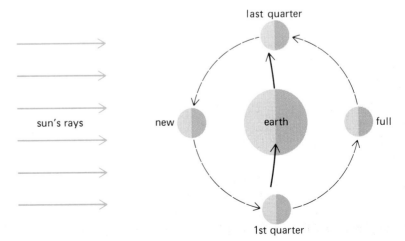

16-18. *The positions of the earth, moon, and sun determine the moon's phases.*

Moon phases: The changing appearance of the moon as more of the sunlit side becomes visible from earth.

It seems logical that the earth's closest neighbor would have been very closely studied. Ancient astronomers kept very detailed records of the changes in the moon's appearance. They found that these changes occurred in a cycle every 29½ days. We call these changes the **phases of the moon.** Later it was realized that the *phases* were due to the moon's path around the earth.

We can only see the moon because it reflects sunlight. How much of its lighted side we can see is determined by where in its orbit around the earth the moon is. At one point, the moon is between the earth and the sun. At this time sunlight is reflected away from the earth. We look at the dark side. This phase is called the *new moon*.

As the moon continues in its orbit, a little more of the lighted side is visible from earth each night. Gradually, the appearance changes. First, a curved strip becomes visible. This is called the *crescent moon*. Night by night, we see more of the lighted side. The crescent grows until half the side facing us is in light. This phase is called the *first quarter* because the moon has gone one quarter of

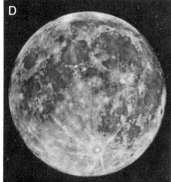

16-19. *The phases of the moon: (A) crescent (B) first quarter (C) gibbous (D) full moon (E) last gibbous (F) last quarter (G) last crescent.*

the way around the earth. The light continues to creep across the moon's surface until most of it is lit. This is called the *gibbous* (**gib**-us) moon. Finally, after about two weeks, the entire side of the moon facing the earth is in light. The moon is on the opposite side of the earth from the sun. This is called the *full moon*.

As the moon's journey continues, its face begins to slip back into darkness. Another gibbous phase (the *last gibbous*) is reached, followed by the *last quarter* and the *last crescent*. Finally the moon is back between the earth and the sun. It is back to the new phase and the cycle begins again. Have you ever noticed that except for the changes in phase, the surface of the moon always looks the same? This is because the moon turns once on its axis during each trip around the earth. Therefore, the same side always faces us. We had no idea what the back side of the moon looked like until spacecraft began to circle it in 1959.

Occasionally at the full moon stage, the moon's path takes it through the earth's shadow. The surface may be partly or totally in shadow. This is called a **lunar eclipse.** The moon is still visible because a little sunlight still manages to get through the earth's atmosphere. This gives the moon a reddish color during *eclipses*.

At least twice during the year, the new moon phase puts the moon directly between the earth and the sun. Then the moon's shadow falls on the earth. Since the moon is much smaller, its shadow sweeps only a narrow path across the earth. People directly in the shadow see a total **solar eclipse.** Those at the edges of the shadow see a partial eclipse.

The moon is the largest satellite in the solar system when compared to the size of its planet. The moon's

Lunar eclipse: The moon passing through the earth's shadow in space.

Solar eclipse: The moon passing between the sun and the earth, casting its shadow on the earth.

16-20. (left) Eclipses are caused when the moon or the earth move through the other's shadow. (center) A total solar eclipse. (right) A total lunar eclipse.

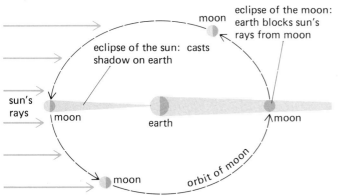

16-21. *The near side of the moon.*

diameter is about one quarter that of earth's. Because the moon's size and mass are less than the earth's, its gravity pull is one sixth that of the earth. With a moon that large only about 380,000 km away, it is reasonable to expect it to affect the earth somehow. You probably know that the pull of the moon's gravity is the main cause of the daily tides in the ocean. Scientists have also discovered "tides" in the atmosphere and a rising and falling in the earth's crust due to the moon's pull. For centuries, people have believed that the moon also had an effect on plants and animals. In some places, they were so sure of the relationship that the people worshipped the moon as a goddess.

When you look at the full moon, can you locate the features that some people say make a face? It is easy to see that there are light and dark areas. Early astronomers thought that the dark areas were seas. Therefore, they named them *maria* (**mahr**-ee-uh, singular: *mare* **mahr**-ay) which is Latin for "sea." Maria are wide, smooth areas with few *craters*. They vary in size from about 300 km to 1,100 km wide. When the back side of the moon was photographed, no maria were found. No one knows why yet. The lighter areas are called *highlands*. Some of the ridges and mountains here are 5,000 m high. The highlands seem to have many more craters than the maria. Craters vary from over 300 km wide to a few milli-

16-22. *Telescopes make mountains, craters, and other features easier to see.*

16-23. *The crater Tycho has a large ray system.*

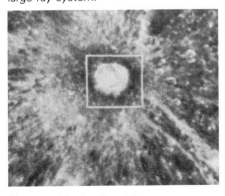

meters in size. Even rock samples from the moon showed tiny craters. Most were formed as material from space slammed into the moon. Most craters have steep inner walls as high as 2.5 km. Many younger craters have ray-like patterns. These rays were formed by material thrown out of the crater when struck by meteorites.

The moon has been studied from earth for ages. However, many questions had not been answered until astronauts finally landed there. For example, the types of rocks on the surface were guessed at, but no one could be sure of what minerals they might contain. Was the moon's interior layered like the earth's? No one knew. Was there any water, anywhere?

Between 1969 and 1972, twelve Americans walked on the moon. They found conditions to be as harsh as expected. There is no atmosphere or water, but small amounts of the gases argon, helium, and neon were found. The temperature in the sunlight reaches over 100°C. During the nights, it drops below −100°C. Because there is no air or water, the rocks do not weather or erode as they do on earth. However, temperature extremes cause the rocks to expand, contract, and break. The impact of meteorites and solar particles also break up the rocks. Therefore, the surface of the moon is covered with a layer of fine dust.

The astronauts set up many instruments and brought

Chapter 16–3 The Moon

16-24. *The astronauts left several scientific instruments operating on the moon.*

Meteorites: Chunks of rock from space that collide with a planet or a moon.

16-25. *Someday sites like this may be common on the moon.*

back samples of rocks and soil. One of the instruments that was left on the surface detects *moonquakes*. This instrument can detect the crash of a meteorite or a satellite into the moon. Data sent back to earth from this instrument has helped scientists determine what the moon's interior is probably like. There seems to be a very thick crust, a mantle and probably a core. The interior of the moon is hot, but there hasn't been any volcanic activity for billions of years. The moon is nowhere near as active as the earth.

This information, combined with data from the types of rocks returned by the astronauts, has allowed scientists to work out a possible history for the moon. It is thought that at one time in the past the moon was a hot molten mass. The heavier elements sank toward its center. The lighter material rose toward the surface and began to cool and harden, forming the crust. During this period, there were many active volcanoes.

Later, the moon, the earth, and the other inner planets were bombarded by billions of **meteorites.** This continued for millions of years and created millions of craters of all sizes. On the earth, weathering processes have destroyed most of these craters. Without air or water to cause weathering, the craters remain on the moon. Some of the larger collisions could have cracked and fractured the moon's crust. This would have allowed tremendous amounts of lava to flow out onto the surface. These flows would fill in many of the large craters with layers and layers of volcanic rock called *basalt*. These areas are the flat maria we see today. The highlands, with their many craters, are parts of the original crust. All of the samples of these rocks brought back from the moon are over 3 billion years old. This indicates that all these events must have happened before that time. Since then, the moon has been very quiet. The fairly smooth surfaces of the maria indicate that few craters have formed since that time.

Humans will return to the moon sometime. The moon is a perfect spot for a scientific laboratory. No atmosphere will interfere with telescopes. Power stations could relay solar energy to earth by microwave. Materials mined from the moon could be used to build space stations or shipped back to the earth. Most of all, we have explored only a very small portion of our nearest neighbor. This fact alone is enough to draw us back.

Materials
Notebook
Pencil

A. Obtain the materials listed in the margin.

B. Look for the moon in the sky around sunset tonight. If the moon is not visible in the evening sky, look for it in the morning.

C. In your notebook, record the date and time. Record the location of the moon (low in the west, high in the south, etc.). Sketch the moon's appearance.

D. Continue these observations for at least two weeks (one month if possible). After completing these observations, answer the following questions.

1. On what date did you see the full moon?
2. On what date was the moon in the new phase?
3. Which phases followed the full moon?
4. Which phases followed the new moon?
5. Arrange the phases in order *starting with the full moon*.
6. Which phases were visible around sunset?
7. Which phases were visible in the early morning?
8. Based on your observations and what you have learned in this lesson, explain the positions of the earth, sun, and moon during each of the phases of the moon.

Scientists have learned much about the moon in the past twenty years. It is no longer looked on as a goddess that controls our lives. Instead, the moon will continue to be a source of information about its history as well as that of the whole solar system. Future generations will explore and use the moon as a stepping stone to space.

1. Compare and contrast the following characteristics of the earth with the moon: atmosphere, diameter, gravity, length of day, mineral composition.
2. Explain two ideas on how the moon formed.
3. Explain how the moon's maria may have formed.
4. During which phases of the moon are the earth, sun, and moon in line?
5. What causes eclipses of the sun and moon?

Chapter 16-3   The Moon

4 THE SOLAR SYSTEM

For centuries, astronomers watched the evening skies and recorded the motions of the five planets they called "the wanderers." Knowing them only as moving points of light, ancient people could only wonder at their nature. They thought these planets were gods. In the 1600's, the telescope was developed. Then these planets were seen as other worlds that could only be studied from afar. Within the short span of twenty years, 1959–1979, all five of these planets were visited by spacecraft from the earth. By 1989, only lonely Pluto of all the nine known planets will not have been visited by these machines. When you finish this section, you will be able to:

- Identify the various types of objects that form the solar system.

- Explain the structure and properties of the sun.

- Compare and contrast the inner and outer planets.

- Construct a model of the solar system.

As the name implies, the sun dominates that part of space called the *solar system*. By our standards, the sun is huge. Its diameter is over 1,400,000 km, more than 100 times that of the earth. Its volume is a million times greater than the earth's. Its mass is 330,000 times greater and contains 99% of all the mass in the solar system. This huge mass creates a gravity force that extends more than 6 billion kilometers out into space. Yet since the sun is made of gases, its density is only one half of the earth's. As huge as it is, this ball of very hot gases is only an average sized star.

The energy given off by the sun travels through space in the form of waves. This type of energy is called *radiant* (**rade**-ee-unt) *energy*. The energy of the sun begins deep inside its core. Here the temperatures may be as high as 15,000,000°C. At these temperatures, the hydrogen atoms are moving at such speeds that when they collide they join to form helium atoms. This process is called nuclear fusion. In the sun, fusion joins four atoms of hydrogen to form one helium nucleus. The mass of the helium is slightly less than the mass of the four hydrogen atoms. The missing mass has been changed into energy. About 4.5 billion tons of hydrogen are changed into helium every second. Even at this rate, there is enough hydrogen in the sun to last another 5 billion years.

The nuclear energy that is released makes its way slowly toward the sun's surface. This surface layer, known as the *photosphere*, absorbs the energy and becomes very hot. Surface temperatures are between 5,000 and 6,000°C. At these temperatures, the gases of the photosphere give off the radiant energy that bathes the solar system.

Above the photosphere is the solar atmosphere. The

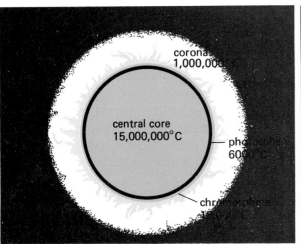

16-26. (left) The sun is made of several layers of very hot gases.

16-27. (right) Occasionally great fountains and arches of gas rise from the sun's surface.

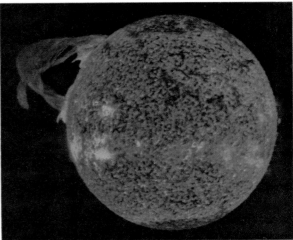

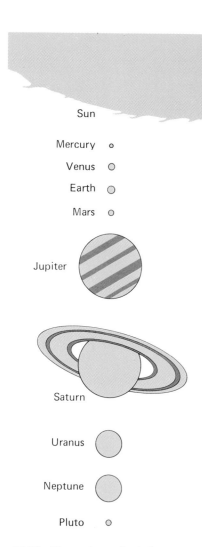

16-28. *The order and relative sizes of the planets.*

16-29. *Olympus Mons is the largest known volcano in the solar system.*

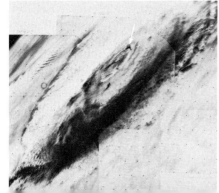

lower layer is the *chromosphere* (**kroe**-muh-sfir). Here great fountains and arches of hot gases have been seen rising from the surface. Temperatures in the chromosphere may reach 15,000°C. The upper part of the sun's atmosphere is called the *corona* (kuh-**roe**-nuh). Temperatures of millions of degrees have been recorded here. The corona sometimes extends millions of kilometers into space.

For years, students have learned the names of the nine planets in order from the sun: Mercury, Venus, Earth, Mars, Jupiter, Saturn, Uranus, Neptune, and Pluto. However, for the twenty years between 1979 and 1999, Pluto's unusual orbit swings it closer to the sun than Neptune. This happens once during Pluto's 248-year-long orbit.

The structure of the planets divides them into two main groups, the *inner planets* and the *outer planets*. See pages 419–421. The *inner planets* include Mercury, Venus, Earth, and Mars. Some astronomers feel distant Pluto is similar enough to belong to this group. However, it is so far away that we are unsure of most of our information about this planet.

The four inner planets are small, solid spheres. They have dense cores made of metallic elements. These cores are surrounded by layers of rocky material. All four show evidence of volcanic activity and earthquakes, although only Earth is still active. The largest known volcano in the solar system is Olympus Mons on Mars. It is 29 km high and 500 km wide. Mt. Everest on Earth is only 9 km high. The planet surfaces are also pitted with craters made by meteorites millions of years ago. Weathering has destroyed most of these craters on Earth. Mercury has so many craters that it looks like our moon.

Early volcanic activity formed atmospheres on each of these planets. However, Mercury's weak gravity could not hold its gases and they were swept away by solar radiation. The composition of the atmosphere on the other three planets varies greatly. Venus is hidden beneath thick clouds of sulfuric acid. Its heavy atmosphere of CO_2 is 100 times more dense than Earth's. It traps so much solar energy that the temperature is an unbearable 480°C.

The atmosphere of Mars is mostly CO_2, but there is water present. Most of it is trapped in small polar ice caps or frozen in the Martian soil. During winter, carbon

Unit IV How Is Our Planet Changing?

The Inner Planets

This planet was heavily cratered by meteorites billions of years ago. *Distance from sun:* 57.9 million km; *Year:* 88 earth days; *Day:* 59 earth days; *Diameter:* 4,800 km; *Temperature:* 430°C to −170°C; No moons.

MERCURY

VENUS

This ultraviolet photo shows some detail of clouds. Venus otherwise appears blank. *Distance from sun:* 108.2 million km; *Year:* 225 earth days; *Day:* 224.7 earth days; *Diameter:* 12,100 km; *Temperature:* 480°C; No moons.

Ours is the only planet with life that we know of. Earth is still geologically active. *Distance from sun:* 149.6 million km; *Year:* 365.3 earth days; *Day:* 23.9 hours; *Diameter:* 12,756 km; *Temperature:* 15°C; One moon.

EARTH

MARS

Mars is the red planet; iron minerals cause the colors. Evidence of water erosion exists. *Distance from sun:* 229.9 million km; *Year:* 687 earth days; *Day:* 24.6 hours; *Diameter:* 6,800 km; *Temperature:* −50°C; Two moons.

The Outer Planets

JUPITER

Largest of the planets at one-tenth the size of the sun, Jupiter has a gaseous makeup with no solid surface. The giant Red Spot seen in the photo may be a storm. *Distance from sun:* 778.3 million km; *Year:* 11.86 earth years; *Day:* 9.9 hours; *Diameter:* 142,800 km; *Temperature:* −130°C at cloud tops: Sixteen known moons (possibly 17). The small photos show two of Jupiter's moons.

SATURN

Saturn's extensive ring system includes hundreds of ringlets. The rings probably are made of ice and rock particles. Saturn's moon Titan is only moon known to have an atmosphere. *Distance from sun:* 1,427 million km; *Year:* 29.46 earth years; *Day:* 10.7 hours; *Diameter:* 120,000 km; *Temperature:* −185°C at cloud tops; Fifteen known moons (possibly 16). The small photos show two of Saturn's moons.

URANUS

Surrounded by nine narrow rings, Uranus shows a blue-green color due to atmospheric gases. Its axis tilts at 98°. *Distance from sun:* 2,870 million km; *Year:* 84 earth years; *Day:* 17-24 hours; *Diameter:* 51,800 km; *Temperature:* −215°C at cloud tops; Five moons.

NEPTUNE

Distance makes observations difficult, but Neptune is similar to Uranus and may have rings. *Distance from sun:* 4,497 million km; *Year:* 165 earth years; *Day:* 18 hours (?); *Diameter:* 49,000 km; *Temperature:* −200°C at cloud tops; Two known moons.

PLUTO

Its size and distance make measurements uncertain. Pluto is currently closer to the sun than Neptune. *Distance from sun:* 5,900 million km; *Year:* 248 earth years; *Day:* 6.4 earth days; *Diameter:* 3,000 km (?); *Temperature:* −230°C; One moon.

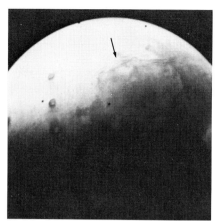

16-30. *Mars shows many huge canyons and evidence of water erosion.*

Asteroids: Chunks of rock and metal that orbit the sun between Mars and Jupiter.

16-31. *Io, one of Jupiter's moons, has several active volcanoes.*

dioxide "snow" makes the ice caps expand. The thin Martian atmosphere is sometimes clouded by dust storms which cover the entire planet. The surface of Mars shows many features that may have been carved by running water in the past. Some canyons are as long as the United States is wide. One is eight times wider and four times deeper than the Grand Canyon. Observers have also found at least one large canyon on Venus.

Between the inner and outer planets is a band of small, rocky objects called **asteroids** (as-tuh-roidz). Some scientists feel these chunks are the remains of a large planet that exploded millions of years ago. Others feel they are lumps that never formed into a planet. They range in size from less than a meter to over 100 km in diameter. Many scientists feel there will come a time when astronauts will mine useful materials from these *asteroids*. These materials could supply stations on the moon or orbiting in space, or be returned to Earth.

Except for tiny Pluto, the *outer planets* are giant balls of gas. Most of the gas is hydrogen and helium. There are also ammonia, methane, and water. All four probably have small cores of heavy elements, but their overall densities are very light. Saturn could actually float on water—if there were an ocean big enough.

Jupiter, Saturn, and Uranus all have rings. Neptune may also have one. Exactly how these rings formed is still unknown. They are probably made of tiny ice particles. Are they leftovers from the formation of these planets or the remains of satellites that have broken apart? No one knows.

Voyager 1 and *Voyager 2* flew by Jupiter in 1979 and by Saturn in 1980 and 1981. We learned more about these planets from these brief visits than was known from all the observations up to that time. Scientists are still studying the data these spacecraft sent back.

The *Voyager* spacecraft discovered Jupiter's ring and a new satellite. They also sent back pictures of the planet's atmosphere and several moons. The moons were startling. Active volcanoes were found on one. Others showed icy crusts fractured by meteorites and the pull of Jupiter's gravity. Close-up studies of Jupiter's churning atmosphere and its giant red spot may help us to understand our own atmosphere better.

Saturn's moons proved no less exciting. At least five, and possibly six, new ones were discovered. Most sur-

16-32. *Icy comets reflect sunlight. They are sometimes visible to the naked eye.*

Meteoroids: Chunks of rock and metal found throughout the solar system. Sometimes they enter the atmosphere or collide with a planet or moon.

Comets: Icy bodies usually found on the edges of the solar system.

Materials
Construction paper
Adding machine tape
Compass
Scissors
Ruler
Tape

prising were the rings. What was thought to be three main rings turned out to be thousands of smaller ringlets. Mysterious spokes appeared across the rings. Tiny moons seemed to act like sheep dogs, chasing particles back into their rings. The moons were no less fascinating than Jupiter's. Huge Titan is the only moon we know of that has an atmosphere. Some of Saturn's moons are made mostly of ice. Others are mostly rocky material. The *Voyagers* raised more questions than they answered. *Voyager 2* is on its way to Uranus in 1986 and Neptune in 1989. Then we will see what surprises await us there.

The remaining members of the sun's family include objects called **meteoroids** (**meet**-ee-uh-roidz) and **comets**. *Meteoroids* are chunks of rock or metal that are found just about everywhere in the solar system. Sometimes they enter Earth's atmosphere where friction causes them to glow. Then they are called *meteors* or shooting stars. If they do not burn up before they reach the ground, they are called *meteorites*. Meteorites are responsible for most of the craters that are seen on Mercury, the Moon, and the satellites of Jupiter and Saturn. It seems there were many more of them billions of years ago when these craters were formed.

Comets are thought to be like giant dirty snowballs that orbit the sun beyond the orbit of Pluto. Now and then the path of a comet brings it within the orbits of the inner planets. As it approaches the sun, the surface layers are changed to gas by energy from the sun. The gas is pushed into a tail behind the comet. Sometimes the comet and its tail reflect enough light to be seen from Earth even during the day. *Halley's comet* is named after the astronomer who realized that it returned every 75 years. This huge, bright comet should be clearly visible when it returns in 1986.

A. Obtain the materials listed in the margin.

B. The diameter of Earth is 12,756 km. Let this distance represent one diameter unit. Table 1 on page 424 gives the diameters of the sun and the planets in terms of Earth's diameter. Thus, if Earth's diameter is assigned a value of 2 cm, the diameter of Mercury would be 2 cm × 0.4 = 0.8 cm.

1. How large will Jupiter be?

2. How large a circle would you need for a model of the sun?

Chapter 16-4 The Solar System

TABLE 1

Object	Diameter (Earth = 1.0)
Sun	110.0
Mercury	0.4
Venus	1.0
Earth	1.0
Mars	0.5
Jupiter	11.2
Saturn	9.4
Uranus	4.0
Neptune	3.9
Pluto	0.2

TABLE 2

Planet	Av. Distance from Sun (Earth = 1.0)
Mercury	0.4 AU
Venus	0.7
Earth	1.0
Mars	1.5
Jupiter	5.2
Saturn	9.5
Uranus	19.2
Neptune	30.1
Pluto	39.5

C. Use a compass to draw circles on paper to represent each planet. Cut out the circles.

D. On your model, it would be impossible to use the same scale to show the size of the planets and their distance from one another. If you tried to do this, you would either have to make some circles so small you could hardly see them or you would have to place Earth 2,300 cm from the sun. For this reason, you will use a different scale for distance here. Let 1 AU = 10 cm.

E. Earth is about 150 million km from the sun. This distance is called an Astronomical (as-truh-**nom**-ih-kul) Unit (AU). Table 2 gives the distances of the planets from the sun in AU's.

  3. How far from the sun would Earth be on this scale?

  4. How far from the sun would Jupiter be?

  5. How long a piece of adding machine tape would you need to stretch from the sun to Pluto?

F. Assume the sun is at one end of this piece of tape. Measure off the distances from the sun to each planet using the information in Table 2. Attach your model of each planet in its proper place on the tape.

  6. Note the large gap between Mars and Jupiter. What objects belong in this gap?

G. Label your model.

Humans have learned more about our solar system in the past twenty years than in all recorded history. We have come to know the planets as other worlds on which many of Earth's processes are also at work. However, there are enough differences and puzzles to keep humans exploring and questioning for centuries to come.

1. Name five different types of objects that form the family of the sun.
2. Sketch and label a cross section of the sun.
3. List two major differences between the inner and outer planets in terms of their structures.
4. List five ways the inner planets are similar to each other.

Unit IV How Is Our Planet Changing?

VOCABULARY REVIEW

Match the description in Column II with the term it describes in Column I.

Column I
1. current
2. salinity
3. continental shelf
4. equinox
5. eclipse
6. comet
7. meteoroid
8. solstice
9. mid-ocean ridge
10. moon phase

Column II
a. An icy body usually found on the edge of the solar system.
b. Earth or moon passing through the other's shadow.
c. Number of grams of salt found in 1,000 grams of sea water.
d. Day having the most hours of daylight.
e. Chunks of rock or metal found in space.
f. A chain of ocean bottom volcanic mountains.
g. Formed by sediment deposited at the top of the continental slope.
h. Water moving in a particular direction.
i. The changing appearance of the moon as seen from the earth.
j. Period of daylight and nighttime equal length.

REVIEW QUESTIONS

Choose the letter of the response that best completes each of the following statements.

1. The boundary between the continents and the ocean floor is always marked by a (a) seamount (b) abyssal plain (c) trench (d) continental slope.
2. The salinity of a 1,000 g sample of water that contains 56 g of salt is (a) 0.56 (b) 5.6 (c) 56 (d) 1,000.
3. Two forces on the earth caused by the sun that make sea water move are (a) temperature changes and turbidity (b) earthquakes and turbidity (c) temperature changes and wind (d) wind and earthquakes.
4. The tilt of the earth on its axis causes (a) a change in the number of daylight hours during the year (b) a change in the amount of solar energy that reaches the earth's surface during the year (c) the seasons (d) all of the above.

5. The outer planets (a) all have rings (b) are large and rocky except for Pluto (c) are large gas balls except for Pluto (d) are all ice balls.
6. Which of the following facts about the moon is true? (a) It once had active volcanoes (b) Most of its craters were formed by meteorites (c) It has a crust, mantle, and core (d) All of these are true.
7. The inner planets are (a) large gas balls (b) small and rocky (c) very cold (d) very hot.
8. The moon's maria are (a) mountains (b) oceans (c) craters (d) lava plains.
9. The sun's energy is released by (a) splitting atoms (b) burning atoms (c) joining atoms (d) friction between atoms.
10. The proper sequence for the following phases of the moon is: (a) new, first quarter, full (b) new, third quarter, full (c) new, gibbous, crescent (d) full, crescent, gibbous.

REVIEW EXERCISES

Give complete but brief answers to each of the following exercises.

1. How can differences in temperature and salinity cause ocean currents?
2. List the following terms in order from the continental coastline to the deep ocean: abyssal plain, continental shelf, continental rise, mid-ocean ridge, continental slope.
3. Define the term *salinity*.
4. Describe three possible ways the moon may have formed.
5. Compare and contrast tides and waves.
6. What happens to day length in the Northern Hemisphere between the summer solstice and the winter solstice?
7. Compare and contrast the atmospheres of the inner planets.
8. How does the tilt of the earth on its axis cause the seasons?
9. Describe the structure, composition, and temperature ranges of the sun.
10. List several possible reasons for humans to return to the moon.

UNIT 5
WHAT MAKES UP OUR LIVING WORLD?

CHAPTER 17

LIFE ON EARTH

1 EARTH: A SPECIAL PLACE

Since the 1960's, over thirty spacecraft have visited other planets in the solar system. Some have landed; others made observations as they flew past. The information they gained was radioed back to scientists on Earth. These spacecraft discovered much about the physical characteristics of these planets. Conditions on these planets have been compared and contrasted with those on Earth. In doing so, we have discovered something we already knew. Earth is a very special place. When you finish this section, you will be able to:

- List and describe some of the conditions necessary for life on Earth.

- Compare and contrast some environmental factors on Earth, Venus, and Mars.

- Define the biosphere and describe its boundaries.

▲ Hypothesize descriptions of organisms that might exist on Venus and Mars.

Organism: A complete living thing.

Environmental factors: The several conditions in an environment an organism needs to survive.

Environment: Everything in its surroundings that affects the way an organism lives and acts.

Biosphere: That area at or near Earth's surface where life is possible.

Over the past ten years we have learned a lot more about our neighboring planets. We have found that conditions on their surfaces are quite different from those on Earth. So different in fact, that most forms of life on Earth could not live on these planets. What is it that makes Earth so special? Why can humans and other **organisms** exist on Earth but not on Venus or Mars?

Organisms that live on Earth need an atmosphere containing certain gases, water, a source of energy, and favorable temperatures. We refer to these conditions as **environmental factors.** Each type of organism has a certain combination of factors that is best for its survival. A particular combination of these factors is called an **environment.**

There are many possible *environments* on earth. The total of all the environments where life is found is called the **biosphere,** or zone of life. On Earth, organisms are found in environments from high in the atmosphere to the bottom of the oceans. Each environment contains

17-1. *The biosphere includes all the places at or near the surface where living things can exist.*

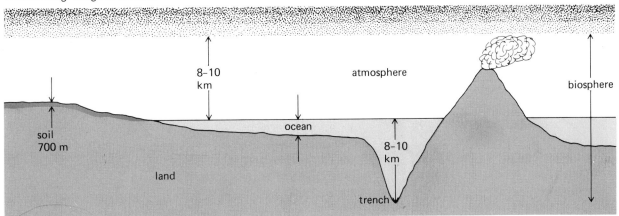

Chapter 17–1 Earth: A Special Place

a certain range of *environmental factors* needed by the organisms found there. The biosphere includes all the places at or near Earth's surface where life can exist.

Earth's atmosphere is mostly nitrogen and oxygen with smaller amounts of carbon dioxide and water vapor. Each of these gases is very important to organisms on Earth. Living matter on Earth is made of compounds containing these molecules. Carbon dioxide is used by plants to make food. The carbon is passed along to animals when they eat the plants. Oxygen is used to release energy from food. Very few organisms can survive without a supply of oxygen. Most of them get their oxygen from the atmosphere. Some organisms take it from the water in which they live.

All living things contain water. Humans are about two-thirds water. Organisms such as jellyfish contain even more water. Water is vital to carry on all of the activities of life. Nitrogen is also part of the molecules that make up living matter. Certain organisms take it from the atmosphere and pass it along to others. Carbon dioxide, oxygen, water, and nitrogen are all necessary to life as we know it.

The atmospheres of the other planets are not like Earth's. The atmosphere of Venus contains about 90% carbon dioxide. It contains small amounts of nitrogen and water vapor and very little oxygen. There are also thick clouds of sulfuric acid. The atmosphere is so dense it would be like trying to walk underwater. Most Earth life could not live here.

The atmosphere of Mars is over 95% carbon dioxide, with traces of nitrogen and oxygen. There is also water

17-2. *The planets (a) Venus (b) Earth, and (c) Mars have different physical conditions.*

vapor in the Martian atmosphere. Mars has permanent ice caps at the poles and more water may be frozen in the soil. However, no trace of liquid water has been found. The Martian atmosphere at the surface is about as dense as Earth's is at 200 km above the surface. Earth life would have a hard time living here too.

Earth has large amounts of liquid water. Some of this water is fresh, but most is salty. The oceans cover about three fourths of the earth's surface. Most of the fresh water is frozen in the polar ice caps. The rest is in ponds, lakes, rivers, and the soil. Large amounts of liquid water are necessary to sustain most forms of Earth life.

The little water Venus has is a gas. Water on Earth may be solid, liquid, or gas. The water on Mars is mostly frozen. The reason for these differences seems to be each planet's distance from the sun. All of these planets receive most of their energy from the sun. Venus is much closer to the sun than Earth so it receives more solar energy. Mars, on the other hand, is much farther away and receives less solar energy. This energy affects the surface temperature of the planets.

Earth has an average temperature of about 15°C. Hot springs can reach 200°C. High in the atmosphere, the temperature is as low as −100°C. Although different forms of life can exist at these extremes, most organisms favor a range between 0° and 50°C. Thus, on Earth water can exist as a solid, a liquid, and a gas. Water vapor and other gases hold heat energy and distribute it around the Earth. Without this effect, Earth's temperature would be about 100°C lower than it is. Earth's atmosphere also protects us from harmful solar radiation.

The large amounts of carbon dioxide in Venus' atmosphere absorb most of the sun's energy. As a result, the temperature averages higher than 750°C. No Earth organism could live in that heat. Scientists are concerned that Earth's atmosphere could also heat up. Each year increasing amounts of carbon dioxide are released into our atmosphere by the burning of fuels. A small change in the average temperature could cause major changes in our climate.

The average temperature in the atmosphere of Mars is about −200°C. However, daily highs can reach 250°C in the sunlight. Thus, the water on Mars is either vapor in the form of clouds or frozen as ice.

Earth's organisms use solar energy in other ways.

17-3. Earth life can survive in many harsh locations such as this hot spring.

17-4. *These ocean bottom organisms obtain energy from the earth's interior heat instead of the sun.*

Plants use it to make their food. The energy trapped in plant matter is passed to animals in the food they eat. Recently, bacteria have been found on the ocean bottom that do not get their energy from the sun. These bacteria use heat energy from within the earth to make food. Other organisms then feed on the bacteria. Whole communities of living things have been found at places where heat is escaping through the ocean crust. It is possible that life on other planets could also obtain energy from similar chemical reactions instead of the sun.

A new branch of biology called *Exobiology* (**ek**-soe-bie-**ol**-uh-jee) has developed during the space age. Exobiologists study other planets and the universe for the possible existence of life. To do this, they must first understand the conditions that make life possible on Earth. They then compare Earth conditions with those on other planets. Finally, they decide where life could possibly exist and try to find it. The experiments performed by the *Viking Lander* on Mars in 1976 were to search for

17-5. *Exobiologists (and science fiction authors) often describe what organisms on other planets may look like.*

Unit V What Makes Up Our Living World?

Martian life. These experiments were designed by exobiologists.

There may be places on other planets or their moons where conditions are favorable for some Earth-type organisms. Other forms of life may exist under conditions Earth life cannot. Attached to the *Voyager* spacecraft are recorded messages explaining who made them, where they came from, and what life on our planet is like. Will any other form of life ever read them?

Materials
Colored paper
Marking pens or crayons

Exobiologists often try to guess what organisms on other planets might look like. (See Fig. 17-5.) In this lesson, you learned some of the physical conditions on Venus and Mars. More complete descriptions of these planets are found in Chapter 16.

A. Obtain the materials listed in the margin.

B. Make a list of the environmental conditions on Venus. Write a description of what you think an organism suited to live in such an environment would look like.

C. Draw a picture of your organism.

D. Repeat steps B and C for an organism living on Mars.

Earth is a special place. Its particular combination of physical factors is unlike that of our neighboring planets. Organisms that live on Earth could not survive on the other planets. The zone within which life exists is called the biosphere.

1. List three ways Earth's atmosphere is important to life.
2. Contrast Earth's atmosphere with those of Venus and Mars.
3. Contrast Earth's temperature with those of Venus and Mars.
4. In what ways does life use energy from the sun?
5. List four gases found in Earth's atmosphere and explain how each is used by organisms.

Chapter 17–1 Earth: A Special Place

2 NATURE'S RECYCLING BUSINESS

In the city of Baytown, Texas, homes are being flooded by tidewater as the land slowly subsides. In California's San Joaquin (wah-**keen**) Valley, the land has subsided as much as 10 meters. Both of these events are happening because humans are pumping water out of the ground faster than it can be replaced by nature. Even the mighty Colorado River is being drained almost dry before it reaches the ocean. We need more and more water for our homes and farms. We are using our water supplies carelessly. We may soon face a water crisis like the present energy crisis. When you finish this section, you will be able to:

- Diagram and explain the carbon dioxide-oxygen cycle.

- Diagram and explain the water cycle.

- Diagram and explain the nitrogen cycle.

- Trace these cycles through several environments.

You are probably familiar with the water cycle and

the role it plays in the earth's weather. A cycle is a series of steps or events that sooner or later will lead back to where they started. Many materials in nature go through such cycles. Without this constant recycling, the available supplies of many materials would be used up. For example, without the constant release of oxygen from plants during food making, animals could not long survive. The supply of oxygen would be used up in a fairly short time. Carbon dioxide and nitrogen are also recycled in nature. Without the water cycle, there would be no water on land for living things.

Solar energy plays a major part in the water cycle. Energy from the sun causes water to change from a solid or a liquid to a gas called water vapor. This process is called *evaporation*. Water evaporates into the atmosphere from living organisms, lakes, rivers, and mostly from the oceans. When the temperature falls, the water vapor turns back into liquid drops or snow crystals. This is called *condensation*. This is the reverse of evaporation. When the drops or snow crystals are large enough, they fall to the ground. Some fall directly into the oceans, others fall on land. See Fig. 17-6.

Once on land, several things can happen. Sometimes snow becomes ice and stays frozen for thousands of years. If the temperature is warm, rain and snow may evaporate again. You have seen this happen after a storm. Some of the water may run off over the land surface in streams and rivers. Eventually, this water may return to the sea. Or the water may sink into the soil and become part of the ground water supply. This water may remain in the ground for centuries.

17-6. *The water cycle is powered by solar energy.*

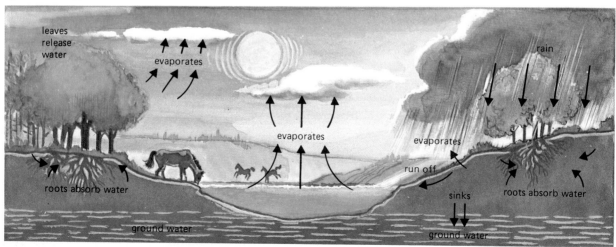

Chapter 17-2 Nature's Recycling Business

Living organisms make use of water in several ways. Plants absorb water from the ground through their roots. Animals need water to drink. For these uses, water may be pumped out of the ground. The water then becomes part of the living matter of organisms. Some of it is later lost by evaporation. More is given off in wastes. The rest is released when the organisms die. All of these pathways are part of the water cycle.

The earth's atmosphere contains much less carbon dioxide and much more oxygen than Venus or Mars. Many scientists feel that life itself is at least part of the reason for this difference. They say that billions of years ago, the atmosphere contained much more carbon dioxide and almost no oxygen. Simple forms of life in the sea, called algae, used the energy of the sun to combine carbon dioxide and water into food and living matter. Oxygen was released as a waste during this process. At the same time, more carbon dioxide became locked up in the rocks of the earth's crust. So, over long periods of time, the amount of carbon dioxide in our atmosphere decreased while the activities of living things caused the oxygen to increase. This did not happen on Venus or Mars because these planets do not have liquid water or life.

The ability to combine carbon dioxide and water into food is common to all green plants, both in the oceans and on land. The plants in turn become food for most animals. In order to use their food, both plants and animals need oxygen from the atmosphere. The oxygen is combined with the food to release the energy needed to carry on the activities of life. During this process, car-

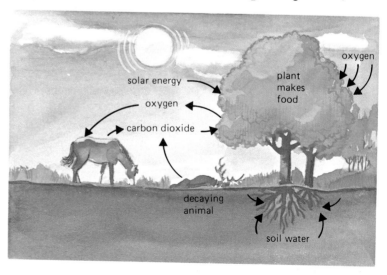

17-7. Carbon dioxide and oxygen are recycled by plants and animals.

bon dioxide is released as a waste gas. Much of the carbon dioxide taken in by plants and animals is used in forming the chemicals that make up living matter. It is released later in wastes or when an organism dies and decays. Thus, both carbon dioxide and oxygen are continually recycled by living organisms. The carbon found in your body cells may have been part of billions of other organisms before becoming part of you. This process is referred to as the carbon dioxide-oxygen cycle (CO_2-O_2 cycle). See Fig. 17-7.

Nitrogen makes up about 78% of the earth's atmosphere. Animals and plants need nitrogen but cannot use it directly from the atmosphere. Bacteria called **nitrogen-fixing bacteria** combine nitrogen with other elements to make compounds called nitrates. These bacteria are found in the soil and on the roots of some plants. The nitrates formed by these bacteria are used by plants. They are taken into the plants through the roots along with water from the soil. The nitrogen then becomes part of the plants' molecules. Animals get nitrogen from the plants they eat. This nitrogen then becomes part of the animals' molecules. See Fig. 17-8.

Nitrogen is returned to the soil in animal wastes and when plants and animal matter decay. Another type of bacteria, called **decay bacteria,** breaks the material down, releasing the nitrogen compounds so they can be used again. Some of the nitrogen does escape into the atmosphere. This will have to be recaptured by *nitrogen-fixing bacteria* before it can be used again.

Farmers have found that some crops can use up the

Nitrogen-fixing bacteria: Tiny organisms that combine nitrogen and oxygen from the air to make nitrogen compounds.

Decay bacteria: Tiny organisms that obtain their food by breaking down dead plants and animals.

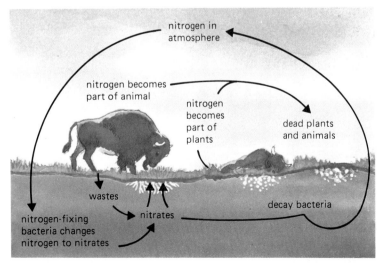

17-8. *All living organisms rely on nitrogen-fixing bacteria to make atmospheric nitrogen usable.*

Chapter 17-2 Nature's Recycling Business

17-9. *Human activities often interfere with natural cycles.*

nitrates faster than they can be replaced. To solve this problem, fertilizers containing nitrates can be added to the soil. Planting crops such as clover or alfalfa for several years can also replace the nitrogen. The fertilizer you might use on a lawn or garden serves the same purpose.

Nature has been in this recycling business for millions of years. Human activities often interfere with these cycles. Our intensive use of fertilizers to produce more food has caused many problems with land and water. Fertilizers are washed into lakes and ponds, causing water plants to grow rapidly. This can choke a lake and cause the animal life to die. Our fossil fuels pour tons of carbon dioxide into the air as they burn. This may cause our atmosphere to warm up. Air pollution and the cutting of forests reduces the amount of CO_2 being recycled for O_2. This may affect the amount available for animals.

The water cycle is a major concern at the present. You are already aware of the problems of water pollution. Some pollutants from burning fuels dissolve in water vapor in the air making the rain acid. This kills life in lakes and rivers. The heavy use of ground water for irrigation and drinking is depleting this resource rapidly. Along the coasts, salt water sinks into the ground to replace it. When too much water is pumped out, the land sinks or collapses forming large sink holes. If rainfall is less than normal, many areas do not have supplies to cover the shortages that result. Water rationing has been used in some places. In the Midwest, the shortage of water for irrigation has forced some farmers to change the type of crop they grow. Others have gone out of business. Water is vital to farming and industry. When it runs short, major problems will be facing us.

Materials
Colored pencils or pens
Pictures of various environments from magazines

A. Obtain the materials listed in the margin.

B. Use the colored pencils to outline and label the CO_2-O_2 cycle on a picture of a natural environment.

C. Answer the following:

1. Would the CO_2-O_2 cycle operate without sunlight?

2. What three things must green plants obtain before they can make their own food?

3. Besides food, what else is produced during the foodmaking process?

4. Would this cycle operate in an aquarium? in a terrarium?

D. Use the colored pencils to outline and label the water cycle on a picture of an environment.

E. Answer the following:
  5. How is the sun involved in the water cycle?
  6. What happens to water that is lost from a plant's leaves?
  7. List three things that may happen to winter's snow as part of this cycle.
  8. Explain how you are a part of this cycle.

F. Use the colored pencils to outline and label the nitrogen cycle on a picture of an environment.

G. Answer the following:
  9. What is the difference between nitrogen-fixing and decay bacteria?
  10. What two things may happen to nitrogen released from decaying organisms?
  11. How might an animal such as a lion, which eats no plants, get the nitrogen it needs?
  12. Many people with home gardens bury the leaves that fall from trees in the fall. Can you suggest why this might be a good idea?

SUMMARY & QUESTIONS

The amounts of many necessary substances on earth are limited. Fortunately, nature has provided a "recycling business" to renew some of these resources. Human activities, however, threaten even nature's ability to renew these resources.

1. Explain the role of the sun in the water cycle.
2. Explain why both plants and animals depend on certain bacteria for the nitrogen they need.
3. Diagram how the CO_2-O_2 cycle might operate in an aquarium.
4. Which of these cycles would not operate without living organisms? Explain why.
5. Plants play a double role in the CO_2-O_2 cycle. Explain how this happens.

Chapter 17-2 Nature's Recycling Business

3 THE ACTIVITIES OF LIFE

Is there life elsewhere in the universe? This question has been asked over and over again. We simply do not know yet. However, some meteorites that strike the earth contain complex carbon compounds. These compounds are identical to some of those in living things on the earth. Recently, these compounds were also found in gas clouds in space. Were these chemicals made by living matter or were they formed by simple chemical processes? Again we do not know. What we do know is that the ingredients for life are found throughout space. When you finish this section, you will be able to:

● Describe the life processes common to all living things.

● Explain the two parts of the cell theory.

▲ List the characteristics that living things and nonliving things have in common.

17-10. *This meteorite is made of stone.*

How do we determine what is alive and not alive on earth? Not everyone agrees on what *life* really means. Most biologists agree that living things can carry on certain activities that nonliving things cannot. What are some of the activities you carry on that a nonliving thing cannot?

Unit V   What Makes Up Our Living World?

ACTIVITY

Materials
pencil
paper

17-11. *Do the students and the car carry on some of the same activities?*

PART I

You are alive. A car is not.

A. Make a list of some of the things you must do in order to stay alive.

1. Does a car appear to do any of these things?

2. Did you list the following activities in question 1? On your paper answer yes or no to each item here.
a. take in food and oxygen
b. release energy from food
c. get rid of wastes
d. grow
e. move
f. respond to changes in the environment
g. give off useful chemical substances
h. reproduce

3. Does a car appear to do any of the things listed in question 3? List the things a car can do.

Sometimes a nonliving object appears to carry on one or more of these activities. For example, a car moves. A car needs gasoline to run and gives off waste through the exhaust. A car can also release energy from the gas.

17-12. *Which of the life processes does this car show? Why is it not alive?*

Chapter 17–3 The Activities of Life

Life Processes: Activities or processes carried on by all living organisms.

Many nonliving things seem to carry on some of the **life processes.** Only living things can carry on all these *life processes.* The life processes are explained in the list below.

a. Food getting: Taking in the materials that provide energy and that are needed for growth.
b. Respiration: Using oxygen to release energy from food. Carbon dioxide is released as waste.
c. Excretion: The removing of wastes.
d. Growth: Increasing the size of the organisms.
e. Movement: Moving of the whole organisms or material inside the organism.
f. Response: Reacting to changes in the environment.
g. Secretion: Making and giving off useful chemical compounds.
h. Reproduction: Producing more of its own kind.

The life processes described above are common to all living organisms. Some of these processes are quite obvious. Others are difficult to observe. Biologists agree that living organisms must carry on all these processes.

17-13. *What life processes are shown in this photograph?*

Besides carrying on all the life processes, living things also have the same basic structure. In the late 1600's, people first observed things even smaller than the naked eye could see. An Englishman named Robert Hooke was examining materials with a microscope. One of these materials was a piece of cork. He cut a very thin piece of cork and examined it with a microscope. He observed tiny, orderly, empty spaces that reminded him of a honeycomb. Because they reminded him of the cells in a honeycomb, he called them **cells.**

Cell: The smallest organized unit of living protoplasm.

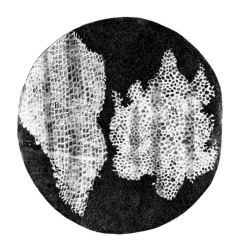

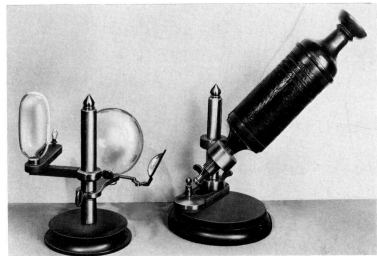

17-14. *Robert Hooke's microscope and drawings of cork cells.*

Ten years later, a Dutchman named Anton von Leeuwenhoek (**lau**-ven-hook) focused his microscope on a drop of water. He saw a new world of life that no one ever knew existed. He called the tiny creatures he saw "wee beasties." As the years went by, more and more microscopic observations of living organisms were made. Investigators found that all living things were made up of the tiny units called *cells*. They also discovered that cells were full of a thick liquid substance they called **protoplasm** (**pro**-toe-plas-um). Another important discovery was that cells could only be reproduced by other living cells. This led to the development of the *Cell Theory*. This theory may be stated in two parts:

A. The cell is the basic unit of all living organisms.
B. Only living cells can produce new living cells.

Protoplasm: All the living material found in a cell capable of carrying on all the life processes.

Materials
microscope
cover slips
microscope slides
medicine dropper
pond water samples
pencil
paper

PART II

A. Obtain the materials listed in the margin.

B. Place a drop of the water from the sample on the center of a clean microscope slide. Include some of the material you see in the water. Carefully place a cover slip over the drop of water. Examine the materials under the microscope, using low power.

1. Describe what you see through the microscope.
2. Which of the life processes is easiest to observe in the material on the slide?
3. What other life processes do you observe?
4. Why can't all the life processes be observed?

C. Leeuwenhoek saw similar organisms when he observed a drop of pond water. In order to

Chapter 17-3 The Activities of Life

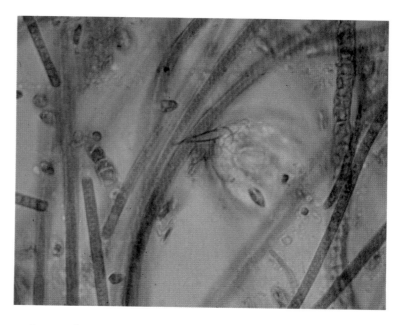

17-15. *How many of these organisms can you find in your water sample?*

make careful observations, he described and made sketches of each organism.

In your notebook, carefully describe as many organisms as you can find. Next to each description, sketch the organism you observed.

5. All of these organisms are made up of the same structure, called a ——.

Living things on earth carry on all the basic life processes. As humans explore other planets, they will be looking for signs of life. If living things are found, will they carry on the same life processes? Will they be made up of the same small units we call cells?

1. All living things carry on the life processes. List and describe each of these life processes.
2. What are some of the life processes that are common to both living and nonliving things?
3. The Cell Theory states (a) all living things are made up of units called elements (b) only living cells can produce new living cells (c) nonliving material can produce living material (d) all living cells are made up of water.

VOCABULARY REVIEW

Match the descripton in Column II with the term it describes in Column I.

Column I
1. cell
2. biosphere
3. organism
4. energy
5. life processes
6. protoplasm
7. atmosphere

Column II
a. All the areas on earth where life can exist.
b. The basic unit of life
c. The living material found in a cell.
d. The activities of life.
e. Ability to cause motion or change.
f. A complete living thing.
g. A layer of gases surrounding a planet.

REVIEW QUESTIONS

Choose the letter of the response which best completes the statement or answers the question.

1. Protoplasm is (a) an inorganic mixture of elements and compounds (b) a thick solution of elements and compounds capable of carrying on the life processes (c) the material that makes up hair (d) a solution of mixtures.
2. Which of the following statements is a part of the Cell Theory? (a) All living things are made up of units called compounds. (b) Living material may come from nonliving material. (c) All living things are made up of units called cells. (d) Only living things contain carbon dioxide.
3. When Robert Hooke examined a slice of cork under his microscope, he was really looking at (a) cells (b) cell membranes (c) cell walls (d) cytoplasm.
4. In order to be considered living, something must (a) carry on all the life processes (b) carry on most of the life processes (c) reproduce (d) be able to move.
5. Which of the following would probably not be an environment for living organisms? (a) deep cave (b) deep ocean trench (c) hot spring (d) an active lava flow.
6. Which of the following statements about sunlight is correct? (a) Sunlight is needed to provide a favorable temperature range. (b) Green plants

need sunlight in order to make food. (c) Some animals need sunlight in order to see. (d) All of these.
7. All of the following are part of the nitrogen-cycle except (a) nitrogen-fixing bacteria (b) decay bacteria (c) green plants (d) carbon dioxide.
8. Which of the following cannot be recycled and used again? (a) nitrogen (b) energy (c) oxygen (d) carbon dioxide.
9. The process of evaporation is a part of the (a) water cycle (b) nitrogen cycle (c) CO_2-O_2 cycle. (d) All three cycles.
10. When an organism uses oxygen to release energy from its food, it is carrying on the life process of (a) growth (b) movement (c) food getting (d) respiration.

REVIEW EXERCISES

Give complete but brief answers to the following.

1. A geranium plant and a mouse are placed in separate, sealed containers with a supply of water. What do you think will eventually happen to each? Explain your answers. What will happen if they are both placed in the same sealed container?
2. The sun plays a major role in the water and CO_2-O_2 cycles and a minor role in the nitrogen cycle. Explain briefly how the sun is involved in each cycle.
3. Each of the following plays an important part in one of the cycles we studied in this chapter. Identify which cycle the process is most closely related to and briefly explain the process.
   a. Taking in and giving off gases b. Evaporation c. Water loss from leaves d. Decay e. Fixing bacteria
4. List four ways in which solar energy affects living things.
5. Examine the definitions of life processes on page 442 carefully. For each process name one nonliving object that *seems* to carry on that process. For example, a car takes in material that provides energy.
6. What are the two parts of the cell theory?
7. What is protoplasm?

Unit V   What Makes Up Our Living World?

CHAPTER 18 ALL LIVING THINGS ARE SIMILAR

1 ANIMAL AND PLANT CELLS

A factory is a place of great activity. Fuel and raw materials are delivered. Fuel is burned to provide energy. Workers put the raw materials together into finished products. Wastes are produced and removed. The finished product is packed and stored until it is shipped. How is a living cell like a factory? When you finish this section, you will be able to:

● Compare the cell with a factory.

● Describe the structure of a typical cell.

● Compare plant and animal cells.

▲ Prepare and observe microscope slides of cells.

Nucleus: The control center of the cell.

Cytoplasm: The protoplasm surrounding the nucleus.

Cell Membrane: The outer boundary of the cell.

Nucleolus: Small body found in the nucleus.

Nucleoplasm: A type of protoplasm found inside the nucleus of a cell.

The structure and functions of our cells could be compared to a factory. The manufacturing processes may be compared to the life processes carried on in a cell. The finished products are the compounds that form the many parts of the cell.

Cells are not all the same. They may differ in size and shape. Many cells have special structures that have special functions. However, all cells are similar in some respects. Most cells have a **nucleus** (new-clee-us) and **cytoplasm** (sy-toe-plas-um). They also have an outer boundary called the **cell membrane**.

The main office and planning department of our factory cell is the *nucleus*. The nucleus is the control center of the cell. It controls everything that goes on inside the cell.

The microscopes you will use are not powerful enough to show much of the inside of the nucleus. You may see a dark spot called the **nucleolus** (new-clee-oh-luss) inside the nucleus. The *nucleolus* seems to have a job related to making proteins. Scientists use very powerful microscopes to observe the structure of the nucleus. The inside of the nucleus is filled with a type of protoplasm called **nucleoplasm**. Throughout the *nucleoplasm* are

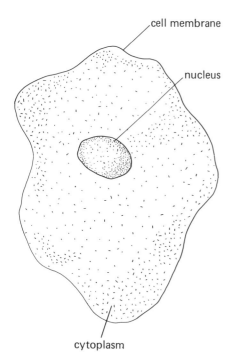

18-1. *(above) The three major parts of an animal cell.*

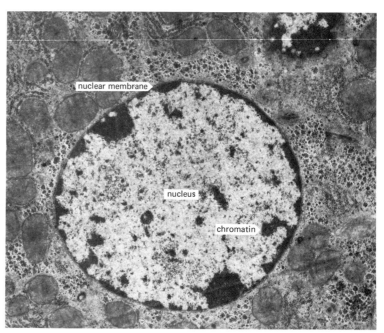

18-2. *(right) Electron microscope photograph of the nucleus of an animal cell.*

Unit V   What Makes Up Our Living World?

Chromatin: Material located in the nucleus which carries the genes.

Genes: Molecules which contain instructions needed to control the activities of the cell.

long, thin fibers called **chromatin** (crow-matt-tin). On the *chromatin* are found small structures called **genes.** The *genes* are the instructions and directions for the manufacture of the finished products of the cell. The chromatin and genes can be compared to the file cabinets and blueprints in the factory's main office. In a factory, the original blueprints never leave the "office." Copies are made and these copies go out of the office into the working area. In a cell, copies of the genes are made and sent out of the nucleus.

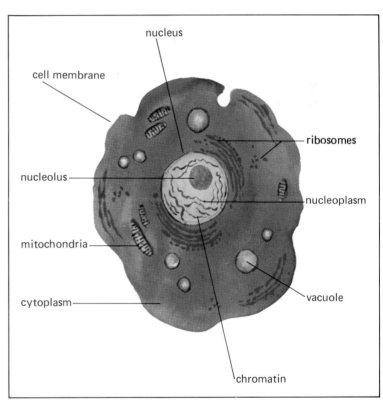

18-3. The structure of a typical animal cell.

The material around the nucleus is called the *cytoplasm*. This is where the cell's manufacturing processes are carried on. The structures that form the "shop area" of the cell are found in the cytoplasm. They include the storage rooms called **vacuoles** (vac-you-oles). Some of these *vacuoles* store food for future use. Some store waste matter until it is removed from the cell. Other vacuoles store fluids. Vacuoles are sometimes large enough to see. They resemble tiny air bubbles.

Vacuoles: Storage areas located in the cytoplasm.

Mitochondria: Structures in the cytoplasm that release energy from food.

The power generators, called the **mitochondria** (my-toe-**con**-dree-ah), are also located in the cytoplasm. The

Chapter 18–1 Animal and Plant Cells

mitochondria release the energy needed to run the cell. The fuels of a cell are the sugars and starches. These are broken down by oxygen in a process called respiration.

Some of this energy is used by structures called **ribosomes** (ry-bow-somes). They put amino acids together to make proteins. The *ribosomes* can be compared to the machines in our factory model. The amino acids are the raw materials of the cell.

Ribosomes: Structures in the cytoplasm that make proteins from amino acids.

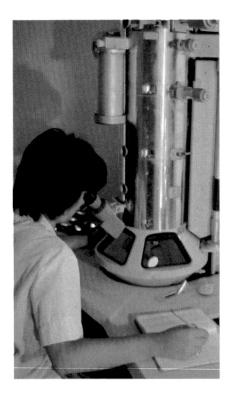

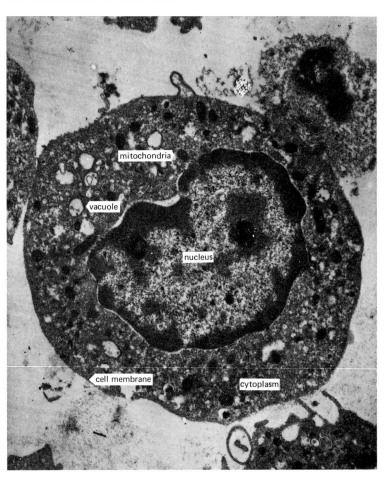

18-4. *An electron microscope helps biologists study the microscopic structures in a cell. (left)*

18-5. *Electron microscope photograph of a typical animal cell (right).*

None of these structures is large enough to be seen with your microscope. Scientists use a very powerful electron microscope to photograph these tiny structures. Electron microscopes can magnify objects over 200,000 times their normal size.

Surrounding the cytoplasm and holding the cell together is the *cell membrane*. The cell membrane acts like the factory wall. It controls what enters and leaves the cell. If the cell membrane is damaged, the cell may die.

ACTIVITY

PART I

Materials
microscope
cover slip
microscope slide
toothpicks
medicine dropper
iodine stain
pencil
paper

A. Obtain the materials listed in the margin.

B. Place a drop of iodine stain on a clean microscope slide. It is used here to make normally colorless protoplasm visible.

C. Gently scrape the inside lining of your cheek with the flat end of a clean toothpick. Stir the material from your cheek into the drop of stain in order to separate the cells.

D. Cover the material with a cover slip. Using low power, examine it under the microscope. See Fig. 18-6.

E. Locate some of the separated cell. Sketch the cell and any parts you can see. Compare this cell with the others on the slide.
  1. Do they all have similar shapes? Are they all about the same size?

F. Focus on one cell with the highest power of your microscope. The dark brown spot in your cheek cell is the nucleus.
  2. Label the nucleus on your sketch. Describe its shape and location.
  3. Does every cell on your slide have a nucleus?
  4. Do they all appear to be located in the same position in the cell?

G. Now carefully examine the cytoplasm around the nucleus of the cheek cell.
  5. Label the cytoplasm on your sketch of the cheek cell.

H. Observe the cell membrane surrounding the cytoplasm. Label the cell membrane on your sketch.

18-6.

Chapter 18–1 Animal and Plant Cells

6. Examine the other cells. Do each of the cells have the same basic structure? Cheek cells are typical of all animal cells and have a nucleus, cytoplasm, and cell membranes.

PART II: PLANT CELLS

Materials
microscope
cover slip
forceps
onion section
microscope slide
paper towel
medicine dropper
iodine stain
pencil
paper

A. Obtain the materials listed in the margin.

B. Slice a raw onion and cut one of the rings into one centimeter sections. Peel the thin layer of cells from the inner curve of the onion section. Place this thin layer in a drop of water on a microscope slide. Make sure this layer is flat and not folded. Now add a cover slip and examine the slide under low power.

1. Are the onion cells easy to see? If not, explain why.
2. How could you make these cells more visible?

C. Place a drop of iodine stain on one edge of the cover slip. Place the edge of a paper towel at the other edge. Observe what happens through your microscope. See Fig. 18-7.

3. Explain what is happening.

D. Sketch a group of several stained cells.

4. How are the onion cells and animal cells alike? Label any parts you recognize.
5. Does each cell have a nucleus? Is it located in the same place as the nucleus in the animal cell?
6. Do onion cells have cytoplasm? Does the cytoplasm fill all of the space outside the nucleus?
7. What do you notice about the vacuole in the onion cell? How does this compare with the size of the vacuoles in an animal cell?

E. Carefully look at the shape of the onion cell. Examine the outer boundary of one cell very closely.

8. How is the shape of the onion cell different from the shape of the cheek cells?
9. Do you notice any differences between the outer edge of this cell and the outer edge of an animal cell?
10. What function do these rigid walls have?

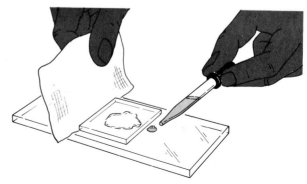

18-7.

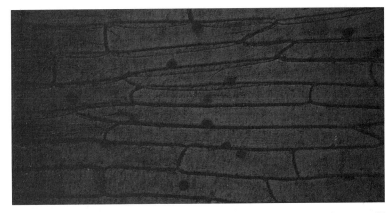

18-8. *Which parts of these onion cells can you identify?*

The most obvious difference between the plant cell and the animal cell is the size of the vacuoles. Vacuoles in the plant cell are very large. The vacuole still acts as a storage area. It is filled with a clear fluid. This fluid is mostly water but also contains sugar, starch and protein molecules.

Plant cells also have a thick, firm outer boundary called the **cell wall**. This outer boundary supports and protects the cell. Animal cells do not have a *cell wall*. These cell walls formed the little "cells" that Robert Hooke saw in the piece of cork. Plant cells also have a cell membrane. The cell membrane is pressed up against the cell wall.

Cell Wall: The rigid protective layer which surrounds the plant cell.

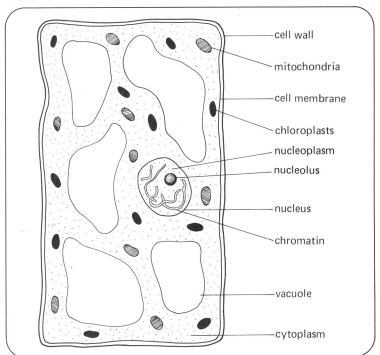

18-9. *Structure of a typical plant cell.*

Chapter 18—1   Animal and Plant Cells

453

These thick and rigid walls will remain long after the plant cell dies. In trees, many of these cell walls are crushed together into layers. Eventually, these layers form what we call wood.

Making food is one of the most important plant life processes. The ability to make food is the major difference between plant and animal cells. In the cytoplasm of the plant cell are many small, green structures called **chloroplasts** (**chlor**-oh-plasts). These *chloroplasts* contain molecules of **chlorophyll** (**chlor**-oh-phil). The *chlorophyll* allows a plant cell to make its own food. Animal cells do not have chloroplasts. Therefore, they cannot make their own food. However, not all the cells of green plants have chlorophyll.

Chloroplasts: Oval shaped structures containing chlorophyll.

Chlorophyll: Green material in green plants that uses sunlight to make food.

Materials:
Microscope
Microscope slide
Cover slip
Medicine dropper
Scissors
Elodea leaf
Pencil
Paper

PART III

Now we will examine the leaf of the elodea plant.

A. Remove a leaf from the tip of the elodea plant. Cut the leaf in half. Place the tip of the leaf in a drop of water on a slide. Carefully add a cover slip so no air bubbles form. Examine under low power.

1. Do you notice any difference between the elodea and the onion cells? What is the difference?

B. Focus on a group of cells near the center of the tip of the leaf. Change to high power. The elodea leaf is only a few layers thick. It is possible to focus on the different layers by adjusting your focus knob very

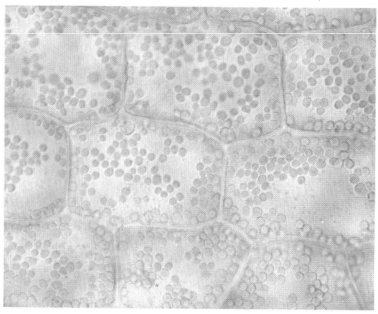

18-10. *How are these elodea cells different from the onion cells?*

carefully. Adjust the focus so you are viewing one layer of cells. What you should notice are the small, green football-shaped chloroplasts.

2. Where are the chloroplasts located?

3. Why didn't the onion cells we examined have chloroplasts?

C. Focus on a cell in which the chloroplasts are arranged around the edges. Allow the microscope light to warm the cells. Now you may be able to see a movement of the chloroplasts in the cytoplasm.

4. Describe the motion you observe. Why is it around the edges of the cell? This motion is an example of how a cell carries on the process of movement. This movement of cytoplasm occurs in all living cells.

D. Sketch a group of elodea cells. Label all the different structures you can identify. Don't forget to include the cell membrane even though it is not visible beside the cell wall.

Plant cells contain all the other structures that are found in the animal cell. There are mitochondria, ribosomes, chromatin, and genes. All these structures have the same functions in the plant cell.

SUMMARY & QUESTIONS

Below you will find unlabeled drawings of typical plant and animal cells. To show your understanding of the structure of these cells, identify the numbered parts. Then state briefly the function of each part. Finally, list two major parts found in plant cells but not in animal cells.

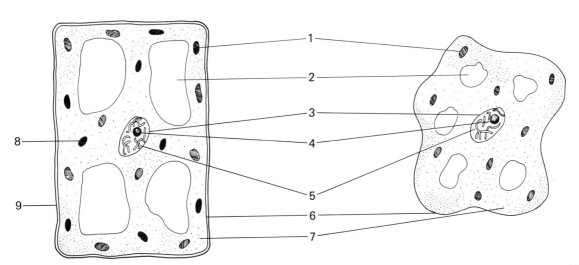

Chapter 18–1 Animal and Plant Cells

2 CLASSIFICATION: KEY TO UNDERSTANDING

Do you collect stamps as a hobby? Millions of people do. Stamps from all over the world tell stories of history, great people, sports, and different types of plants and animals. How would you organize your stamp collection? Would you do it by the country they are from? By the subject shown on the stamp? By the price of the stamp? There are many ways.

There are many different types of plants and animals. How do scientists group living things? When you finish this section, you will be able to:

● Explain how living things are classified.

● Describe the major characteristic of the four types of *protists*.

▲ Use the microscope to observe protozoans.

Classify: To arrange organisms into groups according to similar characteristics.

Species: A group of related organisms that have the same characteristics.

Genus: A group of related species differing only in a few characteristics.

Over two hundred years ago, a Swedish biologist named Carolus Linnaeus (lyn-**nay**-us) developed a way of **classifying** the different plants and animals. His system was based on likenesses in structure. Linnaeus examined thousands of organisms. He found that many organisms had the same characteristics. All the organisms with the same characteristics he put into the same groups. This group he called a **species**. Linnaeus placed similar *species* into a larger group called a **genus**. The scientific name of the organism is made up of the *genus* and species name. The genus name is capitalized.

Today there are many more characteristics used to *classify* an organism. Scientists study the structure and chemical makeup of the cells. They also examine the number and type of genes on these cells. From this information, they can decide which organisms are related and which are not.

The classification system used today has seven groupings or levels. This system begins with the broadest group, called the Kingdom. A kingdom includes the largest number of different organisms. As you move down through the levels, the number of organisms in each gets smaller. The smaller the group, the more alike the organisms in that group are. Eventually, you will end up with a single species.

KINGDOM–includes several related Phyla (plural of phylum)
PHYLUM–includes several related Classes
CLASS–includes several related Orders
ORDER–includes several related Families
FAMILY–includes several related Genera (plural of genus)
GENUS–includes one or more related species
SPECIES–includes all organisms with the same characteristics.

For a long time, all living organisms were placed into either the Animal or Plant Kingdoms. The classification of most plants and animals did not pose a problem. It was easy to see that a redwood was a plant and a whale an animal. However, many microscopic organisms presented a problem. For years, biologists disagreed on the kingdom these organisms belong in. Finally, most biologists agreed to put these organisms in a third kingdom, **Protista** (**pro**-tiss-ta).

Protists: Kingdom of organisms that are neither plants nor animals.

The *Protista* Kingdom is made up of many phyla. In order to study the protists, we will place these phyla into four groups. These four groups are the algae, fungi, protozoans, and bacteria.

Did you ever observe a small pond and notice that the water wasn't crystal clear? The water probably had a greenish color. This color was caused by the green **algae** living in the water. These *algae* often form a messy film on the surface of the pond. You may have wondered how anything could live in such a mess. Actually, algae are vital to the organisms living in the pond. All algae contain chlorophyll and make their own food. During this process, oxygen is released into the water. The oxygen dissolves in the water and is used by fish.

Algae: Protists that contain chlorophyll and make their own food.

18-11. *The algae of this pond serve as food for the fish and produce oxygen.*

Fungi: Plantlike protists that cannot make their own food.

Protozoans: Single-celled, animal-like protists.

In addition to making their own food, algae are food makers for the organisms around them. Many small animals eat the algae. Fish eat the smaller animals. Larger fish eat the smaller fish. Thus the algae directly or indirectly feed the organisms around them.

Another group of protists is the **fungi**. *Fungi* include yeast, mushrooms, molds, and mildews. Fungi do not have chlorophyll and cannot make their own food. They must obtain their food from other organisms. Fungi can live in or on other living or dead organisms.

Some fungi can be very harmful. They cause the destruction of crops and spoil food. A few types are harmful to humans. Athlete's foot and ringworm are caused by a fungus growing under the skin. There are other fungi that attack the lungs.

On the other hand, fungi can be very useful. They break down decaying materials, so that the materials can be used again. In industry, yeast, a fungus, is used in baking and brewing. Other types of fungi are used in making cheese. The mold, penicillium, produces a substance useful in medicine. Many nonpoisonous mushrooms are used for food.

Protozoans (pro-toe-**zoe**-ens) are another group of protists. For many years, biologists thought that these single-celled organisms were simple animals. A few of these organisms have chlorophyll and can make their own food. Thus, some biologists called them plants. Today they are classified as protists.

18-12. *(left) Bracket fungi.*

18-13. *(right) Green mold.*

Unit V What Makes Up Our Living World?

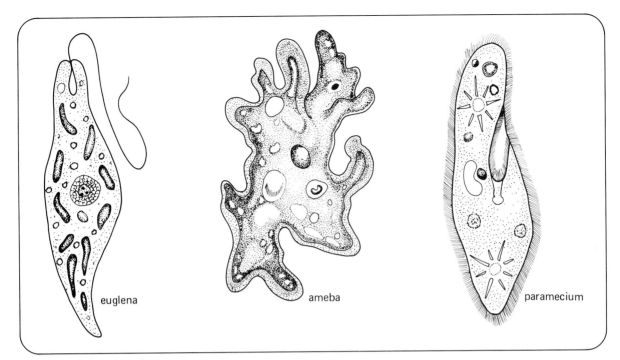

18-14. *Typical single-celled protists.*

Materials
Microscope
Microscope slides
Ameba and paramecium samples
Stained yeast cells
Cover slips
Medicine dropper
Pencil
Paper

Protozoans are very interesting to study because they move freely. Many different kinds of protozoans can be found in a sample of pond water. They may even be found in an aquarium. You will be given two samples of protozoans. One of these samples is the ameba.

A. Place a drop of the ameba sample on a glass slide. Cover with a cover slip. Focus on the specimen with low power. Then move to high power.

1. Does the ameba always have the same shape? Describe its shape.

2. Is the ameba moving? If so, describe how it moves.

B. Make a diagram of the ameba. Label all the cell parts you can identify.

3. Does the ameba have a cell membrane?

4. Does it have a nucleus? If so, where is it located?

5. Do you see any clear, circular structures in the cytoplasm? What do you think these structures are?

C. The ameba cannot make its own food. It must eat other organisms. Place a drop of yeast on the slide. Observe the action of the ameba.

6. Describe how the ameba takes in food.

D. Place a drop of paramecium sample on a clean slide.

Chapter 18–2 Classification: Key to Understanding

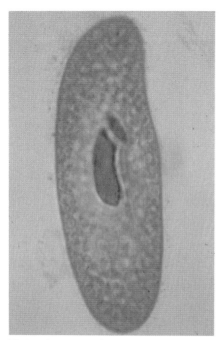

18-15. *A paramecium.*

Cover with a cover slip. Observe it under low power and then high power. Draw a diagram of the paramecium. Label all the parts you can.

7. Does the paramecium change shape?

8. Describe how the paramecium moves.

E. Carefully observe the vacuoles located at each end of the paramecium.

9. What do the vacuoles in the paramecium appear to be doing?

F. Add a drop of yeast to the paramecium slide.

10. Describe how the paramecium takes in food.

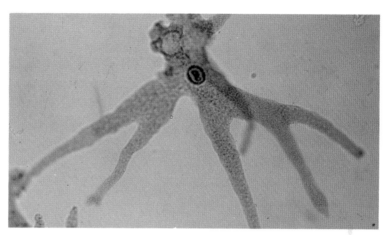

18-16. *An ameba.*

Bacteria: Very small protists with simple cell structure and no nucleus.

The last group of protists are the **bacteria**. *Bacteria* are very small one-celled organisms. They are so small that it would take about 50,000 of them to cover the head of a pin. Bacteria are everywhere. They are found living in soil, in the depths of oceans, in the atmosphere, and in the bodies of animals.

Many diseases of humans are caused by bacteria. Pneumonia, diphtheria, and tuberculosis are such diseases. Other bacteria are used in making dairy products such as yogurt, buttermilk, butter, and cheese. Bacteria are also useful in producing vinegar, tanning leather, and curing tobacco.

Before we leave the classification of living organisms, we should look at a group that is a real puzzle. These are the **viruses** (vie-ris-ses). They are so simple that it is hard to tell whether they are living or nonliving. *Viruses* are not organized into cells. However, when they enter a

Virus: Particles that are not cells but can reproduce in the cells of living organisms.

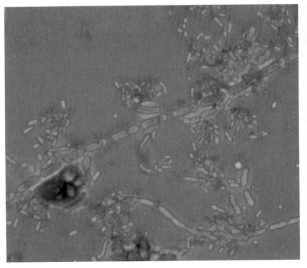
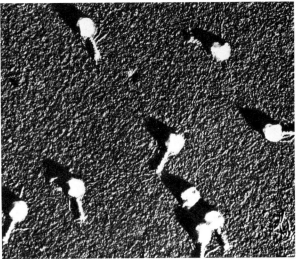

18-17. (left) Typical bacteria organisms

18-18. (right) The viruses in this photograph are magnified 100,000 times.

living cell, they are able to reproduce. In doing so, they damage the cell and cause disease. Polio, mumps, and the common cold are diseases caused by viruses.

SUMMARY & QUESTIONS

Classification is as much a tool of the scientist as is a microscope or a test tube. Classification of organisms is based on the ways they are alike and different. In this section, we considered the algae, fungi, bacteria, and protozoans as protists. However, many biologists feel that these groups should be separated even further. In time, it is possible we may have to develop other classification kingdoms.

1. Arrange the following list of classification groups in their proper order, from largest to smallest: Genus, Phylum, Order, Species, Kingdom, Family, Class.
2. What is the major difference between algae and the other types of protists?
3. List the four groups of protists and describe how each is useful to humans and how each is harmful to humans.

Chapter 18–2 Classification: Key to Understanding

3 CLASSIFYING: PLANTS AND ANIMALS

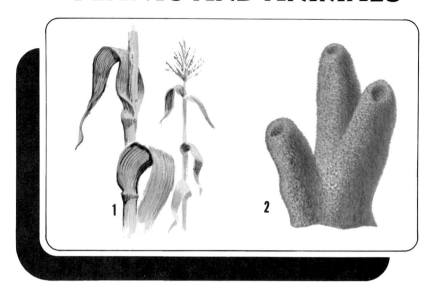

Can you identify the plants and animals shown here and on pages 466 and 467? Most people enjoy looking at plants and animals, but few can identify many of them. In many parks and zoos, plants and animals are identified by small name tags. Unfortunately, the more than a million and a half living things do not come with natural labels. Scientists have classified these organisms in order to simplify their identification. When you finish this section, you will be able to:

● List characteristics of the major animal phyla.

● Describe the characteristics of the major groups of plants.

▲ Classify selected animals and plants.

The animal kingdom may be divided into two major groups: animals with backbones and those without. Animals that have backbones are called *vertebrates* (**vere-tee-braytes**). These all belong to Phylum *Chordata* (core-**dot**-ah). There are five classes of vertebrates in this phylum.

Warm-blooded: An animal that maintains a constant internal body temperature

Cold-blooded: An animal whose internal temperature changes with the temperature of its surroundings.

Mammals have hair or fur on their bodies. They give birth to live young. They are also **warm-blooded**. Mammals may be as different as a whale and a human, but they all have the same general characteristics.

Birds have feathers. They are also *warm-blooded* animals. They lay eggs with brittle shells.

Reptiles have a scaly skin and lay eggs on land. The eggs have a tough shell. Reptiles are **cold-blooded**. They breathe air through their lungs. Snakes, lizards, turtles, and crocodiles are reptiles.

Amphibians have a smooth skin and lay their eggs in water. Their eggs do not have hard shells. Amphibians are also *cold-blooded*. The young breathe oxygen from the water through their gills. The adults develop lungs. The amphibian group includes frogs, toads, and salamanders.

Fish have a scaly skin and lay small, jelly-like eggs in water. They are cold-blooded and breathe oxygen from the water through their gills.

The rest of the major animal phyla do not have backbones. These animals are thus called *invertebrates*. The animals in Phylum *Arthropoda* (are-throw-**poe**-dah) have a hard outside covering made of the same material as your fingernails. Their legs and antennae bend at joints. The animals are commonly called arthropods. The four major groups may be identified by the number of pairs of legs they have.

If there are three pairs of legs the animal is an *insect*. Most insects also have one or two pairs of wings. There are more species of insects than any other type of animal.

If there are four pairs of legs and two sections to the body, it is an *arachnid* (ah-**rack**-nid). Arachnids include spiders and scorpions.

If there are at least five pairs of legs and two body sections, it is a *crustacean* (cruss-**tay**-shun). Crustaceans also have several pairs of feelers, movable mouth parts and breathe through gills. Most crustaceans live in water.

Animals with many body sections with legs on each section are called *many-legged*. If there is one pair per section it is called a *centipede*. If there are two pairs on each section, the animal is a *millipede*.

The Phylum *Echinodermata* (ee-**kine**-oh-der-mat-ah) contains animals with a hard internal skeleton. Some have five to ten arms that spread out from the body like spokes on a wheel. All the animals have spines covering their bodies. Starfish, sea urchins, and sand dollars are all echinoderms.

Mollusks have a hard outer shell protecting a soft body. They are in Phylum *Mollusca* (moe-**luss**-ka). Most mollusks have a one- or two-part shell. The group contains clams, oysters, snails, and even squid and the octopus.

Animals with wormlike bodies with many sections are in Phylum *Annelida* (**ann**-ah-lid-ah). They are also called segmented worms. They are the simplest animals to have complex body systems such as digestive and nervous systems. Earthworms and sandworms are in this phylum.

Animals with long, smooth, threadlike bodies are in Phylum *Nematoda* (knee-ma-**tow**-dah). Many roundworms, which live in or on other animals, are nematodes.

Flatworms are in Phylum *Platyhelminthes* (play-tee-**helm**-ent-these). Flatworms have digestive systems with only one opening. The other systems are poorly developed. The tapeworm is a flatworm that lives in the intestines of many animals.

Coelenterates (coe-**lent**-err-ates) are in Phylum *Coelenterata* and have hollow bodies. There is a ring of tentacles around the mouth opening. The tentacles catch food and stuff it into the mouth. The body walls are only two cell layers thick. Jellyfish, coral, and sea anemones (ann-**em**-oh-knees) are in this phylum.

Phylum *Porifera* (poor-**if**-err-ah) contains the sponges. Sponges are colonies of cells. They take in water through their pores and filter out food. They are even less organized than coelenterates.

There are two major plant phyla. The first, the Phylum *Bryophyta* (brr-**eye**-oh-fight-ah), contains plants that do not have special tissue for transporting water, minerals, and food. They are called **nonvascular** plants. These plants have no true roots, stems, or leaves. Mosses and liverworts are two common types of bryophytes.

The Phylum *Tracheophyta* (tray-**key**-oh-fight-ah) contains the **vascular** plants. Tracheophytes include ferns and seed plants. Ferns can be found growing in damp, shady areas. The big, lacy structure is the leaf of the fern. The stems grow just beneath the surface. From the stem, small roots grow out into the soil.

The most common type of *vascular* plants are the seed plants. These plants reproduce by forming seeds. If conditions are right, the seeds will grow into new plants. Seed plants are divided into two groups. The first group

Nonvascular: Plants without specialized tissue for transporting water, minerals and food.

Vascular: Plants having specialized tissue for transporting water, minerals and food.

Gymnosperms: Vascular plants with seeds that are not covered.

Angiosperms: Vascular plants with seeds covered by a protective tissue.

has uncovered or naked seeds and are called **Gymnosperms**. The second group have covered seeds and are called **Angiosperms** (an-gee-oh-sperms).

Many naked seeds plants, the *Gymnosperms*, have seed cones. They are called conifers or, more commonly, evergreens. The seeds are located on the scales of the cones. The leaves of conifers are needlelike and remain green all year round. You may be familiar with pines, spruce, fir, or cedar which are all conifers.

All remaining plants belong to the group called *Angiosperms*. These plants have covered seeds. The seed is surrounded by protective tissue called fruit. Apples, blueberries, corn kernels, and pea pods are all examples of this protective fruit. Angiosperms are also called flowering plants since they produce flowers. These flowers are the reproductive organs of the plant.

Angiosperms may be either **monocots** or **dicots**. These terms refer to the structure of the seed. A peanut is a *dicot* seed. There are two halves to the seed. However, a corn kernel has only one part. It is a *monocot* seed.

Monocot: A plant that produces seeds with one part.

Dicot: A plant that produces seeds with two halves.

There are other easily noticeable characteristics of monocots and dicots. Monocot leaves have veins that run the length of the leaf, side by side. Their flowers have three petals or multiples of three (6, 9, and so on). Dicot leaves have branching veins. The dicot flowers have four or five petals or multiples of four or five.

Materials
Pencil
Paper

You have read about the important characteristics of the major types of plants and animals. Now look at the drawings on page 462, 466, and 467. Using the descriptions you have read in this section, classify each organism. Make a list of the characteristics each organism possesses.

Chapter 18–3 Classifying: Plants and Animals

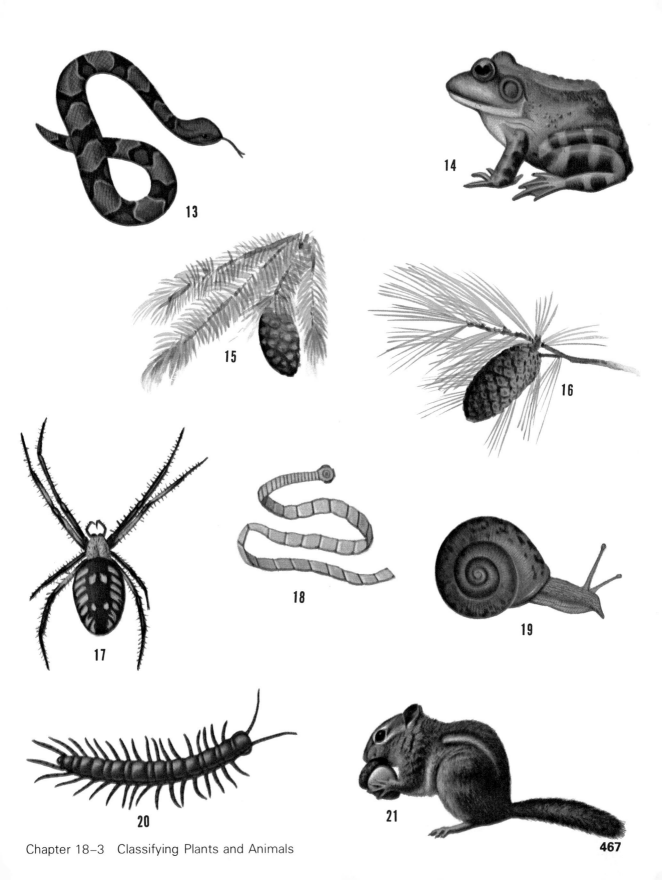

SUMMARY & QUESTIONS

Classification of living things is based on structure, behavior, and complexity of the organism. All organisms in a phylum share certain general characteristics by which they may be identified.

1. The classification key used in this lab is based on the same characteristic used by Linnaeus. What is that characteristic?
2. The animal kingdom is sometimes divided into two general groups, vertebrates and invertebrates. What structural characteristic is the basis of this grouping?
3. What is the function of the vascular tissue in plants?
4. Why are conifers classified as "naked seed" plants?

VOCABULARY REVIEW

Match the number of the term with the letter of the phrase that best explains it.

1. nonvascular
2. fungi
3. species
4. dicot
5. warm-blooded
6. monocot

a. Animals that maintain a constant body temperature.
b. A group of related organisms that have the same characteristics.
c. Plants without specialized tissue for transporting water and food.
d. Plantlike organisms that do not contain chlorophyll.
e. A plant with one-part seeds.
f. A plant with two-part seeds.

REVIEW QUESTIONS

Choose the letter of the phrase that best completes the statement or answers the questions.

1. The Protist Kingdom was developed (a) because the animal kingdom was too big (b) to group together organisms that are not true plants or ani-

mals (c) to simplify the system Linnaeus developed (d) all of the above reasons.
2. The scientific name of an organism consists of its (a) species and kingdom (b) family and class (c) genus and order (d) genus and species.
3. Seed plants that have uncovered seeds are called (a) angiosperms (b) gymnosperms (c) monocot (d) dicot.
4. Each of the following is a characteristic of mammals except (a) they have a backbone (b) they give birth to live young (c) they have hair or fur on their bodies (d) they all live on land.
5. Different species within the same genus are (a) related in many ways (b) unrelated (c) the same organism (d) kingdoms.
6. Dicot plants have (a) leaf veins side by side (b) two seed parts (c) flower parts in three or multiples of three (d) all of these.
7. A starfish, a sponge, and a mosquito have which of the following in common? (a) no supporting skeleton (b) no backbone (c) are warm-blooded (d) have a nervous system.
8. Which of the following groups of protists all contain chlorophyll? (a) fungi (b) algae (c) protozoa (d) bacteria.
9. All of the following are used in classifying organisms except (a) the number of genes in their cells (b) the structure of their body (c) the number of offspring they produce (d) the chemical makeup of their bodies.
10. The fungi include some organisms that (a) cause human diseases (b) are useful to humans (c) damage and destroy human resources (c) all of the above.
11. Each of the following is an anthropod except (a) a spider (b) a snail (c) a crab (d) a dragonfly.
12. What is the proper sequence of these four terms? (a) Kingdom, Phylum, Genus, Species (b) Kingdom, Species, Phylum, Order (c) Species, Family, Kingdom, Class (d) Genus, Phylum, Class, Order.

REVIEW EXERCISES

Give complete but brief answers to the following questions.

1. Explain why biologists found it necessary to use a third Kingdom of living things, the Protists.
2. How are vascular and nonvascular plants different?
3. What are some of the characteristics used to classify living organisms?
4. How are the green algae important to a pond environment?
5. What do the terms *vertebrate* and *invertebrate* mean?
6. Arrange the following classification groups in order, beginning with Kingdom: Kingdom, Order, Species, Phylum, Family, Class, Genus.
7. Why are humans, fish, snakes, eagles, and frogs all in the same Phylum?
8. Is the peanut a monocot or dicot seed? Why?

CAREERS IN LIFE SCIENCE

(left) As a result of technology, relatively few farms and farm workers can produce our food. None the less, there may still be an opportunity to work on a family farm or a farm nearby. To be successful, you must develop certain skills concerning livestock and machinery. You must also have a knowledge of the plants and animals you will be working with.

(below) <u>Marine biologists</u> study the plants and animals that live in the ocean. They investigate the life processes of marine life and determine the effects of pollution on these organisms. The marine biologist in the photograph is collecting samples of sea water. These samples will be tested for the presence of pollutants.

(above) <u>Forest rangers</u> play an important role in protecting the wildlife and forests. In the photograph, a forest ranger observes the surrounding forest for the early signs of fire. A fire discovered early enough can be stopped before it destroys the forest community.

CHAPTER 19 COMMUNITY RELATIONSHIPS

1 ECOSYSTEMS

On February 8, 1975, three astronauts returned to earth from Skylab. For 84 days they had lived and worked 450 km above the earth. Everything needed for survival was contained in a 90 m by 20 m cylinder. Solar batteries provided the needed energy. The space lab was an artificial environment. When you finish this section, you will be able to:

- Describe the living and nonliving parts of an ecosystem.
- Define the term *ecology*.
- ▲ Observe and follow the flow of materials and energy through an ecosystem.

Ecosystem: A group of organisms and their physical environment.

Biotic: The living organisms in an ecosystem.

Abiotic: The nonliving materials and energy in an ecosystem.

A group of organisms and their nonlivng environment is called an **ecosystem** (ee-coe-sis-tem). An *ecosystem* is a lot like the Skylab station. All the materials needed for life are present. Most of these materials are used again and again.

A pond is a complex system of living and nonliving things that affect one another. The living things in any ecosystem are called the **biotic** (by-**ot**-tick) parts. The nonliving things, such as water, soil, light, and temperature, are called the **abiotic** (aye-by-**ot**-tick) parts. A typical

19-1. *What are biotic parts of this ecosystem? What abiotic factors might affect them?*

pond ecosystem is shown in Fig. 19-1. The biotic parts include the animals and plants. Some animals and plants are too small to be seen. The abiotic parts include all the nonliving things that might affect the living things. Light, water, and the muddy bottom are abiotic parts of a pond.

The biotic parts of an ecosystem cannot exist without the abiotic parts. An ecosystem contains everything needed to support life. However, materials are always moving into and out of an ecosystem. The raccoon and frog may leave the pond and move to a nearby woodland.

Water and air are two abiotic parts that move freely between ecosystems They carry many living and non-living materials into and out of the ecosystem. Rain, snow or streams may add water to the pond. Evaporation and streams may carry water away. Sand, silt, carbon dioxide, and oxygen are a few of the abiotic things moved by water and wind. Seed and insects are some of the biotic things wind and water may move in or out of the pond.

19-2. *Describe how wind and water affect this grasslands ecosystem.*

Energy also enters the ecosystem. Most of it is in the form of sunlight. Green plants store this energy during photosynthesis. The energy is then passed on to animals that eat these plants. Other animals eat these animals to get their energy. Plants and animals use the energy to carry on their activities. The energy that they use can no longer be used by other organisms. This energy cannot be recycled. Like Skylab, an ecosystem needs a constant supply of energy.

19-3. *How does energy from the sun affect the plants and animals in this ecosystem?*

Chapter 19–1 Ecosystems

Most of the materials in the ecosystem are re-usable. Water, oxygen, carbon dioxide, wastes, and dead organisms are recycled and used again. This recycling occurs in other ecosystems as well as in ponds. In general, most materials are recycled while energy is lost.

Biologists call the study of the relationships between living things and their environment **ecology** (ee-**kahl**-oh-gee). The word comes from the Greek words OIKOS, which means "house," and LOGOS, which means "to study." Thus, *ecology* is the study of our "house." Our house is the ecosystem in which we live.

Humans live in many different ecosystems. Large cities, small towns, ranches or farms are all ecosystems. Each has a different combination of biotic and abiotic parts. However, each must provide everything that living things need to survive.

Ecology: The study of the relationship between organisms and their environment.

19-4. What abiotic factors in this ecosystem affect the organisms that live here? How?

We have seen that the pond ecosystem is affected by the surrounding ecosystems. So it is with all ecosystems. Each is affected by what happens in its neighboring ecosystems. All the individual ecosystems of the earth make up one large ecosystem. We call it the biosphere.

Materials
pencil
paper

At one time, you may have set up and maintained a small ecosystem. An aquarium is a good example of an ecosystem.

A. Carefully examine a classroom aquarium or a nearby pond. Your teacher may provide other examples of ecosystems for study. The study could also be carried out in your own yard.

B. Answer the following questions about the flow of energy and materials in the ecosystem you are observing.

  1. What biotic parts do you observe in this ecosystem?
  2. What other biotic parts might be present even though they may be too small to see?
  3. What are the abiotic parts of this ecosystem?
  4. What materials may be entering this ecosystem from other ecosystems? How?
  5. What materials may be leaving the system? How?
  6. How does the water cycle work in this ecosystem?
  7. How are the plants and animals in this system involved in the carbon-dioxide-oxygen cycle?
  8. How does energy enter and pass through this system?
  9. What do all ecosystems have in common?

19-5. *Students studying the biotic and abiotic parts of a stream ecosystem.*

Chapter 19–1 Ecosystems

SUMMARY & QUESTIONS

The skylab space station and an ecosystem have much in common. Both contain all the nonliving things needed for the survival of living things except for a source of energy. The sun supplies energy to both systems. Both can recycle necessary materials. But in both, the energy is used and lost, and must be replaced.

1. Which of the following components of a meadow ecosystem is a biotic part? (a) a stream (b) the grass (c) the soil (d) the air
2. Make a list of the biotic and abiotic parts of the ecosystem in which you live.
3. Which of the following abiotic parts of an ecosystem is not recyclable? (a) energy (b) water (c) materials (d) soil
4. How does a large city ecosystem provide the food and energy it needs for its living organisms?
5. Define the term *ecology*.

2 THE CHAIN OF LIFE

A butterfly in the marsh moves from flower to flower in search of food. Suddenly, it is seized in midair by a dragonfly. Later, the dragonfly pauses in its hunting to rest. In an instant, the dragonfly is captured by a hungry bullfrog. Distracted for a moment, the frog fails to see a water snake moving closer. The snake strikes, and the frog becomes a victim. Later in the day, the snake leaves the marsh grass to sun on a nearby beach. Its motion catches the eye of a hawk, high above. There is a sudden swoop and a flurry of wings. The hawk flies off, the snake dangling from its claws. When you finish this section, you will be able to:

- Describe the *food chain* and identify the organisms it contains.

- Identify the difference between an *herbivore*, a *carnivore*, and an *omnivore*.

- Describe a food web.

▲ Construct models of food chains and webs.

Food chain: The passing of energy in the form of food from one organism to another.

The chain of events just described is called a **food chain**. All the organisms in this chain need energy to carry on their activities. The sun is the source of energy for all life. Only green plants can "capture" this energy.

19-6. *Green plants capture the sun's energy to make their own food. This is the start of a food chain.*

Producer: A green plant able to make its own food by photosynthesis.

Consumer: An organism that depends on producers for its food needs.

They store the energy in the food they make during photosynthesis. Because they can produce their own food, green plants are called **producers**. Animals that eat the plants to get energy are called **consumers**. A *consumer* may also eat other animals.

In this food chain, the marsh grass and the other plants are the *producers*. These plants pass energy along to the consumer, the butterfly. Because the butterfly feeds directly on the plants, it is a first-level consumer. The butterfly was eaten by a second-level consumer, the dragonfly. The frog, snake, and hawk are third-level consumers. They eat animals that ate other animals. In another *food chain*, these third-level consumers may be second-level consumers. For example, the hawk is a second-level consumer when it eats a rabbit. The rabbit feeds on plants, so it is a first-level consumer.

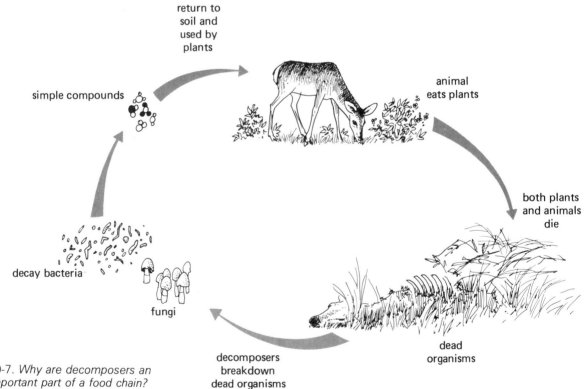

19-7. *Why are decomposers an important part of a food chain?*

Decomposer: An organism that obtains its food from wastes and dead organisms.

Food chains may be longer or shorter than this one. Each chain leads to an animal, such as a hawk or a human, that is not eaten by other animals. But this is not the end of the chain. The wastes and remains of these top consumers are food for the **decomposers**. Decay bacteria and fungi are examples of *decomposers*. They break down wastes and dead organisms all along the food chain. These broken-down materials are returned to the soil. Here they are recycled and used again. A food chain may be summarized as follows:

PRODUCERS → FIRST-LEVEL CONSUMERS → HIGHER-LEVEL CONSUMERS
                          ↓
              → DECOMPOSERS ←

Let's look at another example of a food chain. The forests of North America provide many examples. The green plants such as shrubs and trees are the producers. Feeding on them are the first-level consumers such as rabbits, deer, and mice. These animals are the **herbivores** (**erb**-i-vores), or plant-eaters.

Herbivore: A consumer that eats only plant tissue.

19-8. Organisms that feed directly on plants are called first-level consumers or herbivores.

Carnivore: A consumer that eats only animal tissue.

Animals that feed on other animals are called **carnivores** (**car**-neh-vores), or meat-eaters. In the forest, a fox (meat-eater) might eat the rabbit (plant-eater). A bear may eat both plants and animals. Such an animal is called an **omnivore** (**om**-neh-vore), or "everything-eater." A second way of diagramming the food chain is shown in Fig. 19-10 on the next page.

Omnivore: A consumer that eats both plant and animal tissue.

Chapter 19–2 The Chain of Life

19-9. *The grizzly bear is an example of an omnivore.*

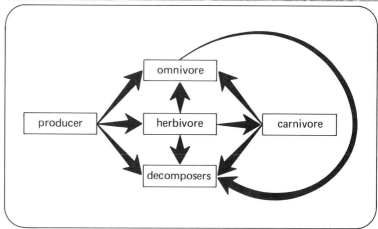

19-10. *The levels in a food chain.*

You don't have to go to a forest or a marsh to observe food chains. Wherever there are living things, you will find food chains. A vacant lot or park may have the following food chains:

WEEDS-(SEEDS) → RAT OR MOUSE → OWL

GRASS → CRICKET → SKUNK → FLEA

OAK (ACORN) → SQUIRREL → CAT → FLEA

Predator: An animal that hunts other animals for food.

Prey: Animals that are hunted by other animals for food.

There is another relationship between organisms in a food chain. It is called a **predator/prey** relationship. Two animals are always involved. The animal that does the catching is the *predator*. The animal that is eaten is the *prey*. For example, the snake in the marsh was a predator. The frog was the prey of the snake. A predator may become the prey of another predator. In the marsh food chain, the snake became the prey of the hawk. In your backyard, a cricket may be prey for a mouse. The mouse may become prey for a cat.

19-11. *Predator/prey relationship on the African plains. Which organism is the predator? the prey?*

One point that was mentioned before should be repeated. A food chain is really an energy chain. Energy from the sun is captured by the producers and passed on to the consumers. Through these chains, living things get the energy they need to live.

An ecosystem may have many food chains. Most animals in a food chain eat a variety of foods. An animal in one chain often eats animals from other chains. As a result, food chains overlap. This network of overlapping food chains is called a **food web**. Fig. 19-13 shows how two simple food chains may overlap to form a *food web*.

Food web: All the feeding patterns in an ecosystem.

Food webs exist in every ecosystem. The African plains is a good example. The lion prefers to eat zebra and wildebeest. However, it will also eat giraffes, buffalo, and baboons. In the same ecosystem, leopards, eagles, vultures, and hyena eat many of the same animals. The resulting food chains form a tangled food web.

Chapter 19–2   The Chain of Life

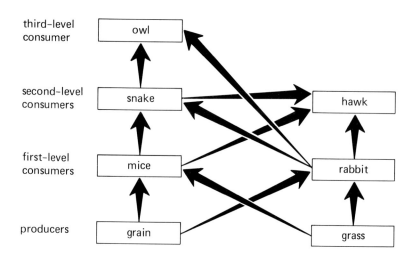

19-12. Identify the producers, and first-, second-, and third-level consumers in this web.

Materials
Paper and pencil
Cardboard patterns (square, circle, triangle, diamond)
Scissors

Table 19-1 lists some of the organisms found in two different ecosystems. One ecosystem may be a park or yard in a city or town. The other is a pond. In each list,

TABLE 19-1

□	○	△	◇
PRODUCERS	FIRST-LEVEL CONSUMERS	SECOND-LEVEL CONSUMERS	THIRD-LEVEL CONSUMERS

CITY ECOSYSTEM

Grasses	Squirrel	Robin	Cat
Weeds	Beetles	Starling	
Vegetables	Mice		
Nuts & Seeds	Grasshoppers		
	Other Insects		

POND ECOSYSTEM

Elodea	Minnows	Turtle	Human
Algae	Insects	Frog	
Other Water	Small Water	Sunfish	
Plants	Animals	Perch	
		Bass	

Unit V What Makes Up Our Living World?

there are producers and three levels of consumers. Each of these groups has a different shape.

A. Draw two different food chains for each ecosystem.

B. Draw four squares on a piece of paper. Write the name of one producer from the city ecosystem in each square.

   1. What is the source of energy for these producers?
   2. What other producers might be found in this ecosystem?

C. Use the patterns to draw 5 circles, 2 triangles, and 1 diamond on the paper. Cut them out. Scatter the various shapes around the producers. Write the name of each consumer in its proper shape.

D. From each organism, draw arrows to all the other organisms that may eat it. Some organisms may not be eaten by any others. Some may be eaten by several others.

   3. Which organisms are predators?
   4. Which may be prey?
   5. Squirrels and mice sometimes eat insects. What level consumer are they then?
   6. Why is this drawing called a food web?
   7. Which animals might be affected if the yard or garden were sprayed with an insecticide? How?

E. Using the same procedure, draw a food web for the pond ecosystem. Using this drawing, answer Q. 1-4.

Earlier, you studied the different levels of a food chain. Now you will examine these levels a little closer. There are usually more producers in a food chain than other organisms. As you move to the next level, the number of organisms decreases. A grass field shows this decrease in organisms. The grasses feed thousands of plant-eating insects. These insects feed a few hundred sparrows. Finally, the sparrows feed only a few hawks.

Only about 1 percent of the energy from the sun is used by green plants. Most of this energy is used to carry on the plant's life activities. Only a small part of the

19-13. *As the level of consumer increases, the size of the organism also increases. Why do the numbers decrease?*

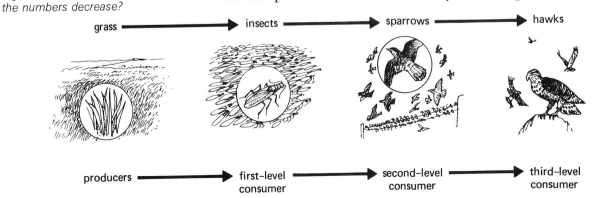

Chapter 19-2 The Chain of Life

energy is left to be stored in the plant. This stored energy is passed on as food to the next level. At each level, most of the energy is used and lost as heat. The rest of the energy is passed on to the next level. As a result, each level has less energy available to it than the previous level. This explains why there are fewer organisms at the end of the chain.

SUMMARY & QUESTIONS

Living organisms need to obtain energy. Only green plants can obtain energy directly from sunlight. All other organisms must obtain their energy from the green plants or from animals that ate the plants. These relationships are shown in food chains.

1. In the food chain, GRASS → MOUSE → WEASEL, which organism is the predator? (a) grass (b) mouse (c) weasel (d) both mouse and weasel
2. Why are food chains often called energy chains?
3. Identify each of the following organisms as a producer, herbivore, carnivore, or omnivore.
   a. antelope   e. sheep   i. pond lily
   b. chicken   f. grass/grain   j. human
   c. coyote   g. mountain lion
   d. corn   h. frog
4. From the above organisms, construct at least four food chains. Include humans in at least two of them.

3 THE LIVING COMMUNITY

Reston, Virginia, is a new town outside of Washington, D.C. The developers specifically planned the town to provide for the needs of the citizens. Each of the basic needs people have—food, jobs, housing, services, recreation, and open spaces—was considered. The town was also designed to preserve the natural environment. When you finish this secton, you will be able to:

● Describe and give examples of a *community*.

- Describe an organisms's *habitat* and *niche* in a *community*.

- Describe three different types of human communities.

- Examine and describe the community in which you live.

Community: All the plants and animals living in a particular area or ecosystem.

All ecosystems are made up of living organisms and their nonliving environment. Very few of these organisms live alone. The survival of each depends on the other organisms in the area. All the plants and animals living in a certain area make up a **community**. All the animals and plants living in a forest are called a forest *community*.

Many communities are made up of smaller communities. As you walk through the forest, you may come upon a stream. The organisms living here make up a stream community. Farther along, the stream may enter a pond. This would be a pond community. A seashore may have a sandy beach, a rock breakwater, a salt marsh or a tidepool. Each is a community. A nearby park may contain pond, open field, and flowerbed communities. A single tree can be a community. It may be home to many insects, birds, and squirrels.

19-14. *Name several different communities that might be found in this forest ecosystem.*

Chapter 19–3 The Living Community

Conditions never stay the same in a community. They change with the seasons. The changing physical factors cause some organisms to leave the community. Others become inactive in cold weather. Many communities change drastically over long periods of time. Some even change from day to night.

Each organism in a community has a certain place in which it lives. This place is called its **habitat** (**hab**-i-tat). It is sometimes called the organism's address. An organism's *habitat* supplies all the conditions it needs to survive.

Habitat: Where an organism is usually found in a community. Its address.

Within the forest community are many different habitats. The soil is the habitat of earthworms and many small insects. Salamanders may find a home in a rotten log. Deer live among the trees and shrubs. The earthworm, deer, and salamander have different habitats. However, they all live in the same community.

Organisms may share the same habitat. The green plants of the forest grow in the soil where the earthworm lives. Insects share the rotten log with the salamander. Many birds live among the shrubs with the deer.

19-15. *What is the habitat of this spotted salamander? What other organisms might share this habitat?*

Organisms that live in the same habitat depend on each other in many ways. Earthworms eat the decaying plants in the soil. The plants, in turn, use the wastes released by the worms. Birds spread the seeds of berry bushes.

Every organism plays a role in its community. In a grassy field, the plants are producers. The insects, rabbits, and cattle that eat the grass are consumers. The role each organism plays is called its **ecological niche** (**nitch**). It is sometimes called its occupation. A field community

Ecological niche: The role an organism plays in a community. Its occupation.

19-16. *What ecological niche does this red-tailed hawk occupy in this community?*

may have many other *niches*. Some insects may be second-level consumers. Birds and shrews may fill third-level consumer niches.

Each community has its own food chains and webs. Herbivores, carnivores, and omnivores are usually present. Some larger organisms may be part of food webs in several communities. Animals such as hunting birds, deer, and bobcats are examples.

Two different organisms may live in the same habitat and have the same niche. They are said to be in *competition* with each other. The one that is better adapted to the habitat will crowd the other out. For example, sheep and cattle are both herbivores. However, sheep bite the grass close to the ground. As a result, they make the grassland unfit for cattle.

Humans live in communities, too. When we study human communities, we really study two different communities. First, there is the natural community. This includes humans and the other organisms around them. It includes the food webs of which humans are a part. Some of these food webs go far beyond the location in which people live. City dwellers get their food from all over the world. People who live on a farm may produce much of their own food.

In a large city, the natural community would include dogs, cats, rats, birds, insects, grass, flowers, trees, and weeds. Habitats might include parks, yards, vacant lots, sidewalk cracks, gardens, and window boxes. Each organism would fill a certain niche in the community. There are many examples of food chains and webs for such a community.

19-17. *A park may contain many different habitats.*

The other type of community is a social community. In general, there are three types: urban, suburban, and rural. In large urban communities, neighborhoods can be compared with habitats. People in these habitats have niches to fill. Grocers, waitresses, utility workers—all fill niches necessary to maintain the community.

19-18. *Three types of social communities: urban (left), suburban (middle), and rural (right).*

A difference among these communities is the size of the population. As the population increases, other problems increase, too. Food and water supplies must be obtained from outside the community. Energy demands, air and water pollution, traffic, and transportation become problems. Other social problems such as poverty, slums, crime, and disease are also affected.

Materials
Pencil and paper

In this activity, you will examine the social community in which you live.

A. On a piece of paper, draw a very general map of your community. Include major roads, public buildings, and natural features such as rivers.

1. Do you live in an urban, suburban, or rural community?

2. Describe the nonliving parts of your community.

3. Which of these factors do you think have an effect on the people in the community? How?

4. How are the nonliving parts of your community different from those of other community types?

B. On your map, outline the various habitats in which people live and work.

5. How do these habitats differ?

C. Communities consist of the organisms that live in a certain area. Often the organisms depend on each other.

6. How does an ecological niche compare with the occupations of the people in your community?

7. List several "niches" in your community and show how you depend on them.

8. What are some of the problems facing your community? Are any related to population size?

An ecosystem consists of a community and its environment. Each organism lives in one or more habitats and has a niche within its community. Humans live in habitats and have niches in both natural and social communities.

Chapter 19–3 The Living Community

1. What is a community? List several examples.
2. Describe the ecological niche of each organism in the food chain: GRASS → MOUSE → SNAKE → HAWK → DECOMPOSERS.
3. What is a habitat? List at least five different habitats that organisms could occupy in your neighborhood.
4. Describe some ways the population size would affect human natural and social communities.

4 CHANGES IN POPULATIONS

Isle Royale National Park is a large island in Lake Superior. It is a long, narrow island lying 25 km from the shore of Canada. Because of its separation from the mainland, there were no large grazing animals such as deer on the island.

Around the beginning of this century, a few moose reached the island. They either swam from shore or crossed on the winter ice. What they found was a paradise. There were no other large herbivores to compete with them for food. There were no natural enemies. As a result, the number of moose increased rapidly. Within 20 years, over 2000 moose were living on the island. Most organisms rapidly increase in numbers when conditions are favorable. When you finish this section, you will be able to:

- Describe the factors that affect a *population* of organisms.
- Explain how one population can affect another population of organisms.
- Apply these ideas to human populations.
- Examine graphs showing changes in a population.

Population: A group of one kind of organism in a particular community.

Isle Royale is a forest community. All the organisms of the same species within a community are called a **population** (pop-you-lay-shun). The moose of Isle Royale are such a *population*. The island also has populations of beaver, fir trees, birches, and foxes.

Scientists are always observing what happens to populations in these special cases. It gives them a better picture of what normally goes on in a community. They try to find out what populations are present. There may be hundreds of different ones. They try to find out the size of each population and if the numbers are changing. They also try to find out which species is the **dominant species**.

Dominant species: Organism present in the greatest numbers.

19-19. *The increase in the moose population on Isle Royale caused a food shortage.*

Chapter 19—4 Changes in Populations

Density: The number of organisms found in a certain area at a given time.

Communities are often named for the most *dominant* species present. An oak-hickory community would have large populations of these trees. The dominant species also determines the type and numbers of animals that live in such a community.

By the 1900's there were four moose for each square kilometer of land on Isle Royale. This is called the **density** of the population. The *density* of the moose population was so great that it caused a food shortage. In the next few years, all but a few hundred moose died. Most of them died of starvation. Such a sudden decrease in numbers is called a population crash. Scientists feared the entire herd would die of starvation.

In the summer of 1936, a forest fire burned over a quarter of the island. It looked like the end for the herd. Over the next several years, the burned area began to fill with lichen and moss. Young birch and aspen trees soon appeared. The herd began to thrive on this food source. The number of moose steadily increased again. By the 1940's, park rangers again began to find dead moose. It looked like another crash was about to occur.

Migration: The movement of animals from one place to another.

Moose are not the only animals that have extreme population changes. A well-known example is a tiny mouselike rodent, the lemming. Lemmings are found in Norway, Sweden, and Canada. Every three or four years, the population of lemmings gets very large. As a result, they **migrate** to find new food sources. Lemmings can swim across rivers and lakes, but they cannot tell a lake from the sea. During these *migrations*, thousands plunge into the ocean and drown.

19-20. *The lemming is a small rodent. What limiting factors may cause a migration of lemmings?*

Limiting factor: A condition in the environment that may cause a change in the size of a population.

The density of the moose and lemming populations became a **limiting factor.** It affected the growth rate of the populations. In both cases, the high density resulted in a shortage of food. This caused the increasing death rate of the moose and the migration of the lemmings. There are other *limiting factors* involved with population changes. Higher population densities also cause an increase in disease. These diseases may kill large numbers of organisms. Others become weakened and are easy prey for predators. Lemmings are eaten by foxes, weasels, wolves, bears, and birds. Even large trout eat them as they cross streams.

As the number of an organism increases, so does the number of its predators. The increased number of predators will cause a drop in the number of prey. Then the predators begin to starve to death. This decrease in predators allows the prey to increase again. This is called the predator/prey cycle.

Materials
pencil and paper

PART I

In this activity, you will study an example of this predator/prey cycle. You will see the changes in the populations of snowshoe hare and Canadian lynx. The data used comes from the numbers of furs received by the Hudson's Bay Company of Canada.

A. Examine the graph carefully. Answer the following

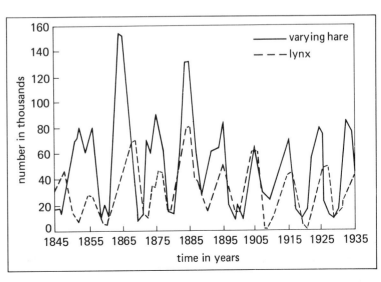

19-21. *The graph shows changes in snowshoe hare and lynx populations over a period of time.*

Chapter 19–3 Changes in Population

questions based on the graph.

1. About how many years apart are the peaks in the snowshoe hare population?
2. About how many years apart are the peaks in the lynx population?
3. When do the peaks in the lynx population occur (just before, at the same time as, or just after the peaks in the snowshoe hares)? Explain why.
4. Why is the line for the lynx almost always lower than the line for the snowshoe hares?
5. The lynx causes a change in the snowshoe hare population. Name two other limiting factors that may also cause a change.

There is a pattern to a population cycle in nature. At first, there is a slow increase. Then, the population begins to grow rapidly. This is due to an increase in the birth rate under favorable conditions. Soon, natural limiting factors cause the population to level off. These factors include density of population, a decrease in food, diseases, and predators.

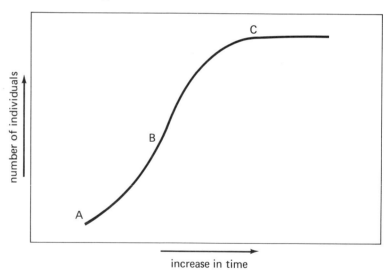

19-22. The "S" curve shows a typical population cycle. At first there is a slow increase (A). Then the population grows rapidly (B). Soon the population levels off (C).

A typical population cycle can be represented by an s-shaped curve. What happens to the curve in the future depends upon these limiting factors. If the food supply increases or predators decrease, the population curve will increase. If not, the curve may decrease.

What happened to the moose on Isle Royale? The problem there was the lack of a large predator. Thus, the moose population became too large for the food supply.

In the late 1940's, a pack of wolves crossed the winter ice to the island. Many people feared that the wolves would increase and destroy the moose herd. This has not happened. The wolves kill mainly the old, the sick, or the injured moose. The populations of moose and wolves have leveled off. There are about a thousand moose and two dozen wolves. This seems to be the number of both animals the island can support.

19-23. *Why are predators such as these wolves a necessary part of any ecosystem?*

However, will this balance continue? Will some unexpected natural factor cause a change in either population? Would this upset the balance and create chaos again? Only time will tell.

Changes in human population have also been studied. There are two major differences between human and animal populations. First, there are no large-scale predators of humans. Second, humans can make large-scale changes in their environments. Both of these factors affect human populations.

The only limiting factors affecting our population are density, food, disease, ... and war. The population of the earth for the past 2000 years is shown in Fig. 19-24 on the next page. Around 1 A.D., the population is estimated to have been about 250 million. By about 1500 A.D., the population had reached 500 million.

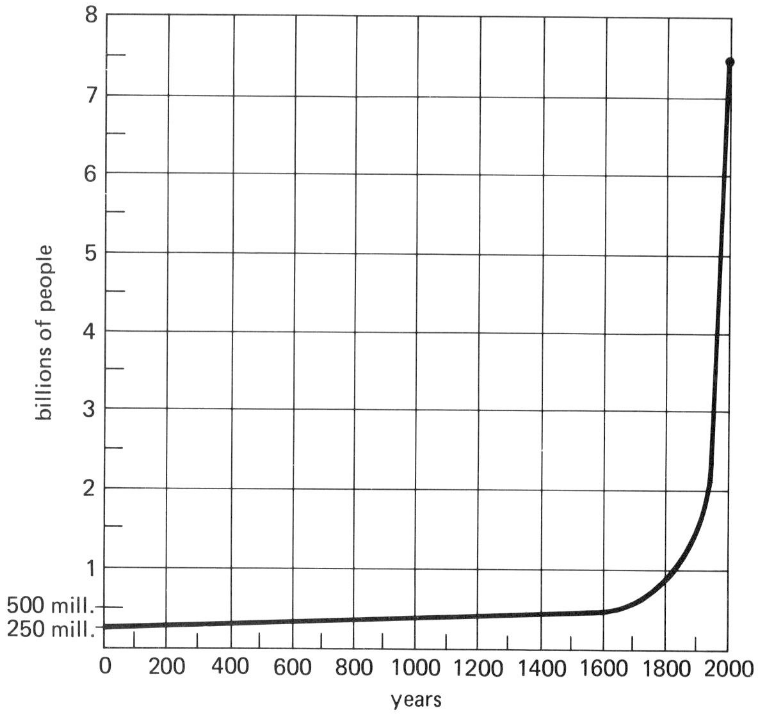

19-24. Estimated world population changes since 0 A.D. What factors may limit human populations?

PART II

A. Answer the following questions based on the graph in Fig. 19-24.

6. How many years did it take for the population to double from 250 million to 500 million?

7. When did the population reach 1 billion people?

8. How long did it take to double the population from 500 million to 1 billion?

9. When did the population reach 2 billion? How long did this doubling take?

10. When did the population reach 4 billion? How long did this doubling take?

11. What has been happening to the length of time necessary to double the human population?

12. If this rate were to continue, what would the world's population be in the year 2000?

13. What problems would such an increase cause?

People have made great advances in producing more food and controlling diseases. As a result, people live longer. This causes an increase in the population. Will food production continue to keep pace? If the population doubles in the next 25 years, will the food supply also double? If it does not, do people face the same type of population crashes as the Isle Royale moose? More people will need more room. What effect will this have on land we need to grow food? Will increased population density increase diseases? Will it cause other problems?

19-25. *What problems might continued increases in human populations cause in the future?*

SUMMARY & QUESTIONS

The population of any species in a particular area is affected by several factors. The food available, numbers of predators, disease, and overcrowding all limit a population. Scientists are always studying changes in populations. This will help them to better understand the future of the human population.

1. What do the terms *population* and *density* mean?
2. What are some of the limiting factors that might affect the population of insects in a vegetable garden?
3. Bounties were often paid for shooting mountain lions. What effect might this have on the local deer population? Why?
4. Why is there so much concern over the continued rise in the world's human population?
5. What would happen to a population of 1000 pairs of organisms after 5 generations if each pair produced:
   a. 4 offspring?
   b. 2 offspring?
   c. 1 offspring?

   What do you predict might eventually happen to each population?

Chapter 19-4 Changes in Populations

VOCABULARY REVIEW

Match the number of the term with the letter of the phrase that best explains it.

1. consumer
2. food chain
3. ecosystem
4. food web
5. producer
6. community
7. population
8. niche
9. habitat
10. decomposer

a. All possible food pathways in an ecosystem.
b. Green plant that carries on photosynthesis.
c. Passing of food energy from one organism to another.
d. A group of organisms and their physical surroundings.
e. Organism that depends on green plants for its food needs.
f. Organism that gets its food from wastes and dead organisms.
g. The place where an organism is usually found (organism's address).
h. All the different organisms in a particular area.
i. All the organisms of the same species found in a particular area.
j. The role an organism plays in a community.

REVIEW QUESTIONS

Choose the letter of the phrase that best completes the statement or answers the question.

1. The study of the relationships between organisms and their environments is called (a) biology (b) ecology (c) zoology (d) botany.
2. All the possible feeding patterns in an ecosystem are called a (a) food chain (b) food pyramid (c) food level (d) food web.
3. The final level in all food chains and food webs is the (a) producers (b) herbivores (c) carnivores (d) decomposers.
4. Which of the following would be a limiting factor in a population of animals? (a) food (b) water (c) predators (d) all of these
5. The annual journey of Canada geese southward in the fall and northward in the spring is called (a) population (b) migration (c) pollination (d) mutation.
6. Which of the following ecological niches could not be occupied by a person? (a) consumer (b) carnivore (c) producer (d) herbivore
7. The crossovers between food chains that produce food webs are due to the fact that (a) many consumers may feed on the same producers (b) second-level consumers feed on first-level consumers (c) consumers usually eat many different organisms (d) all of the above statements.

8. Which of the following are in the correct order? (a) decomposer, producer, carnivore, herbivore (b) producer, herbivore, carnivore, decomposer (c) producer, carnivore, herbivore, decomposer (d) carnivore, herbivore, producer, decomposer.
9. Which of the following parts of a lawn ecosystem is an abiotic part? (a) soil (b) grass (c) worms (d) insects.
10. Which of the following statements about ecosystems is correct? (a) materials are cycled, energy is lost (b) energy is cycled, materials are lost (c) both materials and energy are cycled (d) none of the above.

REVIEW EXERCISES

Give complete but brief answers to the following questions.
1. The food web of a large city may extend far beyond the city limits. It may even be worldwide. Explain why.
2. Explain why humans are called omnivores. Draw two food chains that will help explain your answer.
3. Why does the available energy in a food chain decrease as the feeding level increases?
4. How might the abiotic factors of temperature, water, air, and soil be different from one ecosystem to another?
5. Draw a simple food web using the following organisms: grass, weeds, worms, grubs, beetles, robin, starling, cat.
6. Identify the producers and first-, second-, and third-level consumers in the food web you drew in question 5.
7. Describe some of the effects a high population density might have on a human population.
8. Why do scientists study populations or organisms in communities?
9. Imagine that a small herbivore such as the rabbit escaped into a community. Here it had no natural enemies and plenty of food. What might happen?
10. Organisms may occupy more than one niche in a community. Explain how and give an example.

CAREERS IN LIFE SCIENCE

Nursing aides help the nursing staff of a hospital care for the patients. Some of their tasks are serving meals, feeding patients, and helping patients dress or bathe. The nursing aide in the photograph is helping one of the younger patients of the hospital.

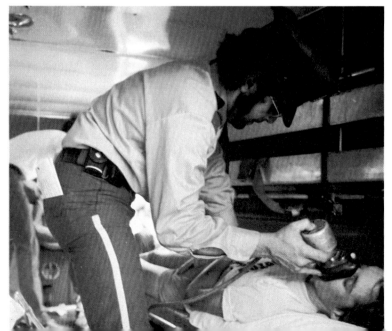

Paramedics are trained to give emergency treatment to victims at the scene of an accident. They begin treatment for emergencies such as burns, broken bones, and heart attacks. In the photograph, a paramedic is giving oxygen to an exhausted marathon runner.

CHAPTER 20
CONTINUING THE SPECIES

1 LIFE GOES ON

Decaying meat produces worms. Sour wine produces vinegar eels. Dirt produces insects. Do these statements sound ridiculous? Before the late 1800's many people believed these statements. They thought that living things could come from nonliving things. Today we know that living things come only from other living things. Worms come from other worms. Insects come from other insects. People come from other people. Even bacteria must come from other bacteria. When you finish this section, you will be able to:

● Compare *asexual* and *sexual* reproduction.

• Define *pollination* and *fertilization*.
▲ Examine the parts of a flower.

The process by which living things produce more of their own kind is called *reproduction*. Organisms reproduce in many different ways. However, there are two main ways: **asexual reproduction** and **sexual reproduction**.

Asexual reproduction (ay-**sex**-you-al ree-pro-duc-shun) requires only one parent. *Sexual reproduction* requires two parents. There are many forms of asexual and sexual reproduction. Most single-celled organisms reproduce

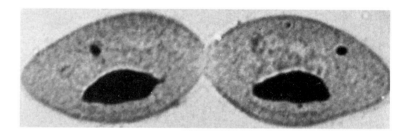

by asexual reproduction. However, sexual reproduction is the most common method of reproduction.

Sexual reproduction requires two special reproductive cells, one from each parent. The reproductive cell from the male parent is called the **sperm**. The reproductive cell from the female parent is called the **egg**. Both contain half the number of **chromosomes** of the parent cells.

Chromosomes are found in pairs in the nucleus. Chromosomes are the blueprints of the cell. They direct and control the cell's activities. Cells contain a certain num-

Asexual reproduction: Reproduction that requires only one parent.

Sexual reproduction: Reproduction that requires two parents.

20-1. *A paramecium undergoing asexual reproduction. The parent cell divides into two approximately equal parts.*

Sperm: A reproductive cell from a male.

Egg: A reproductive cell from a female.

Chromosomes: Long rod-shaped bodies in the nucleus that control a cell's activities.

20-2. *Why do reproductive cells have only half the normal chromosome number?*

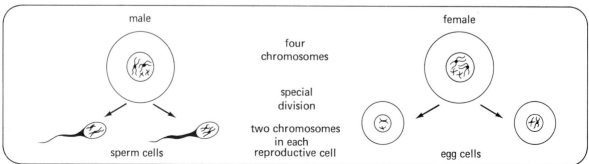

502 Unit V What Makes Up Our Living World?

Fertilization: The process during which a sperm cell and egg cell join.

Trait: A characteristic that is passed from parent to offspring.

Pollination: The process in which pollen is transferred from an anther to a stigma.

20-3. *A flower is the reproductive structure of a flowering plant.*

20-4. *The structure of a flower.*

ber of pairs of chromosomes. The reproductive cell contains one of each pair of chromosomes. See Fig. 20-2.

In sexual reproduction, the *sperm* cell and the *egg* cell join. This process is called **fertilization** (fur-till-lie-**zay**-shun). When *fertilization* occurs, a new cell is formed. This new cell has a combinaton of chromosomes from both parents. It receives half from the female egg and half from the male sperm. The new cell is the beginning of a new offspring. The offspring will have a combination of **traits** from both parents. It will not be identical to either parent.

All flowering plants produce seeds by sexual reproduction. The reproductive structure of a flowering plant is the flower. Flowers come in many different shapes and colors. However, most flowers have the same general structure. The *petals* and *sepals* protect the reproductive organs of the flower.

In the center of the flower is the *pistil*, the female reproductive organ. Inside the ovary at the base of the pistil are one or more *ovules*. The ovules later become seeds. Some flowers have hundreds of ovules.

The pistil is surrounded by male reproductive organs called *stamens*. The *anther*, at the top of the stamen, produces *pollen grains*. They contain the sperm cells.

For fertilization to occur, pollen must be transferred from the anther to the *stigma*. This process is called **pollination**.

When a grain of pollen lands on the pistil, it begins to grow a tube. This pollen tube grows until it reaches

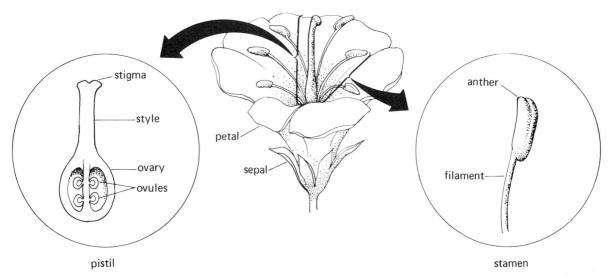

Chapter 20-1 Life Goes On

20-5. Before fertilization can occur, the pollen grain must grow a tube down the pistil.

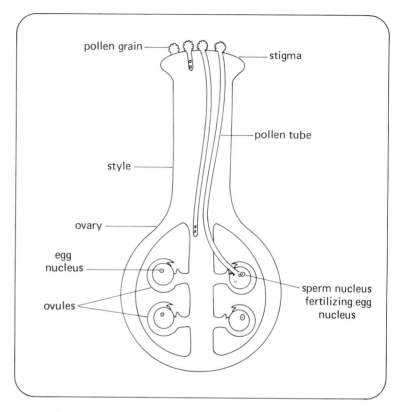

20-6. The structure of a typical seed.

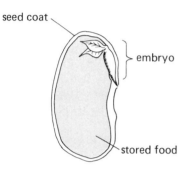

Embryo: A developing organism in its earliest stages of development.

Seed: A ripened ovule that contains the embryo, a stored food supply, and is protected by a seed coat.

Fruit: A ripened ovary that contains seeds.

an ovule. A sperm nucleus from the pollen grain moves down the tube into the ovule. Within the ovule, the sperm nucleus fertilizes the egg nucleus.

Seed development now begins. The fertilized egg divides many times to form an **embryo** or young plant. At the same time, other parts of the ovule form the stored food supply and a protective coat. A **seed** is a developed ovule. While the *seed* is developing, the ovary gets bigger and develops into a **fruit**. A *fruit* is the ripened ovary of a flower. Fruit helps protect and nourish the seed.

20-7. Different types of fruit and their seeds.

504

Unit V   What Makes Up Our Living World?

Materials
Flower
Hand lens
Scalpel
Ruler, 15 cm
Paper
Pencil

A. Obtain the materials listed in the margin. Examine the flower using the hand lens. Locate the sepals, petals, stamens, and pistil.

1. How are the sepals different from the petals?

B. Carefully remove the sepals and petals from the flower.

2. What is a function of the sepals?
3. What is a function of the petals?

C. Remove a stamen and examine it.

4. What is the function of the stamen?
5. Draw and label the parts of the stamen.

D. Closely examine the pistil. Look at the top of the pistil. This is the stigma.

6. What is the advantage of the stigma being sticky?

E. Slice the pistil in half lengthwise along the style.

CAUTION! DO NOT CUT YOUR FINGERS. Measure the length of the style.

7. What is the length of the style?
8. What structures are inside the ovary?
9. After fertilization, what do these become?
10. After fertilization, what does the ovary become?
11. How long must a pollen tube grow to reach one of the ovules?
12. What is the function of the pollen tube?

Living things come from other living things as a result of reproduction. Asexual reproduction involves one parent. Sexual reproduction requires two parents. A fertilized egg results from sexual reproduction. In flowering plants, pollination must take place before fertilization. After fertilization, a seed develops and a fruit forms to help nourish and protect it.

1. What is the difference between asexual and sexual reproduction?
2. Why must pollination occur before fertilization?
3. Describe how a pollen grain reaches an ovule.
4. Name the parts of a seed.

Chapter 20-1 Life Goes On

2 PATTERNS OF INHERITANCE

When you look in a mirror, what do you see? Hopefully you see a special person. You are special because you have a unique combination of traits. Others may have some of the traits you have. However, no one is exactly like you. In addition, there is a very small chance that the same combination of traits will ever occur again. In other words, there will probably never be another you. You really are a special person. When you finish this section, you will be able to:

- Describe the general patterns of inheritance that apply to all organisms.

- Compare and contrast present knowledge of inheritance with Mendel's conclusions.

▲ Make a survey of some inherited traits.

Genes: Structures on chromosomes that determine inherited traits.

Generation: A level in the succession of a family.

Genetics: The science that studies the passing of traits from parent to offspring.

The phrases "related by blood" or "blood relative" have very little meaning scientifically. These phrases reflect a time when people believed that children had exactly the same blood as their parents. It was also believed that a person's characteristics were determined by their blood. Today we know this is not true. Characteristics or *traits* are determined by certain other factors inherited from a person's parents. These factors are called **genes.** In other words, *genes* are structures that determine inherited traits. Genes are located on chromosomes the way beads are on a string.

The person who first began to study how characteristics are passed from one **generation** to another was Gregor Mendel. His experiments on ordinary garden peas were published in 1866. The scientific community of Mendel's day did not accept his ideas, so they were soon forgotten. Not until long after Mendel's death was his work rediscovered. Then it formed the basis for a new branch of science called **genetics.** *Genetics* is the study of how traits are inherited.

Mendel studied several different traits of pea plants.

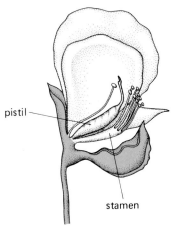

20-8. The garden pea was an excellent plant for Mendel's studies. Its flower may be hand-pollinated if the stamens are removed before natural pollination occurs.

Pure: Refers to an organism or cell in which the paired genes for a trait are identical.

Dominant: Refers to the stronger of a pair of traits.

Recessive: Refers to the weaker of a pair of traits.

Some of these traits included stem length, seed color, seed shape, and pod color. Each of these traits have two forms. For example, stem length was either tall or short. Seed and pod color were either yellow or green. Seed shape was either round or wrinkled. From his experiments Mendel saw that the inheritance of these traits followed a pattern.

To do his experiments, Mendel first made sure that he had plants that were **pure** for each of the traits he was studying. *Pure* means that the same trait showed up generation after generation. Mendel did this by growing plants that he pollinated himself. Then Mendel crossed two plants that showed different pure forms of the same trait. For example, he took pollen from pure short plants and used it to pollinate pure tall plants. He also took pollen from pure tall plants and used it to pollinate pure short plants. All the offspring that were produced from these crosses were tall. Mendel called tallness the **dominant** trait.

Mendel then crossed these tall offspring with each other. Of the plants that were produced from these crosses, about three fourths showed the dominant characteristic of tallness. The remaining one fourth were short. Mendel called shortness the **recessive** characteristic. He said a *recessive* characteristic was one that could be "hidden" in one generation and show up again in a later generation. Thus, after crossing all the tall offspring, three fourths of the plants that grew were tall and one fourth were short.

This pattern, shown in Fig. 20-9 on page 508, occurred for each of the traits Mendel studied. From the patterns of inheritance he observed, he concluded the following.

1. Each pea plant has two factors for each trait.
2. When a sperm or egg cell is formed, the two factors separate.
3. Each factor passes into separate sperm or egg cells. Thus, each sperm or egg cell contains one factor for each trait.
4. A new plant receives one factor for each trait from each parent.
5. When two factors are different, one may be dominant and hide the effects of the other factor. The "hidden" factor would be called the recessive factor.
6. All traits are inherited in the same way but independently of each other.

Chapter 20–2 Patterns of Inheritance

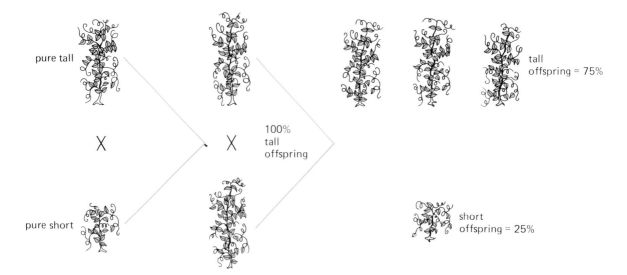

20-9. *Some results of Mendel's work with pea plants.*

Mendel did not know anything about genes, chromosomes, or how reproductive cells are formed. Yet his conclusions were amazingly correct.

Today we know that Mendel's conclusions and the general patterns of inheritance he found apply to all organisms. However, Mendel's ideas need to be clarified using today's knowledge of cells. Cells contain a certain number of paired chromosomes. Human body cells each contain twenty-three pairs of chromosomes. Each chromosome of a pair has factors for the same traits. These factors are now called genes. There are many genes on each chromosome. This means that each pair of chromosomes controls many different traits. For example, one pair of chromosomes may have the genes for eyelash length, dimples, eye color, and many other traits.

When reproductive cells are formed, they receive one member of each pair of chromosomes. See Fig. 20-10. As a result, they also receive the genes that are on those chromosomes. During sexual reproduction, the reproductive cells from each parent join. When these cells join, their chromosomes form pairs again. Thus, the offspring receive one gene for a trait from each parent.

The genes on a pair of chromosomes that control the same trait do not have to carry the same information. If they do carry the same information, the organism is *pure* for that trait. For example, there is a pair of genes that controls whether a person will have straight or curly hair. If a person has two genes for curly hair, that person is pure for curly hair. If a person has two genes for straight hair,

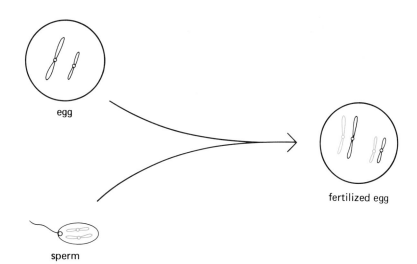

20-10. *During fertilization, the chromosomes of the reproductive cells combine. This restores the proper paired number of chromosomes.*

that person is pure for straight hair. If a person has one gene for straight hair and one gene for curly hair, that person is **hybrid** (**hie**-brud) for that trait. If a person is *hybrid* for a particular trait, the dominant gene will "hide" the recessive gene. In this case, curly hair is dominant. Therefore, a person who is hybrid for this trait will have curly hair. The information for straight hair will be "hidden." If a person is pure for straight hair, that person's hair will be straight. The recessive trait shows up because there is no dominant gene to "hide" it. The only way people know whether they are pure or hybrid for a dominant trait is if they know the type of genes they inherited from their parents. How can a person tell if they are pure for a recessive trait?

Hybrid: Refers to an organism or cell in which the paired genes for a trait are different.

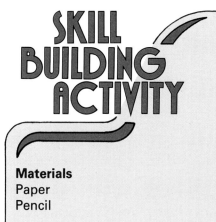

SKILL BUILDING ACTIVITY

Materials
Paper
Pencil

MAKING A SAMPLE SURVEY

In this activity, you will try to determine whether a trait is dominant or recessive.

A. Copy Table 20-1 from page 510 into your notebook.

B. Count the number of students in your class who possess each of the traits listed. Write this number in the column to the right of each particular trait.

C. Use your completed table to answer the following questions.
　　1. Which trait of each pair is more common?
　　2. Which trait of each pair do you think is dominant? Why?

Chapter 20–2 Patterns of Inheritance

3. Can you tell from your table whether a person is pure or hybrid for a trait? Why or why not?

D. Find out from your teacher which traits are dominant and circle them.

4. Are all the dominant traits the most common traits?

5. Do you think taking a sample like this is a good way to decide which traits are dominant? Explain.

6. Now that you know which traits are dominant and which are recessive, can you tell whether a person is pure or hybrid for a trait? Explain your answer.

7. If there are only a limited number of genes people can inherit, why is each person a unique individual? Use your knowledge of genetics to explain your answer.

TABLE 20-1

Can Roll Tongue into U		Cannot Roll Tongue into U	
Cleft Chin		No Cleft Chin	
Attached Ear Lobes		Free Ear Lobes	
Straight Hair Line		Widow's peak	
White Forelock of Hair		No White Forelock of Hair	

SUMMARY & QUESTIONS

The work of Gregor Mendel was rediscovered in 1900. At that time a new branch of science, called genetics, began. Mendel showed that there was a pattern to the way traits are inherited. We now know that heredity is controlled by genes. Genes are located on the chromosomes.

1. Choose a trait Mendel studied and describe the patterns he found.
2. Explain what the terms *dominant* and *recessive* mean and give examples.
3. What is a hybrid? Give an example.
4. How do Mendel's conclusions relate to recent knowledge of how traits are inherited?

3 HUMAN INHERITANCE

The two people in the photograph above are brother and sister. Some of the characteristics they inherited from their parents are the same and some are different. For this reason, they are not the same sex and do not look exactly alike. Even some of the same genes will act differently in each of them. In this section, you will learn why this is so. When you finish this section, you will be able to:

● Describe the inheritance of *sex-influenced* and *sex-linked* traits.

● Explain and give examples of *incomplete dominance*.

● Compare and contrast the *sex-chromosomes* and explain their function.

▲ Apply knowledge of genetics to problem solving.

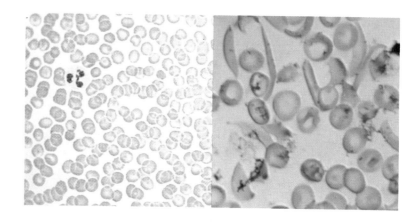

20-11. *(left) Normal red blood cells; (right) Abnormal red blood cells of sickle-cell anemia.*

Incomplete dominance: A condition in which neither of the two genes that cause the appearance of a trait is dominant.

Not all genes are dominant or recessive. Sometimes when plants or animals with opposite traits are crossed, the offspring do not look like either parent. Instead, the offspring show a blending of traits from both parents. For example, when a red four o'clock flower is crossed with a white four o'clock flower, the offspring are all pink. Neither the red nor the white color is dominant. This type of inheritance is called **incomplete dominance**.

The gene for a blood disease called sickle-cell anemia (uh-**nee**-mee-uh) also shows *incomplete dominance*. People who have two genes for sickle-cell anemia have circulation problems. When their blood is low in oxygen, their red blood cells become a long, thin half-moon or sickle shape. See Fig. 20-13. As a result, the sickled cells tend to get stuck in the tiny blood vessels. This prevents the blood from carrying oxygen to all parts of the body as it should. A person who is hybrid for sickle-cell anemia has some normal red blood cells and some red blood cells that become sickled. These people do not suffer the effects of the disease because there are not enough abnormal red blood cells to clog the blood vessels.

Another inherited trait that shows some incomplete dominance is that of human blood types. The blood types found in humans are O, A, B, and AB. There are three different forms of the gene for blood type. These are O, A, and B. Each person inherits two genes for blood type. Since there are three different genes for blood type, there are six possible combinations of these genes a person can have. These combinations and the blood types they produce are shown in Table 20-2.

TABLE 20-2

Gene Combinations	Blood Type
OO	O
AA or AO	A
BB or BO	B
AB	AB

Notice that people can only have O type blood if they are pure for that trait. The O type gene is the recessive gene. If A or B type genes occur with an O type gene, the blood type is either A or B. The gene for O is "hidden". Therefore, the A and B type genes are dominant to the O gene. However, if people inherit both an A type gene and a B type gene, their blood type is AB. One is not dominant to the other. This is a type of incomplete dominance.

Some genes act differently in males and females. The traits that result from these genes are called **sex-influenced** traits. Baldness is a *sex-influenced* trait. It is dominant in males and recessive in females. This means that one gene for baldness is enough to make a man bald. However, for a woman to be bald she must have two genes for baldness. The reason for this seems to involve chemical differences between the male and the female.

Sex-influenced trait: A trait that is dominant in one sex and recessive in the other.

Are all inherited differences between the sexes due to sex-influenced traits? The answer is no. So far you have learned that individuals have two of each type of chromosome. You have also learned that each member of a pair of chromosomes contains genes that control the same traits. These statements are true for both males and females for all except one pair of chromosomes. These are called the **sex chromosomes.** The *sex chromosomes* determine whether a person will be male or female.

Sex-chromosomes: A pair of chromosomes that determine the sex of an individual.

There are two types of sex chromosomes. One is larger and has many genes. This is called the *X chromosome*. The other sex chromosome is smaller and has only a few genes. This is called the *Y chromosome*. In humans, females have two X chromosomes and no Y chromosome. Males have one X chromosome and one Y chromosome.

There are several traits controlled by the genes found on the X chromosome. These traits are called **sex-linked traits** because their genes are located on a sex chromosome. Since females have two X chromosomes, they also have two of each gene found on the X chromosome. However, males only have one X chromosome. Therefore,

Sex-linked trait: A trait controlled by genes on the X chromosome.

Chapter 20–3 Human Inheritance

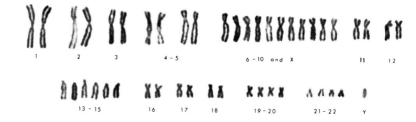

20-12. *The chromosomes of a normal human male.*

males only inherit one of each gene found on the X chromosome. This means that males have only one gene for some traits instead of a pair. For this reason, *sex-linked* traits act differently in males than in females.

Color blindness is an example of a sex-linked trait. People with color blindness cannot see the colors red or green. To them, red and green appear gray. This trait is caused by a recessive gene on the X chromosome. A female can inherit a gene for color blindness from one parent and a gene for normal color vision from the other parent. If this happens, the recessive gene for color blindness is "hidden" by the dominant gene for normal color vision. This female would not be color blind. In order for a female to be color blind, she must inherit a gene for color blindness from *each* of her parents. A male may receive a gene for color blindness from his mother. Since he does not receive an X chromosome from his father, he cannot also have a gene for normal color vision. This man would be color blind because there is no dominant gene to "hide" the effects of the recessive gene. Other sex-linked traits are inherited the same way as color blindness. Therefore, sex-linked traits appear more frequently in males than in females.

Materials
Paper
Pencil

PROBLEM SOLVING

Obtain the materials listed in the margin. Read the following problem, then answer the questions.

The father of a family was reported lost at sea when his children were very young. All of the pictures of him were destroyed in a fire. As a result, the children do not remember what he looks like. The children, a boy and a girl, are now young adults. Both are bald. The boy has blood type A and is color blind. The girl has blood type B and is not color blind. The mother, who was not bald and was not color blind, died and left the children a large amount of money. The mother had type O blood.

Recently, several men have contacted the children claiming to be their long lost father. The first man was not bald, had blood type A, and was color blind. The second man was bald, had blood type B, and was color blind. The third man was bald, had blood type AB, and was not color blind. The fourth man was not bald, had blood type O, and was not color blind.

1. How many genes for baldness could the boy have?
2. How many genes for baldness must the girl have?
3. Would the father most likely be bald? Explain.
4. How many genes for color blindness does the boy have? From whom was his color blindness inherited?
5. Would the father have to be color blind? Explain.
6. What genes for blood type must the mother have?
7. What genes for blood type must the girl have?
8. What genes for blood type must the boy have?
9. What genes for blood type must the real father have?
10. Which man is the real father of the children?

SUMMARY & QUESTIONS

Some genes are neither dominant nor recessive. In this case, a hybrid offspring shows a blending of traits. This type of inheritance is called incomplete dominance. Some genes act differently depending upon the sex of an individual. This is true for sex-influenced and sex-linked traits. The sex of an individual is determined by the sex-chromosomes.

1. What is incomplete dominance? Give an example.
2. In humans, what combination of sex chromosomes does a female possess? A male?
3. Why are some traits dominant in males and recessive in females? What is such a trait called?
4. What type of trait is color blindness? Why are there more color-blind males than females?

4 ADAPTATION AND SURVIVAL

Someday in the future, you may go to the supermarket and buy "sunbeans" (part sunflower and part bean) for dinner. You may also purchase silk produced by bacteria instead of worms. Nitrogen fertilizers may no longer be necessary because plants may be able to take nitrogen directly from the air. These are just some of the developments being studied through genetic research. Would some of these developments eventually occur naturally? We do not know the answer to that question. However, as you will see in this section, genetic changes in nature usually take a long time. When you finish this section, you will be able to:

● Identify two ways differences can occur in a group of organisms.

● Explain why some harmful genes continue to be passed from generation to generation.

● Describe how organisms become adapted to their environment.

▲ Investigate characteristics that adapt organisms to their environment.

Fossil evidence suggests that organisms may have undergone changes over thousands of years. One possible example of an organism that has changed this way is the horse. See Fig. 20-16. It is thought that over the years some horses had different combinations of traits. These came from different combinations of chromosomes and genes as a result of sexual reproduction. Changes in traits may also have been caused by changes in specific genes.

In nature, genes occasionally change. These changes happen by chance. Such sudden changes in genes are called **mutations** (myoo-**tay**-shun). The seedless grape and fuzzless peach (nectarine) resulted from *mutations*. Insects that are resistant to insect poisons and fruit flies born without wings are also results of mutations. Muta-

Mutation: A change in one or more genes that results in a new trait.

about 50,000 years ago

about 35,000 years ago

about 20,000 years ago

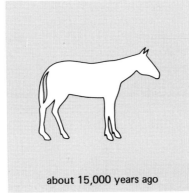
about 15,000 years ago

Modern horse

20-13. *Fossil evidence suggests that the modern horse developed from an ancestor that had four toes on each foot and was about the size of a fox terrier.*

Adaptation: A trait that enables an organism to survive better than similar organisms.

tions can be caused by X rays, nuclear radiation, and some chemicals. Only mutations in the sex cells (the eggs and sperm) can be passed on to the offspring.

Some mutations help an organism to survive and reproduce. Such mutations become **adaptations**. *Adaptations* enable an organism to survive better than similar organisms. An insect that is resistant to insect poisons can survive better than other insects.

However, many known mutations are harmful. They upset an organism's delicate body balance. They also cause traits that can make an organism less adapted to its environment. For example, fruit flies born without wings cannot fly to get food. Often harmful mutations are not passed on for more than a few generations. This is because the effected organisms do not survive to reproduce. Some mutations such as seedlessness in fruits can continue to be passed on generation after generation. You might think that the grapes would not reproduce without seeds. However, humans grow seedless grapes by asexual reproduction. In this way, the seedless trait can be passed on.

Some *mutant* genes are harmful to the organism yet continue to be passed from generation to generation. This usually happens because the mutant gene is helpful when it occurs in the hybrid combination. Thus, in the hybrid combination the mutant gene provides an adap-

Chapter 20-4 Adaptation and Survival

WIDE RANGE

Dandelions will grow and thrive in all types of soils. They are native to Europe and Asia but are found in almost all parts of the world.

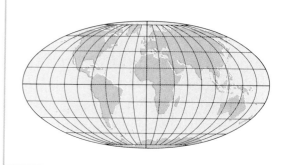

BROAD FLAT LEAVES

Dandelion leaves measure approximately 23 cm long by 5 cm wide. They spread out from the base of the plant. As a result, they crowd and shade many other plants out of existence. Dandelions have a bitter taste, so grazing animals do not like to eat them.

LONG TAPROOTS

The coarse, thick root may be well over 23 cm long. These can get moisture and food not reached by other plants. As a result, the plant is able to survive periods of drought while other plants cannot.

SHORT STEMS

The leaves come almost directly from the tap-root. This makes it difficult to cut or pull out the plant. Even the flower stalk is very often so short that a lawn mower cannot cut off the flowerhead.

Special Adaptations of the Dandelion

LONG GROWING SEASON

In regions where it does not snow dandelions blossom all year round. In regions where it does snow dandelions blossom early in the spring and continue until the snow falls. Thus, they have a long season in which to produce seeds.

PRODUCES MANY SEEDS

Although the plant makes pollen, reproduction is asexual. Therefore the plant does not have to rely on chance pollination or fertilization for reproduction to take place. This allows many seeds to be easily and efficiently produced.

SEEDS EASILY SCATTERED

The feathery propeller-like strands attached to the seeds are easily carried over large areas by the wind. This enables dandelions to grow and become established in new areas.

SURVIVES FROM YEAR TO YEAR — PERENNIAL

In areas where it snows and it is very cold in the winter, dandelions form broad leaf bases in the fall. As a result, they are able to begin growth early in the spring.

tation for the organism. One example is the gene for sickle-cell anemia. This recessive gene is common in many black populations in Africa. The disease can cause death at an early age. If people die early from sickle-cell anemia, how is the gene passed on to the offspring? To answer this question we must look at the people who are hybrid for this trait. These people show either mild symptoms or none at all. They have also been found to have a higher than normal resistance to another disease called malaria. Malaria is very common in many parts of Africa. Thus, people who are hybrid for sickle-cell anemia are better adapted to living in areas where malaria is common. This means that these people are better able to survive and reproduce. As a result, the sickle-cell gene is passed on to the offspring.

One of the most studied examples of adaptation is the dark- and light-colored peppered moth in England. These moths are active during the night and rest on tree trunks during the day. Before England had many factories, most of the tree trunks and moths were light-colored. Shortly after the factories began pouring out their polluting smoke, the tree trunks were covered with soot and became dark. In time, most of the moths developed a naturally dark color. How could such a change occur so quickly?

In the mid-1950's, a scientist performed a series of experiments to answer this question. He found that when the color of the moths matched the trees, the moths could not be seen as easily. Thus, they were eaten less often by birds. Therefore, when the trees were light, more of the dark-colored moths were eaten. When the trees became dark, more of the light-colored moths were eaten. The surviving moths passed their traits along to their offspring. So, when the trees became dark, more of the dark moths survived to produce dark-colored offspring.

20-14. *Peppered moths on a tree trunk. How does their color affect their survival?*

Another example of adaptation can easily be seen with weeds. Weeds are adapted to live in many different environments. A weed is a plant that persists in growing where we wish it would not grow. Therefore, a plant can be a weed in some locations but not in others. A common weed is the dandelion. Some of the characteristics of dandelions that make them so successful are discussed on pages 518 and 519. Many of these characteristics also apply to other weeds.

SKILL BUILDING ACTIVITY

Materials
Paper
Pencil

READING FOR INFORMATION

Dandelions continue to grow in spite of efforts to get rid of them. In this activity, you will investigate why they are so successful.

Refer to pages 518 and 519 as you answer the following questions.

1. Where do you find dandelions growing? What advantage would this be to the dandelion population?
2. How does having a short flower stalk help dandelions to survive?
3. How does the dandelion's root adapt it for survival?
4. How are the leaves arranged around the root? How does this help the dandelion and hinder the other plants?
5. Describe a dandelion head full of seeds.
6. How do dandelions produce seed so efficiently? Why is this an advantage?
7. How are the seeds scattered? How is this an advantage?
8. How late into the fall do dandelions blossom? Why is this an advantage?
9. Write a brief paragraph explaining why dandelions are so difficult to eliminate.

SUMMARY & QUESTIONS

Different combinations of genes cause a variety of traits within each species. Mutations also cause organisms to be different. Some traits allow organisms to be better adapted to their environment. These organisms survive better. When they reproduce, they pass these traits along to their offspring.

1. What are two ways in which new traits occur?
2. Define and give examples of mutation.
3. Explain the dark-colored peppered moth population increase in size in England.
4. Why have seedless fruits survived?
5. What advantage do people who are hybrid for sickle-cell anemia have?
6. Why is the dandelion considered a weed? What are some characteristics that make it so successful?

VOCABULARY REVIEW

Match the description in Column II with the term it describes in Column I.

Column I
1. sex-influenced
2. sexual reproduction
3. egg
4. sperm
5. chromosome
6. gene
7. asexual reproduction
8. hybrid
9. mutation
10. seed
11. embryo
12. adaptation
13. sex-linked

Column II
a. A structure that controls one inherited trait.
b. A developing organism in its earliest stages of development.
c. An organism in which the paired genes for a trait carry different information.
d. A female reproductive cell.
e. A change in one or more genes.
f. A trait controlled by genes on the X chromosome.
g. Reproduction requiring two parents.
h. Long rod-shaped bodies in the cell's nucleus that control a cell's activities.
i. A male reproductive cell.
j. Reproduction requiring one parent.
k. A ripened ovule that contains the embryo, a stored food supply, and is protected by a coat.
l. A trait that enables an organism to survive better than similar organisms.
m. A trait that is dominant in one sex but recessive in the other.

REVIEW QUESTIONS

Choose the letter of the response that best completes the statement or answers the question.

1. The female reproductive organ of a flower is the (a) stamen (b) sepal (c) petal (d) pistil.
2. What type of trait is color blindness? (a) dominant (b) sex-linked (c) incompletely dominant (d) sex-influenced.
3. For a recessive trait to show in the offspring, the offspring must have (a) two dominant genes (b) two recessive genes (c) one dominant and one recessive gene (d) two X chromosomes.
4. Fruit is formed by the (a) pollen grains (b) sperm (c) ovary (d) fertilized ovule.
5. The process of a sperm joining with an egg is called (a) mutation (b) fertilization (c) pollination (d) adaptation.

6. In shorthorn cattle, a red bull crossed with a white cow gives offspring that are all *roan*, a shade between red and white. This is an example of (a) mutation (b) incomplete dominance (c) dominance (d) sex-linked trait.

REVIEW EXERCISES

Give complete but brief answers to each of the following.

1. In humans, brown eyes are usually dominant over blue eyes. Suppose a blue-eyed man marries a brown-eyed woman. The woman's father has blue eyes.
   (a) What are the genes of the husband?
   (b) What are the genes of the woman's father?
   (c) What are the genes of the wife?
   (d) Would the couple be able to have a brown-eyed child? Explain.
   (e) Would they be able to have a blue-eyed child? Explain.
2. Both Mrs. Smith and Mrs. Jones had babies the same day in the hospital. Mrs. Smith took home a baby boy whom she named Scott. Mrs. Jones took home a baby boy whom she named James. Mrs. Jones began to question whether her child had been accidentally switched with the Smith baby in the nursery. To help solve the problem, blood tests were done. Mr. Smith was found to have type A blood and Mrs. Smith had type AB blood. Mr. and Mrs. Jones both had type A blood. Scott had type O blood and James had type B blood. Could a mixup have been made? Explain your answer.
3. Describe how pollination and fertilization occur in plants.
4. One reason King Henry VIII had one of his wives executed was because she had not given him a son. What should he have known about the genetic inheritance of an individual's sex?
5. A woman has a mother who is bald, but her father is not. Her brother is rapidly losing his hair. She is a circus performer who hangs by her hair as part of her act. Should she train for a new act or not? In your answer, explain how baldness is inherited.
6. Bananas are a seedless fruit. How can they reproduce without seeds and why have they survived?
7. What types of adaptations would you expect desert plants to have?
8. Explain how the peppered moth population in England changed as a result of the growth of that country's industry.
9. How might a harmful gene continue to be passed on to future generations?
10. Why are weeds such successful plants?
11. Why are more men than women color blind?

CHAPTER 21 THE HUMAN ORGANISM

1 SUPPORT AND MOVEMENT

Running, bicycling, dancing, swimming, playing sports, or working out at a gym are some ways that help a person keep physically fit. Physical fitness involves more than strong muscles. Exercising is a way to relax and release tension. It helps a person maintain a suitable body weight and it increases blood flow. This brings oxygen to all parts of the body and carries away wastes. In other words, being physically fit makes a person feel good and helps a person stay healthy. When you finish this section, you will be able to:

- Explain how the human body is organized.
- Describe the structure and functions of the human skeleton.
- Discuss the structure and movement of two types of *joints*.
- Compare and contrast three types of muscles.
- ▲ Observe the structure and the relationship of muscles to bones at a *joint*.

The human body is made up of a large number of cells and the products of these cells. The cell is the basic unit of structure and function in a living organism. Many cells are specialized to perform certain functions. Groups of similar cells that perform the same function make up a body **tissue.** Some of the functions of *tissues* include protection, support, attachment and movement.

In some parts of the body, groups of tissues work together and form **organs.** Your eyes, ears, heart, lungs, stomach, and brain are all *organs*. Some organs also work together. Organs working together make up **systems.** The human body is made up of several *systems*. Two of these systems are the skeletal system and the muscular system.

The skeletal system gives shape and support to the body. It also helps to protect some body parts. For ex-

Tissue: A group of cells that are similar in structure and function.

Organ: Different tissues that work together to perform a function or functions.

System: A group of organs working together and forming a functional unit.

21-1. *The organization of the structures that make up an organism.*

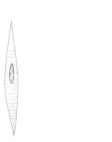

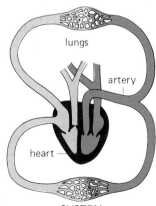

CELL TISSUE ORGAN SYSTEM

Bone: A hard, living tissue made up of bone cells and deposits of calcium and phosphorous compounds.

Cartilage: A firm but flexible tissue that gives shape and support to many parts of the body.

Ligaments: Tough strands of elastic tissue that connect bones at movable joints.

Joint: A place where two bones meet.

ample, the backbone protects the spinal cord. It also supports the upper body and head. In addition, the skeletal system serves as a framework for the attachment of muscles.

The human skeleton is made up of **bone** and **cartilage**. *Bone* and *cartilage* are tissues that are very similar. One difference between the two is that cartilage does not contain the calcium or phosphorus (fos-fuh-rus) compounds that bone contains. This makes cartilage more flexible than bone. You can move the tip of your nose because it is made of cartilage. The bridge of your nose is made of bone and is rigid.

There are 206 bones in the human skeleton. Some of these bones are connected to each other by tough, elastic strands of tissue called **ligaments**. The point where two bones meet is called a **joint**. Since *ligaments* stretch easily, they allow the bones to move freely. This forms what is called a *movable joint*.

Joints can allow movement in different directions. A *hinge* joint allows back and forth movement like the hinge of a door. The knee is an example of a hinge joint.

A *ball and socket* joint allows rotational movement. The joint at the shoulder and upper arm is an example of a ball and socket joint. The end of the upper arm is round like a ball and fits into the shoulder socket.

The inside surface of most joints is covered with cartilage. This acts as a cushion between the bones. Joints also contain a special fluid that lubricates them so they do not wear each other away.

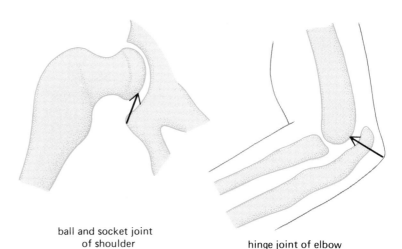

21-2. *Can you think of any other examples of these joints?*

ball and socket joint of shoulder

hinge joint of elbow

Movement at the joints and other parts of the body is caused by the muscles. There are more than 600 muscles in the human body. Men and women have the same number and kind of muscles. However, muscle tissue makes up more of a man's body weight than a woman's. This means that men have more muscle tissue than women. It does not necessarily mean that their muscles are stronger than women's muscles.

In order for muscles to be strong and firm, they must be exercised regularly. If they are not, they become weak, soft, and flabby. When muscles are strong, they can do more work and not get tired as easily.

The muscles of the arms and legs are examples of muscles that aid us in movement. These are called **voluntary muscles** because a person can control them at will. There are some muscles like the ones found in the digestive, respiratory, and circulatory systems that cannot be consciously controlled. These muscles are called **involuntary.** Some muscles such as those that move your eyelids and help you breathe are both *involuntary* and *voluntary*. At times, you can consciously control these muscles, but usually you do not.

In addition to voluntary and involuntary, muscles can be divided into three groups as shown on page 528. Smooth muscle is found in the walls of the blood vessels, digestive system, and other internal organs. *Smooth muscle tissue is involuntary.* The heart is made up of a second type of muscle called *cardiac* (**kahrd**-ee-ak) muscle. This type of muscle is also involuntary.

Skeletal (**skel**-ut-'l) muscle is the third type. Skeletal muscles are voluntary. Most skeletal muscles are attached to bones by structures called **tendons.** Tendons are made of tough, nonelastic tissue. When a muscle contracts, it pulls on the tendon. This causes the tendon to pull on the bone that is attached to its other end. This, in turn, causes movement at the joints.

All muscles work only by contracting. When a muscle contracts, it becomes shorter and thicker. To understand the reason for this, we must look at the structure of muscles. Muscles are made up of individual fibers that are like threads in a rope. These fibers are made up of even smaller structures called *filaments*. When a muscle contracts, these filaments move together by sliding over each other. This is like sliding the fingers of your hands between each other. The sliding together of the filaments

Voluntary muscles: Muscles that can be controlled at will.

Involuntary muscles: Muscles that cannot be controlled at will.

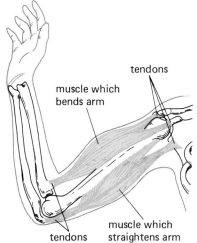

21-3. *Skeletal muscles are attached to bones by tendons. These muscles must work in pairs to bend and extend a joint.*

Tendons: Tough nonelastic tissue that attaches some skeletal muscles to bones.

Types of Muscles

SKELETAL MUSCLE

Skeletal muscles cause the movement of the arms, legs, face, jaws, eyeballs, and so on. These are the most abundant tissues in the human body. Skeletal muscle is so named because it is attached to the bones of the skeleton. Under the microscope, each skeletal muscle cell is cylinder-shaped, contains many nuclei, and looks striped. The skeletal muscle cells are bound together in bundles to form a muscle. Therefore, a skeletal muscle is made up of many bundles of muscle cells.

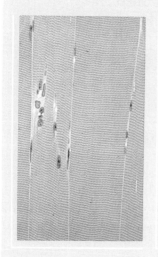 

SMOOTH MUSCLE

Smooth muscle contracts more slowly than other types of muscle. Under the microscope, smooth muscle cells look long, thin, and are usually pointed at each end. They each have one nucleus and do not appear striped. Smooth muscle cells are joined together to form sheets instead of bundles.

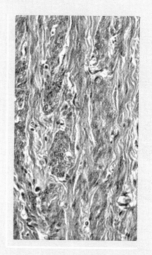 

Cardiac muscle is only found in the heart. This muscle shows some characteristics of both skeletal and smooth muscle. Under the microscope, its cells look striped like skeletal muscle, but it is involuntary like smooth muscle. The muscle cells are joined in bundles that are so closely connected that for years it was not known whether cardiac muscle was made up of individual cells or not.

The heart muscle begins working before birth and continues for twenty-four hours each day of your life. The only time it rests is between beats. Exercise helps strengthen the heart so it can pump more blood with each beat. This makes the heart more efficient.

CARDIAC MUSCLE

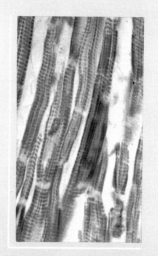 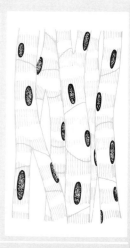

is the reason the muscle gets shorter and thicker. In skeletal muscles, the filaments are alternately thick and thin. This causes skeletal muscle to appear striped.

Since muscles only work by contracting, they can only pull. They cannot push. This is the reason many skeletal muscles must work in opposing pairs. If one set of muscles pulls on a tendon to bend a joint, another set of muscles must pull on a different tendon to straighten the same joint. Combinations of muscles permit movement in several directions. For example, in Fig. 21-3 on page 527, the muscle that bends the arm is opposite the muscle that straightens the arm. Both muscles only contract. When the top muscle contracts, the outer part of the arm bends toward the upper arm. When the bottom muscle contracts, the outer part of the arm moves away from the upper arm.

Materials
Raw chicken wing
Forceps, scalpel
Dissecting tray
Paper towels
Dissecting needles
Microscope, slides, coverslip
Eyedropper
Methylene blue stain (dilute)
Water

A. Obtain the materials listed in the margin.

B. Carefully remove the skin on the chicken wing using the forceps and scalpel. As you are doing this, be careful not to damage the tissues below the skin.

C. Once the skin is removed, observe the structure of the chicken wing and draw a diagram of it. Label the muscles, tendons, bones, ligaments, and cartilage. See Fig. 21-4.

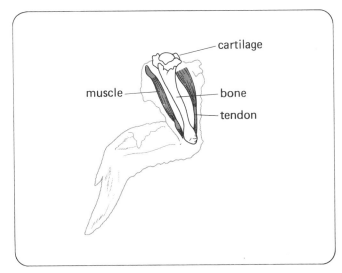

21-4. *The structure of muscles and bones in a chicken wing.*

Chapter 21-1 Support and Movement

D. Use your diagram to help you answer the following questions.
  1. Which muscles extend the wing outward?
  2. Which muscles bend the wing inward?
  3. What is the function of the tendons?
  4. Where is the cartilage located? What is its function?
  5. Where are the ligaments located? What is their function?

E. Place a drop of dilute methylene blue stain on a slide and let it dry.

F. In a drop of water on another clean slide, pull apart a very small piece of muscle. Use dissecting needles to separate the fibers. Then transfer a few fibers to the slide you prepared with the drop of stain on it. Add a small drop of water and a coverslip.

G. Place the slide under the microscope and examine it under low power. Center the slide and carefully switch to high power. Draw a diagram of what you see.
  6. What type of muscle is this? How do you know?

SUMMARY & QUESTIONS

Two of the human body systems are the skeletal and muscular systems. These work together to give our bodies shape, support, protection, and allow movement. The muscular system includes muscles that can and cannot be consciously controlled. These are called voluntary and involuntary muscles. Skeletal muscles are voluntary. Smooth and cardiac muscles are involuntary. All muscles only work by contracting.

1. Give a definition for each of the following: tissue, system, organ, cell, organism.
2. What are the functions of the skeleton?
3. Describe a hinge joint and a ball and socket joint. Give examples of each.
4. Compare and contrast smooth, skeletal, and cardiac muscles.
5. Why must skeletal muscles often work in pairs?

2 DIGESTION

Can you rub your stomach? Seems like a silly question, doesn't it? However, if you think your stomach is near your waist, you're wrong. In this section, you will learn the location of your stomach and other facts about digestion. When you finish this section, you will be able to:

● Describe how the digestive system functions.

● Compare and contrast *mechanical* and *chemical digestion*.

● Explain the functions of the large intestine.

▲ Observe what happens during digestion.

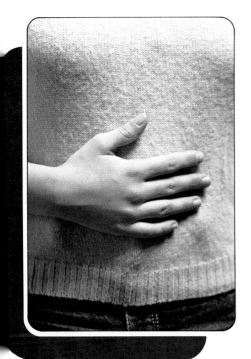

Mechanical digestion: The process by which food is physically broken down into very small pieces.

Chemical digestion: The process by which foods are changed into more simple chemical compounds.

Saliva: A liquid produced by the salivary glands. It contains special starch-digesting chemicals.

Gland: An organ that makes and releases materials having special functions in the body.

The digestive system has several functions. Before your cells can use food, it must be broken down into smaller pieces. This process is called **mechanical digestion.** *Mechanical digestion* begins when you chew your food. Your teeth break up and tear the food. It is important to thoroughly chew your food so that you digest it efficiently.

The stomach also plays a role in mechanical digestion. The muscles in the stomach wall contract in a wavelike motion. This churns the food and helps break up the larger pieces.

Food is a complex chemical substance that must be changed into simpler substances. In both the mouth and the stomach, food is mixed with liquids that contain special chemicals. These chemicals help break food down into simpler compounds. The chemical breakdown of food is called **chemical digestion.**

The watery liquid in the mouth that begins *chemical digestion* is called **saliva** (suh-**lie**-vuh). *Saliva* is produced by the salivary (**sal**-uh-ver-ee) **glands.** *Glands* are organs that make and release materials having special functions in the body. Saliva makes food easier to chew and

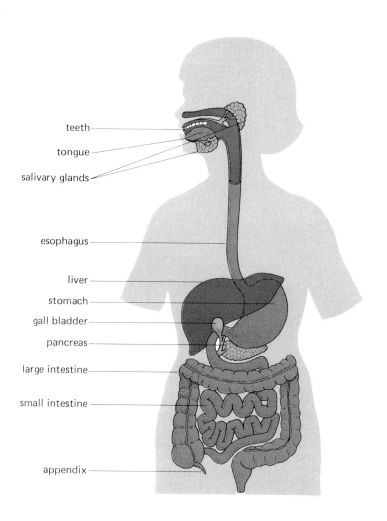

21–5. *The organs of the human digestive system.*

swallow by moistening it. Saliva also begins the chemical digestion of *starches*. Starches are large complex molecules. Through digestion, they are eventually broken down into smaller, simpler sugar molecules called *glucose*. Bread, cereal, and noodles are examples of foods that contain starches.

When food is swallowed, it enters the *esophagus* (ih-**sof**-uh-gus) or food tube. This is a muscular tube that leads to the *stomach*. As the muscles squeeze the walls of the esophagus together, the food is pushed down to the stomach.

The stomach is a j-shaped, muscular storage pouch. The inside wall of the stomach is coated with a layer of *mucus* (**myoo**-kus). Mucus is a thick, slippery liquid that protects the stomach. The stomach contains a special liquid called *gastric* (**gas**-trik) juice. Gastric juice is made by glands lining the stomach and contains an acid. This acid is so strong that it could dissolve the stomach walls if they were not lined with mucus. Gastric juice begins the chemical digestion of *proteins*. These compounds are found in meat, fish, cheese, eggs, and other foods. As the stomach muscles squeeze and churn, gastric juice is mixed with the food. As a result, the food is turned into a semi-solid mass. Next, the food is released into the *small intestine*. This is the main digestive organ. Most chemical digestion takes place here.

The small intestine is a coiled tube about three centimeters thick and seven meters long. Digestive juices also flow into the small intestine. Some of these juices are made by large glands such as the *pancreas* (**pang**-kree-us) and the *liver*. The liver is the largest gland in the body. It is located just under the lower ribs, partially covering the stomach. See Fig. 21-6. A little storage pouch called the *gall bladder* is attached beneath the liver. The pancreas is a smaller gland located below the stomach and above the small intestine. The juices from these glands enter the small intestine through small tubes or **ducts**. Other digestive juice is produced by the lining of the small intestine. Together these juices can chemically break down almost anything you eat.

By the time the food is ready to leave the small intestine, it has been changed from very large complex substances to much smaller and simpler molecules. This has been accomplished by the processes of mechanical and chemical digestion. When chemical digestion is completed, the molecules of digested material move across the membrane of the small intestine into tiny, thin-walled blood vessels. These blood vessels run all along the sides of the small intestine. See Fig. 21-7. Once inside the blood vessels, the **blood** carries the digested food to every cell in the body. The digested food is then either used immediately or stored for later use.

Some food may pass through the system and not be digested. This food moves to the large intestine as a watery mixture. The water is absorbed into the *bloodstream* from the large intestine and is sent to all the body cells. This leaves behind solid wastes called **feces**. The *feces* are eventually eliminated from the body through the opening called the *anus*. This completes the process of digestion and the elimination of solid wastes.

Duct: A tube through which materials can flow from a gland.

Blood: Specialized tissue that transports materials to and from the body cells.

Feces: The solid wastes formed from undigested food materials in the large intestine.

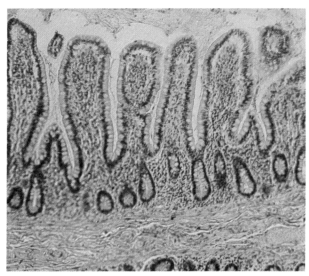
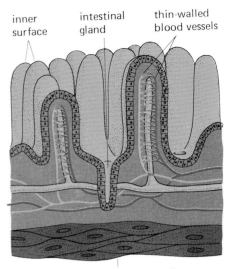

21-6. *The photo at left and the diagram at right show the internal structure of the small intestine.*

Materials
Dialysis tubing (1 pc.)
Tape
2 glass containers
Eyedropper
Starch solution
Iodine
Water
Grease pencil
Stirring rod
Scissors
Paper towels

In this activity, you will see why digestion is necessary.

A. Obtain the materials listed in the margin.

B. The tubing is made of a special plastic that acts like a membrane when it is wet. Thoroughly wet a piece of tubing. Rub it between your fingers to open it. Now tie a string about 3 cm from one end.

C. Fill the tubing with water and dry the outside. Make sure the tube doesn't leak. Tie another string about 3 cm from the open end.

D. Label a glass container with your name. Fill the container 3/4 full of starch solution. Set the tube in the container, taping the ends to the outside. See Fig. 21-7. Only the tube should be in the starch solution.

E. Set your container in a safe place overnight. Answer these questions before the next day.

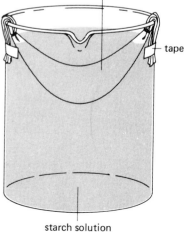

21-7. *Activity set-up.*

Unit V   What Makes Up Our Living World?

1. If the container with the starch solution represents the small intestine, what role do you think the tube is playing?

2. What do you predict will happen to the set-up overnight?

Next day:
F. Get your container.

3. What changes do you observe?

G. Iodine is used to test for the presence of starch. It turns black when it comes in contact with starch. Remove the tube from the starch solution and rinse it off with water. To test if any starch got into the tube, cut off an end of the tube and empty the contents into a clean glass container. Add a few drops of iodine and stir. (**CAUTION: IODINE STAINS! BE CAREFUL NOT TO GET IT ON YOUR HANDS OR CLOTHING.**)

4. Did any starch get into the tube? Why or why not?

H. Look at the predictions you made about the results.

5. How do they compare with what actually happened?

6. Explain how this activity demonstrates the importance of digestion.

SUMMARY & QUESTIONS

The digestive system changes food into a form that can be used by the body cells. As food moves through the digestive system, it is broken down mechanically and chemically. Mechanical digestion takes place by chewing and muscular contractions of the stomach. Chemical digestion takes place by the action of digestive juices. These digestive juices are produced by digestive glands. Digested food enters the bloodstream from the small intestine and is brought to all the cells of the body. Solid wastes are eliminated through the anus.

1. Compare and contrast the two types of digestion. Why are these both necessary?
2. In what organ does most digestion take place? How does the structure of this organ aid in this function?
3. Explain how digested food gets to every cell in the body.
4. What happens to the undigested food?

3 ENERGY RELEASE AND TRANSPORT

Power plants provide energy so people can have light, run appliances, and heat their homes. In many ways, people are like little power plants. Their bodies release energy for work, learning, recreation, body growth and repair, and even heat. When you finish this section, you will be able to:

- Compare and contrast the processes of burning fuel and *respiration*.

- Identify the organs of the human respiratory system and explain their functions.

- Describe the process of *breathing* in humans.

▲ Measure your breathing capacity.

Let's take a look at what happens in a power plant. Fuel such as oil or coal is burned. To burn, the fuel must combine with oxygen. When this happens, energy is given off. Water and carbon dioxide are also formed. This reaction can be shown in a simple **equation**:

FUEL + OXYGEN → CARBON DIOXIDE + WATER + ENERGY + WASTES

The energy from the burning fuel runs special generators that make a more usable energy called electricity.

The cells in your body are like small power plants. The fuel, which is digested food, is brought to the cells by the blood. In the cells, it is "burned" in special generators called *mitochondria* (mite-uh-**kon**-dree-uh). When oxygen combines with glucose, energy is formed. This process is called **respiration** (res-puh-**ray**-shun).

The energy released from glucose is stored in the compound **ATP**. Your cells use the stored energy for many life activities. Like electricity, *ATP* is a more usable form of energy.

Equation: Simple ways to show chemical changes.

Respiration: Process in which oxygen combines with glucose to release useful energy for life activities.

ATP: A compound that stores energy and makes it available to the cell.

Unit V What Makes Up Our Living World?

The process of *respiration* can be shown in the following *equation*:

GLUCOSE + OXYGEN ⟶ CARBON DIOXIDE + WATER + ENERGY (ATP) + HEAT + WASTE

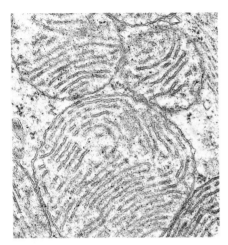

21-9. These large round structures are mitochondria in a muscle cell.

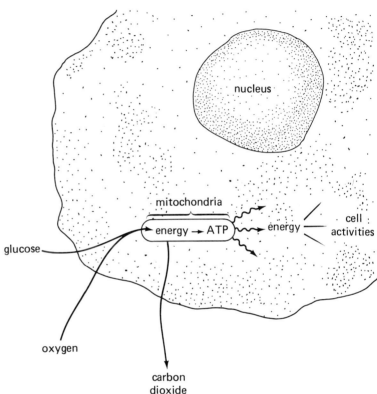

21-10. The mitochondria in a cell change the energy in glucose to energy stored in ATP's.

This equation is similar to the equation for burning fuel in a power plant. However, respiration occurs as a series of many steps. This results in a slower release of energy. Burning in a power plant occurs as one simple step. The release of energy by burning is uncontrolled and rapid.

A power plant cannot change all its fuel energy into usable energy. Likewise, cells cannot change all the energy from glucose into stored energy. What happens to the rest of this energy? A lot of it is lost as heat. This lost heat helps you keep a constant body temperature.

How do each of your cells get the oxygen needed for respiration? In humans, air enters through the nose or mouth. It then passes down the *trachea* (**tray**-khee-ah).

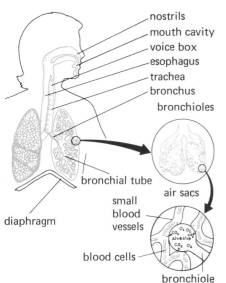

21-11. *The organs in the human respiratory system. Oxygen and carbon dioxide are exchanged in each alveolus.*

Breathing: The process of bringing air into and out of the lungs.

Diaphragm: A sheet of muscle that separates the chest cavity from the stomach cavity.

The trachea branches into the *bronchi* (**bron**-khee). The bronchi lead into each lung and branch into the *bronchial tubes*. In the lungs, the bronchial tubes divide into many smaller tubes, the *bronchioles* (**bron**-khee-ols). These small tubes lead into tiny *air sacs*. The lungs are made up of about 600 million of these tiny sacs. The wall of each air sac has many small pouches called *alveoli* (al-**vee**-oh-lee). These thin moist pouches are surrounded by many small blood vessels. Oxygen entering the alveoli travels through its walls into the *blood*. The blood transports oxygen to every cell in the body. It also picks up the carbon dioxide released by the cells. The carbon dioxide travels from the blood into the alveoli.

How does the oxygen and carbon dioxide get into and out of the lungs? Air moves in and out of the lungs as a result of **breathing**. We usually *breathe* without thinking about it. Your ribs and **diaphragm** (**die**-ah-fram) help you to breathe. Place your hands on your ribs and breathe in deeply (inhale). Now breathe out (exhale). When you inhaled, your ribs moved up and out. At the same time, your *diaphragm* moved downward. The action of both the ribs and the diaphragm increased the size of the chest cavity. This caused the air to rush into your lungs.

When you exhaled, your ribs and diaphragm moved back into place. The size of your chest cavity decreased. This pushed the air back out of the lungs. While the air was in the lungs, the exchange of oxygen and carbon dioxide took place. See Fig. 21-11.

CHART 21-1

	NORMAL EXHALE	FORCED EXHALE	DIFFERENCE
VOLUME			

Materials
Gallon jug marked off in 100 ml.
Water
Rubber tubing
1 soda straw
Pan
Grease pencil
Pencil
Paper

In this activity, you will measure your breathing capacity. Breathing capacity is the amount of air you can forcibly inhale and exhale. This can be done by measuring the amount of air you exhale.

A. Copy chart 21-1 on a sheet of paper. Set up the equipment as shown in Fig. 21-12. Mark the water level with a grease pencil. Insert a straw into the tubing labeled A. Be sure to keep the opening of the jug under the water. Take a normal breath without exhaling. Place your mouth on the straw and exhale normally into the

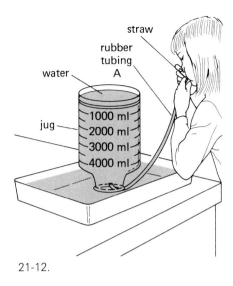

21-12.

straw. Hold your nose closed when you do this. The amount of water forced out of the bottle will equal the amount of air that you exhale. Now mark the level of the water.

1. Why must you hold your nose?

B. Record the volume of air in milliliters you normally exhale. This volume is the difference between the first and second marks.

C. Set up the equipment again. This time, take a deep breath without exhaling. Now place your mouth on the straw, hold your nose, and exhale into the straw. Keep exhaling until you cannot exhale any longer. Mark and measure the amount of water that was forced out of the bottle.

D. Record the amount of air you forcibly exhaled. Now record the difference between the amount of air you normally exhale and the amount of air you forcibly exhale.

2. Why do you think there is such a difference?

3. What type of person would you expect to have a large breathing capacity? Why?

Even after you forcibly exhaled, there is still air in your lungs. This is a reserve amount of air that is always in the lungs. This air is warm and helps to quickly warm the incoming air. It also prevents your lungs from collapsing.

You get energy for daily activities from your body cells. Each of your cells is like a little power plant. Glucose is "burned" in the mitochondria. The mitochondria store the energy from glucose in ATP. Oxygen needed for respiration is brought to each body cell by the respiratory system and the blood. Carbon dioxide is released from each body cell as a result of respiration. Carbon dioxide is also transported by the blood and released from the body through the respiratory system.

1. Explain how the cells in your body are like small power plants. What are these small power plants called?
2. How does the structure of the human lungs allow gas exchange?
3. Describe the pathway that air travels to and from your body cells.
4. How does the movement of the ribs and diaphragm help in breathing?

4 WASTES AND EXCRETION

What do the cars and the people in this photograph have in common? They both need fuel. Gasoline is used by cars, and food is used by people. Also, they both produce wastes. These wastes are given off into the environment. When you finish this section, you will be able to:

● Explain the importance of *excretion*.

● Name and locate the structures of the human urinary system.

● Describe the process of excretion as it occurs in the urinary system.

▲ Trace the pathway that *urine* travels.

Excretion: The process by which an organism gets rid of metabolic wastes.

Kidneys: Excretory organs which remove wastes from the blood.

As a result of the organism's activities, cells release wastes. If *wastes* build up, they become poisonous to the organism. Too many wastes prevent food and oxygen from entering the tissues. Therefore, organisms must get rid of these harmful wastes. The process in which organisms get rid of wastes is called **excretion** (ex-**skree**-shun).

In humans, the main excretory organs are the **kidneys**. Other excretory organs include the lungs and the skin. The *kidneys* are bean-shaped organs. Each kidney is about the size of your fist. They are located in back of the stomach area on either side of the spine. See Fig. 21-13. In the kidney, the blood passes through tiny blood vessels. The tiny blood vessels come in close contact with millions of tiny tubes in the kidneys. The liquid part of the blood (plasma) is *filtered* (separated) from the blood. Wastes, water, glucose, and minerals pass into the kidney. Blood cells and large protein molecules remain in the blood.

If all the materials filtered by the kidney were excreted, many essential substances would be lost. If all

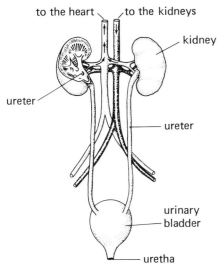

21-13. *The human excretory system.*

Urine: Liquid waste removed from the blood by the kidneys.

the water filtered by the kidney was excreted, a person would have to drink about 170 liters of water per day to replace it. Therefore, after the blood is filtered, another process must take place. In the kidney needed water, glucose, and minerals are returned to the blood. Only wastes are left to be excreted. The remaining wastes are called **urine**. *Urine* contains nitrogen wastes (urea), excess water, and inorganic salts. Urine also contains other substances that are in excess in the blood.

The cleaned blood passes out of the kidney through special blood vessels that connect the kidneys to a larger vessel leading back to the heart. The clean blood is carried back to the heart for circulation again.

The urine passes from the kidney to a tube called the *ureter* (**you**-rhe-ter). The ureter carries the urine to a muscular sac. This sac is called the *urinary bladder.* When the bladder is full it contracts. This pushes the urine into another tube called the *urethra* (yur-**reeth**-rah). The urethra carries the urine out of the body. The kidneys are always filtering the blood to remove wastes. The kidneys, ureters, urinary bladder, and the urethra make up the urinary system. See Fig. 21-13.

Materials
Pencil
Paper

The kidneys, ureters, urinary bladder, and urethra together make up the *urinary system*.

A. Draw and label a model of the human urinary system. Include all the structures that you have studied in this section.

B. Then trace the path that urine travels. Start with the kidney. Use arrows to show the direction the urine travels.

Living things give off wastes. These wastes are the results of the chemical activities that take place in the organisms's body. Organisms get rid of wastes by excretion. The human kidneys remove wastes from the blood. If not removed from the body, these wastes would soon cause death.

Match each structure in column I with its description in column II.

Column I	Column II
1. kidney	a. carries urine from kidney to urinary bladder
2. ureter	b. muscular sac for storage of urine
3. urinary bladder	c. bean-shaped organ that purifies blood
4. urethra	d. carries urine from urinary bladder out of the body

5 CONTROL SYSTEMS

Picture a busy TV newsroom. It receives news stories from all over the world. The news editors decide which stories are important. The news writers rewrite the stories for broadcasting. The news stories are then presented to the public.

In some ways, living organisms are like newsrooms. Organisms receive all kinds of information each day. The organism processes the information it detects and then reacts in a certain way. When you finish this section, you will be able to:

- Identify the structure and function of three types of *neurons*.

- Describe the path an impulse travels in a *reflex*.

- Describe the structure and functions of the human *endocrine system*.

- ▲ Demonstrate that a reflex is an automatic behavior.

Stimulus: Any change in the environment which causes a response in an organism. Stimuli is plural for more than one stimulus.

The ability to respond to a **stimulus** (stim-you-lus) is common to all living things. Responding to a *stimulus* requires three steps: (1) detecting a stimulus, (2) transferring the information, and (3) responding to the information.

Any change in the environment could be a stimulus. But, to become a stimulus, the change must be detected by the organism. Not all changes in the environment are detected.

Humans have a *nervous system*. This system allows us to detect and transfer information and respond quickly to stimuli. There are three different types of cells that work with our nervous system. One cell type receives the stimulus. These cells are usually located in the sense organs. Eyes, ears, nose, mouth, and skin are examples of sense organs. A second cell type transfers the information obtained from the stimulus. These are the nerve cells of the nervous system. The third cell type responds to the stimulus. Muscle cells are usually the third type of cell. These three types of cells work together.

Nerve cells are called **neurons** (kner-rons). A typical *neuron* has an enlarged area called the *cell body*. See Fig. 21-14. The cell body has one or more thread-like branches. These branches are called *nerve fibers*. Some nerve fibers are very long. The nerve fibers that run from your fingertips to your spinal cord are very long.

Neuron: The basic unit of the nervous system; a complete nerve cell.

Chapter 21–5 Control Systems

Impulse: Information from a stimulus; it travels along a nerve fiber.

Information, called **impulses**, travels along the nerve fibers. This is like your voice traveling along telephone cables.

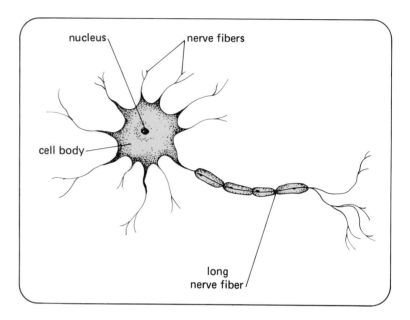

21-14. *A typical neuron.*

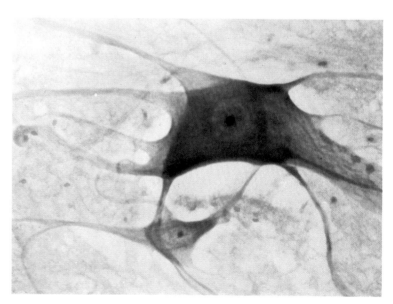

21-15. *A photograph of nerve cells taken through microscope.*

A whole nerve is a bundle of many separate nerve fibers. This is similar to a telephone cable containing many separate telephone wires. See Fig. 21-16. There may be thousands of fibers in a single nerve. Each of these fibers carries a separate *impulse*.

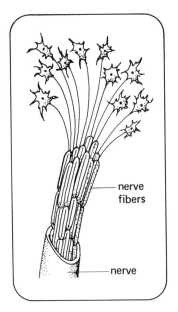

21-16. *How is a nerve like a telephone cable?*

Sensory neuron: A neuron that carries impulses from sense organs to the spinal cord or brain.

Motor neuron: A neuron that carries impulses from the brain or spinal cord to a muscle or gland.

Association neuron: A neuron that carries impulses from sensory neurons to motor neurons.

Reflex: An automatic response to a stimulus in which the brain is not directly involved.

In humans, impulses travel in one direction along definite pathways. These pathways include three special types of neurons: sensory, motor, and association. **Sensory neurons** transfer impulses from the sense organs to the spinal cord or brain. The nerve fibers from your fingertips to the spinal cord are some of the *sensory neurons*. **Motor neurons** transfer impulses to muscles and glands. Impulses that travel along *motor neurons* cause a response to the stimulus. **Association neurons** connect sensory neurons and motor neurons. In humans, *association neurons* are located in the brain and spinal cord. They transfer impulses between sensory and motor neurons. This is like the switchboard operator who transfers telephone calls.

One of the simplest pathways is a **reflex**. A *reflex* is an automatic response an organism is born with. A reflex happens so quickly that you do not think about it until after it happens. Reflexes include sneezing, coughing, and blinking. Pulling away from hot objects and jumping when frightened are also reflexes.

Let's trace the path an impulse travels in a reflex. Suppose you touch a hot object. The sensory nerve endings in your fingers receive the stimulus. The impulse travels along a sensory neuron to the spinal cord. In the spinal cord, association neurons transfer the impulse to a motor neuron. The impulse travels along the motor neuron to muscles in the arm. This causes the muscle in

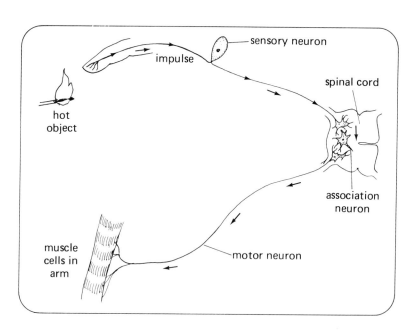

21-17. *Diagram of the pathway of a reflex impulse. How do the three types of neurons work together?*

your arm to contract. Your hand pulls away from the hot object. This happens so fast that you react before you feel pain. While you reacted, an impulse was sent to the brain. When the brain receives this impulse, you feel pain and realize what happened.

Materials
Pencil
Paper

Reflexes happen quickly because the impulse travels to and from the spinal cord. The brain is not directly involved. It receives impulses after the reflex action has occurred. You later realize that the reflex happened.

The knee jerk is also an example of a reflex. In this activity, you and a partner will demonstrate this reflex.

A. One person should sit relaxed with one leg crossed over the other. The other person will use the back edge of her or his hand to tap about 3-4 cm below the knee cap. This tap should be done firmly and quickly. The tapping must be done in the right place with the right amount of force. Otherwise, the reflex will not occur. You may have to try this a few times to get it right.

1. What is the stimulus for the knee jerk reflex?

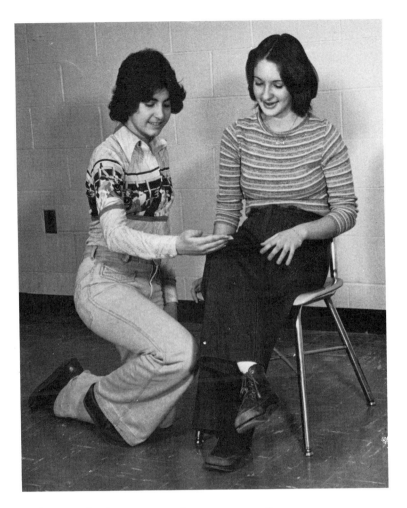

21-18. *The knee jerk reflex occurs when a certain area below the knee is tapped. Can you locate this area?*

   2. What is the response?

B. Draw a sketch of the pathway for the knee-jerk reflex. Label the stimulus, response, spinal cord, and types of neurons. Use arrows to show the direction the impulse travels.

   3. What neuron detects the stimulus?

   4. Where does this neuron carry the impulse?

   5. What neuron is the impulse transferred to next? What is the job of this neuron?

   6. What neuron carries the impulse that causes a response?

   7. Where does this neuron carry the impulse?

Reflexes happen much more quickly than the movements you have to think about. Can you remember when you first tried to hit a ball? It probably took some time before you could even swing the bat at the right time.

   8. What advantages do reflex reactions have over "thought reactions"?

   9. Would it be better if all our reactions were reflexes? Why or why not?

Chapter 21-5 Control Systems

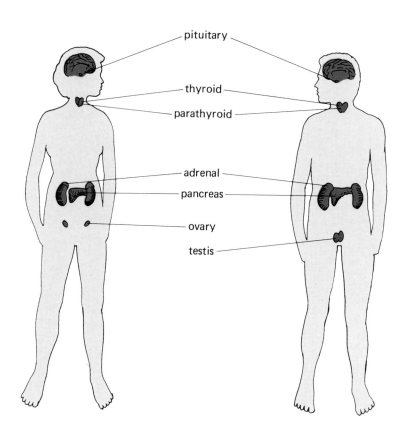

21-19. *The location of the human endocrine glands.*

Hormone: A chemical messenger that regulates and balances body functions.

Endocrine glands: Ductless glands that release hormones into the blood stream.

Feedback: An automatic "turn on" or "turn off" process that occurs when a certain level of hormone is reached.

Our body functions are not controlled only by the nervous system. Many are controlled by special chemicals called **hormones**. *Hormones* regulate life activities such as growth, behavior, and body development.

In humans, hormones are made by special glands. These glands are called **endocrine glands** (**en**-do-krin). *Endocrine glands* are different from digestive glands. They do not have a duct (tube) through which the hormone is released. Instead, hormones are released directly into the bloodstream. Hormones often work far from where they are made. Hormones are transported by the blood.

Only small amounts of hormones are needed to regulate body functions and maintain a balance between these functions. Too much or too little of a hormone can upset this balance. A pituitary hormone regulates the growth of the skeleton. If too much of this hormone is produced, a giant will result. Too little causes a midget.

A mechanism in the body, called **feedback**, regulates the amount of hormones produced. *Feedback* mechanisms are very complex. The signal to turn a gland on or off differs for different glands. Sometimes the signal is the amount of hormone itself. Sometimes the signal is the amount of the substance the hormone regulates. The nervous system also helps regulate the amount of hormones produced.

TABLE 21-2

GLAND	HORMONE	FUNCTION
Thyroid	thyroxine	Regulates release of energy in body. Iodine is needed to make thyroxine. A lack of iodine causes the thyroid gland to enlarge. This is called *goiter*.
Parathyroid	parathyroid hormone	Controls the body use of calcium.
Pituitary	growth hormone other pituitary hormones	Regulates growth of skeleton; controls the release of hormones from other glands; controls kidney function; regulates blood pressure.
Adrenals	adrenalin (epinephrine)	Prepares the body for emergencies by increasing heartbeat, blood flow, and the amount of sugar in the blood; activates the nervous system.
	cortisone	Regulates water and mineral balance in body tissues.
Pancreas	insulin	Allows liver to store sugar, and regulates body use of sugar. Too little insulin results in *diabetes*.
Ovaries (female)	female sex hormone (estrogen)	Controls development of reproductive organs and female characteristics.
Testes (male)	male sex hormone (testosterone)	Controls the development of reproductive organs and male characteristics.

SUMMARY & QUESTIONS

All living things detect and respond to certain stimuli. They are like newsrooms that receive news stories and report them. Humans have two systems that detect and respond to stimuli. The basic unit of the nervous system is the neuron. There are three types of neurons that carry information through the pathways of the nervous system. One of the simplest pathways is the reflex. Reflexes allow us to respond quickly and often protect us from serious injury. The endocrine system releases chemical messengers called hormones. Hormones help regulate and balance body functions.

Chapter 21-5 Control Systems

1. What is a neuron? Name the three types of neurons.
2. Suppose you touch a sharp object and you quickly pull your hand away.
   a. What is the stimulus and response in this situation?
   b. Describe the nervous pathway the impulse travels when this happens. Include the specific types of neurons and their functions.
3. Describe the structure and function of the human endocrine system. List 10 functions that hormones have in the human body.
4. How is the function of a hormone like the function of a building thermostat?

VOCABULARY REVIEW

Match the number of the word with the letter of the phrase that best explains it.

1. endocrine glands
2. mechanical digestion
3. breathing
4. excretion
5. chemical digestion
6. respiration
7. joint
8. ligament
9. stimulus
10. hormone
11. tendon
12. neuron

a. A chemical messenger.
b. A complete nerve cell.
c. Ductless.
d. A place where two bones meet.
e. A tough nonelastic tissue that attaches skeletal muscles to bone.
f. The process by which an organism gets rid of wastes.
g. Any change in the environment which causes a response in an organism.
h. Tough strands of elastic tissue that connect bone.
i. Exhaling and inhaling are part of this process.
j. The chemical process which simplifies food so it can be used by the body cells.
k. The process which uses oxygen to release useful energy from glucose.
l. The process in which food is broken down into small pieces.

REVIEW QUESTIONS

Choose the letter of the response which best completes the statement or answers the question.

1. All of the following have something to do with respiration except (a) exchange of gases (b) the release of energy from glucose (c) digestion (d) oxygen.
2. In the human lungs, gas exchange takes place in the (a) trachea (b) bronchi (c) bronchioles (d) alveoli.
3. A skeletal muscle moves a part of the body by (a) contracting (b) relaxing (c) pushing (d) becoming longer.
4. All of the following are endocrine glands *except* the (a) liver (b) ovary (c) pancreas (d) thyroid.
5. Movement of muscles is caused by (a) association neurons (b) glands (c) sensory neurons (d) motor neurons.
6. Impulses from the sense organs are carried to the brain along (a) association neurons (b) glands (c) sensory neurons (d) motor neurons.
7. Endocrine glands release their hormones into (a) special tubes (b) ducts (c) the blood (d) the urine.
8. The muscles of the digestive tract are (a) skeletal (b) cardiac (c) voluntary (d) smooth.
9. The human spinal cord is protected by the (a) skull bones (b) ribs (c) breastbone (d) backbone.
10. The elbow is an example of which kind of joint? (a) ball and socket (b) hinge (c) immovable (d) partially movable.

REVIEW EXERCISES

Give complete but brief answers to each of the following.

1. Why is digestion necessary? In what two ways are foods changed during digestion?
2. What are two functions of the large intestine?
3. Explain how energy is released in your cells.
4. Why is an artificial kidney important to someone who has kidney disease?
5. What are the organs of the urinary system? What are the functions of each?
6. How are the tendons of a person like the strings of a puppet?
7. Why do some muscles only work in pairs to move parts of the body?

CHAPTER

THE QUALITY OF OUR ENVIRONMENT

1 POLLUTION: OUR PROBLEM

Would you like to live in an environment that looked like this? No one wants to live in a place with dirty water and air, or with trash not picked up. Unfortunately, many people do not have a choice. How would you feel if all the places people lived looked like this? In this century, people have been polluting their environment faster than it can cleanse itself. All too often, efforts to clean up the mess have failed.

When you finish this section, you will be able to:

- Describe some of the causes of air, water, and solid waste pollution.

- Discuss some of the effects these types of pollution may have on living things.

- Suggest some steps that might be taken to reduce these types of pollution.

- Investigate the cause of a fish kill in the Sippiwisset River.

Has your city or town ever had a **smog** alert? There are some communities where warnings like this have to be issued often. On these days, people are warned to stay inside. Students are not allowed out during recess. This is done to keep people from breathing too much polluted air.

Smog is a serious problem in cities surrounded by hills or mountains. The city is in a basin. Sometimes, cold air puts a "lid" on the basin. This traps the warm air below. Auto and factory pollutants are kept in the basin polluting the air. Sunlight then causes reactions that make smog. At times, smog has become so bad that it has caused deaths. In London, New York City, and other places, "killer smogs" have occurred. Many thousands of people have become ill because of smog. It also causes a burning feeling in your eyes.

Pollutants come from many sources. Dust, sand, and salt from ocean spray are carried by the wind. The biggest source is the burning of fuels and wastes. Factories, automobiles, and even cigarettes add to the problem. These are sources of the gases sulfur dioxide, carbon monoxide, and nitrogen oxides.

Animals and plants are affected by these *pollutants*. The growth of many trees is slowed. Vegetable leaves wilt and the plant dies. Many pine trees in the California mountains are dying because of the Los Angeles smog. Air pollution causes respiratory infections, lung cancer, heart problems, and allergies in people.

Can anything be done about air pollution? Estimates show that over 60 percent of all air pollution is due to the automobile. Anti-pollution devices are now required on cars. These controls and mass transportation seem to be part of the answer. Stricter control on pollutants given off by factories and power plants is also

22-1. *More and more cities face smog problems like this. Laws controlling exhausts and burning of trash will help.*

Smog: An irritating mixture of gases, particles, chemicals, and fog.

Pollutants: Any materials such as gases, particles, and chemicals released into air.

22-2. The automobile is the biggest source of air pollution.

22-3. Oil spills such as this endanger fishing grounds and damage beaches and wildlife.

needed. These controls will cost money. That cost will probably result in higher prices for the goods the factories produce.

Did you ever see a river burn? Does this question seem ridiculous? It is not. In 1970, the Cuyahoga River in Ohio, brown from industrial wastes, actually did burn. Industrial wastes, pesticides, chemicals from mines, and sewage poison water organisms. Sometimes the organisms die immediately. Sometimes the poisons build up in their bodies over a long period of time. Eventually, a fatal level is reached. Chemicals and poisons can be passed on through food chains. Higher level consumers may gradually build up harmful levels from the organisms they eat.

A number of oil tankers have collided and sunk near our coasts. The oil they contain spills out and pollutes the oceans. Fish are killed and their breeding grounds destroyed. Clam and oyster beds are ruined and our beautiful seashores damaged.

Recently a new kind of pollution has come into being. Power plants and many factories use water as a coolant. Water is taken from rivers to cool generators. Hot water is then dumped back into naturally cool rivers.

What effects does this temperature change have on living organisms? Many fish are killed by the warm water. Others survive but don't reproduce. In addition, warm water holds less dissolved oxygen than cold. Thus there is less oxygen for the fish to use. Temperature barriers in some rivers prevent salmon from returning upstream to spawn.

Materials
pencil and paper

It is late summer. The water level in the Sippiwisset River is at its lowest level in five years. The river water has reached its highest annual temperature of about 30°C. The Nanatuc Power Plant uses about 1,134,000 liters of river water every minute to cool its generators. This water is then dumped back into the river.

Three days earlier, thousands of dead fish were found downstream from the plant. Most of the fish were herring and perch. There were a few salmon, bass, and shad. Imagine you are a biologist for the state Wildlife Commission. You have been assigned to investigate the fish kill. As part of your research, you have taken temperature readings on the river near the power plant. The results are shown in Fig. 22-4.

A. Trace the diagram on a piece of paper.

B. Shade in the parts of the river which have temperature ranges of 37°, 36°, and 35°. Use a different color for each temperature zone. The temperature zones in the cooling canal have already been shaded. Using your sketch and the information in Table 22-1, answer the following questions.

1. What appears to be the probable cause of the fish kill?

2. Why were most of the dead fish herring and perch?

3. What temperature range would be the most deadly? What part of the river would be deadly to bass and shad?

4. Why would this present a problem to the fish? Which of the fish are least sensitive to temperature changes?

22-4. All temperatures are in °C.

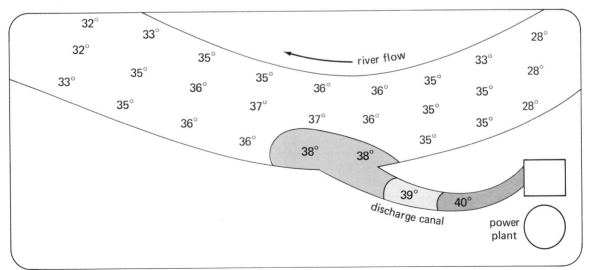

Chapter 22–1   Pollution: Our Problem

TABLE 22-1: **Limiting factors for certain fish.**

FISH	MAXIMUM TEMPERATURES		PREFERRED DISSOLVED OXYGEN RANGE
	PREFERRED RANGE	DEADLY TEMP.	
HERRING	19–22°C	33°C	10+ ppm*
PERCH	20–22°C	33°C	10+ ppm*
ATLANTIC SALMON	13–16°C	34°C	10+ ppm*
BASS	28–31°C	37°C	10+ ppm*
SHAD	23–25°C	38°C	7+ ppm*
CARP	31–33°C	40°C	2+ ppm*
SHINERS	19–21°C	34°C	10+ ppm*
KILLIFISH	16–28°C	34°C	8+ ppm*

*ppm = parts oxygen in a million parts of water

5. Examine Table 22-1, carefully. As the temperature of the river rises, what happens to the level of dissolved oxygen?

6. What types of fish would find the dissolved oxygen level around the mouth of the canal a problem?

7. Would any of the fish listed prefer to live in this part of the river?

TABLE 22-2
Amount of oxygen that dissolves in water

TEMP. °C	MAX. AMT.
0	14.6
5	12.7
10	11.3
15	10.1
20	9.1
25	8.3
30	7.5
35	6.9
40	6.5

SUMMARY & QUESTIONS

People have not always taken care of the environment in which they live. As population and industries grow, pollution becomes more of a problem. Much has been done to solve these problems. Much is left to do.

1. Make a list of the causes of air, water, and solid waste pollution in your community.
2. List several ways you could reduce the amount of paper waste your family throws away.
3. What are some of the effects air and water pollution have on living things?
4. Pesticides sprayed on fields have been found in ocean fish. Explain how this could happen.
5. How can pollution be reduced? What could you as an individual do to reduce pollution in your community?

2 THE SIZE OF THE PROBLEM

In the last section, we studied some of the causes and effects of pollution. We also examined some possible solutions. What about your environment? Does it have its own pollution problems? Is your school troubled by litter, dust, or other pollutants? In this section, we will examine some of these problems. When you finish this section, you will be able to:

● Describe the problem of waste pollution.

▲ Calculate the amount of trash produced by your school community.

▲ Collect samples of particles found in the school's air.

▲ Perform several tests on water samples.

One of the most serious problems today is what to do with the wastes. It is estimated that in 1920, each American produced 6.6 kg of solid wastes per day. By 1970, that estimate had risen to 13.2 kg per day. But in 1920, there were only 100 million people in the United States. As of January 1, 1977, the population had risen to almost 217 million. The amount of trash people throw away each day is fantastic. Think of how much that would be in a year.

About 10 percent of this waste is burned in incinerators. Ash particles produced by burning can seriously increase air pollution. Most of the remaining trash goes into dumps. But there is less and less space to dump this waste. Large cities have serious problems in finding places to put their waste. Many cities have been using the oceans as dumps. This is not the answer. These wastes are damaging the ocean ecosystems.

What can be done to reduce the amount of trash? The most logical answer seems to be to reuse as much material as possible. We will never completely eliminate waste, but it can be reduced. Most glass and metal containers can be made to be reused or recycled. Burnable trash can be used to generate electrical power. Small scale experiments show this to be practical.

Some materials are reused now. The steel in cars, lead in storage batteries, and some plastic and rubber are recycled. About one third of the paper used today is made from recycled waste paper. Some trash can be converted to paving or building materials. Presently, the problem with recycling is that it costs too much. Until it becomes less expensive or until we face critical shortages, we will be a "throw-away society."

22-5. (left) Trash is used as landfill to make room for housing and recreation areas. Unfortunately the habitats of many organisms may be destroyed.

22-6. (right) Recycling of most materials is not yet profitable. These cans will be collected and reused.

Materials
Several large trash bags
Pencil
Paper

PART I

A. Collect all the trash that is left in your room at the end of each day. Place everything you collect into a large plastic bag. At the end of a week, weigh the trash.

1. How many kilograms of trash did you collect in the week?

2. At that rate, how much would accumulate in a year?

B. Assume that your whole school wasn't cleaned for a year.

3. How much trash would accumulate in all the classrooms in that time?

C. Find out from your teacher how many students there are in school. Divide the number of students into the answer for 3.

4. What is the average amount of litter per student per year?

22-7. *How much of the trash you collected is yours?*

Your figures are only an estimate. They give you an idea of the "size of the problem." What happens to the trash after it is collected? Your school is only one small part of your community. How does your community dispose of its trash? Is it burned, buried, or recycled?

D. Carefully examine the trash you collected. Sort the trash into the following groups: paper, wood, metal, plastic, glass, other. Weigh each group.

5. Which group weighs the most?

6. Which types of trash do you think could be recycled for future use?

E. Examine the waste paper.

7. Were most of the sheets of paper completely used?

8. How could you reduce the amount of paper thrown away in your school?

PART II

Let's continue to examine the classroom for evidence of other types of pollution.

A. Look carefully at the windows, windowsills, and any other spot where dust might settle. Do any places seem to be dustier than others?

B. Cover 1/4 of each of three glass slides with masking tape. Label each with your initials and the date. Label one slide FLOOR, another 1 METER, and the third, 2 METERS. Cover the rest of each slide with tape that is sticky on both sides.

C. Place the slides in a quiet corner of the room at the stated height. After three days, examine the slides with a magnifier.

9. Which slide has the most particles on it? Which has least?

10. Which has the largest particles? Which the smallest?

11. Try to identify the types of particles. They may include sand, dust, hairs, fibers, metal shavings.

D. Place several slides throughout the room at the 1 meter level. Make sure the location is marked on each slide.

Materials
3 glass slides
magnifier
double-sided tape
masking tape
pencil
paper

Chapter 22–2 The Size of the Problem

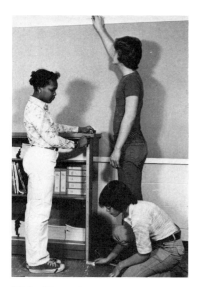

22-8. *Place the slides where they will be exposed but will not be accidentally broken.*

Materials
pH paper and color scale
microscope slides
medicine dropper
ring stand and ring
filter paper
toothpicks
microscope
2 beakers
funnel
water samples
pencil
paper

12. Which part(s) of the classroom seem to have received the most particle pollution?
13. Can you determine why?
14. Can you determine the source(s) of the particles?

PART III

Most types of water pollution require chemical tests to detect. There are several simple tests you can perform on water samples.

A. Obtain the materials listed in the margin.

B. Make a copy of TABLE 22-3. You could extend the table for additional water samples. Perform the tests indicated on each sample. Do not write in this book.

C. Determining pH is a way to measure how acid or basic a water sample is. The pH scale runs from 0 (strong acid) through 7 (pure water) to 14 (a strong base). A reading of 7 means the water is neutral. The closer to 0 the number is, the more acid the water is. The closer to 14 the stronger the base is.

D. With a toothpick, place a drop of the water sample on the pH paper. Compare the color produced with the scale on the dispenser. Record the pH in the table.

TABLE 22-3 **Water Pollution Data Table**

SAMPLE #	pH	COLOR	ODOR	MICROSCOPIC OBSERVATIONS	FILTER OBSERVATIONS
1					
2					
3					
4					
5					

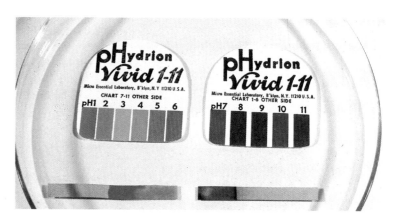

22-9. *Match your test paper to the dispenser scale as accurately as possible.*

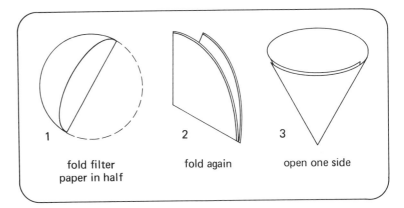

22-10. *(left) Do not fill the funnel over the top of the filter paper.*

22-11. *(right) Fold the paper in half, then in half again. Open so that 3 folds are on one side, one fold on the other side.*

E. Hold the sample against a white background. Record its color and shade as either light, medium or dark.

F. Smell the sample. Record your observations.

G. Prepare a microscope slide of the water. Describe any organisms or materials you see.

H. Filter the water sample. Figure 22-10 shows how to arrange the materials. Figure 22-11 shows how to fold the filter properly. Pour enough water into the filter paper cone to fill it ⅔ full. Allow the water to drain through before refilling. Describe the color of the filter paper and any material on it.

America's solid waste pollution problem is growing. More people produce more solid waste each year. Methods used to get rid of these waste materials pollute the environment. Recycling may be one way to reduce the solid waste problem.

1. How did the amount of solid waste produced in this country increase between 1920 and 1977?
2. What materials are now being recycled?
3. How can you reduce the amount of trash you throw away each day?

Chapter 22-2 The Size of the Problem

3 HANGING IN THE BALANCE

How is your EQ? EQ stands for Environmental Quality. The National Wildlife Federation uses the EQ Index as a way of measuring the quality of the environment. Fig. 22-14 shows the EQ statistics through 1976. A perfect environment would rate 100 in every category. The index is based on the opinions of environmental experts. When you finish this section, you will be able to:

● Describe and list several examples of renewable and nonrenewable resources.

● Explain the need for *conservation* of natural resources.

● Give examples of the use and misuse of *natural resources*.

▲ Locate and display articles relating to conservation of natural resources.

22-12 and 22-13. *Compare these scenes. Which one is most like the roadsides in your community?*

Unit V What Makes Up Our Living World?

Natural resource: Any natural substance of use to humans.

All organisms take matter and energy from their environments. Sooner or later, everything that is taken is returned—everything except energy. Any substance useful to humans in the environment is called a **natural resource**. Animals, plants, soil, air, water, coal, and minerals are all natural resources.

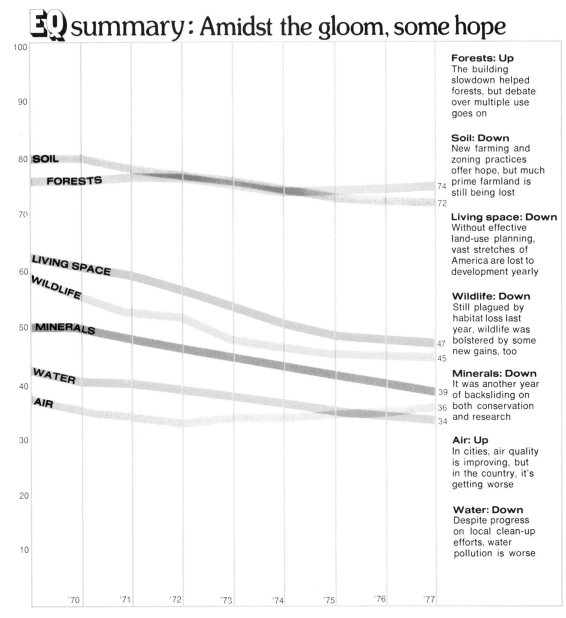

22-14. *Environmental Quality Index.*

Chapter 22-3 Hanging in the Balance

Many *natural resources* are renewable. Renewable resources are those that can be replaced in a reasonable length of time. They include plants, animals, air, and water. Plants and animals reproduce to replace their own kind. Air and water are replaced every time they are recycled.

Nonrenewable resources are those that cannot be replaced. They include metals, minerals, oil, and coal. Some natural resources take a very long time to be replaced. As a result, they can only be used once. These resources are also called nonrenewable. It takes over 300 years for one centimeter of topsoil to be replaced. It is obvious that soil is a nonrenewable resource.

Our natural resources are not always used wisely. For many years, they have been wasted. We have destroyed forests and wildlife. We have polluted the air and water. Our increasing numbers demand more and more resources. The **conservation** (cahn-ser-**vay**-shun) of these resources is up to everyone.

Conservation: The protection and wise use of natural resources.

Earlier sections dealt with the problems of air and water quality. The EQ Index indicates that air quality shows a slight improvement. However, water quality is still declining.

Since the colonial days, over 60 species of animals have become **extinct** in the United States. There are currently about 170 animals now listed as **endangered** species. See Table 22-4. Once an organism is *extinct*, it is gone forever. It cannot be renewed.

Extinct: An organism that no longer exists on earth.

Endangered: An organism that is in danger of becoming extinct.

Of course, many species of plants are also *endangered*. There are about 1700 species that are now considered endangered. This means that the species is protected by

TABLE 22-4

SOME EXTINCT NORTH AMERICAN ANIMALS	
passenger pigeon	ivory-billed woodpecker
heath hen	Carolina parakeet
great auk	Labrador duck
SOME ENDANGERED NORTH AMERICAN ANIMALS	
whooping crane	puma
American bald eagle	musk ox
American crocodile	trumpeter swan
California condor	key deer
grizzly bear	blue whale

22-15. *Wildlife refuges provide secure, protected natural environments for many organisms.*

Note: 1 hectare = 2.47 acres

22-16. *It is difficult to reach a balance between living space for people and other land uses.*

law. Under such protection, the number of some species has increased. As a result of this increase, the chances of survival have also increased.

Human activities affect wildlife in many ways. Hunting and pollution take their toll. One of the biggest problems is the loss of the organism's habitat. As cities, towns, and farms expand, the land available to the wildlife decreases. So does the wildlife.

Fish and game laws restrict the numbers of animals that are hunted. Wildlife refuges provide homes for many animals. These and other *conservation* programs must be expanded to protect our wildlife.

People need land, too. They need it for homes, growing food, roads, and recreation. There is a limited amount of land available. Thus, when one use increases another decreases. In 1976, urban areas increased by more than 300,000 hectares. Other projects, including roads, gobbled up over 100,000 hectares of rural land. Most of the land being lost is good farmland. New laws are aimed at preserving this vital resource. Hopefully, these laws will show results soon.

Unfortunately, laws cannot stop the erosion of topsoil from the land. Each year about 4 billion tons of good soil are carried away by wind and water. However, soil is always being gradually replaced. Rocks exposed to the weather eventually break down to form new soil. Left alone, the removal and replacement of soil just about balances out.

Many human activities increase the rate at which soil is lost. The clearing of grasslands and forests can be very harmful. Natural ground cover, which keeps the soil in place, is removed. This allows rain and wind to rapidly wear away the good topsoil. In the past 300 years, about one-third of the soil in the United States has been lost.

22-17. Scientists estimate about ⅓ of the topsoil has been lost to erosion. What can be done?

22-18. (left) Open-pit copper mine, Bingham Canyon, Utah. Minerals, such as copper, are nonrenewable and cannot be replaced.

22-19. (right) Increased use and waste have led to an energy shortage.

Better soil conservation will help prevent such soil losses. Farming practices such as contour plowing, strip farming and terracing help slow erosion. These systems are not usable on all farms. New methods must be found.

Minerals in the earth's crust are nonrenewable resources. For example, the earth cannot replace the iron, sodium, copper, and other materials we use. Some metals can be recycled. Other materials cannot be reused. At present, not enough recycling is being done. It simply costs too much.

Coal and oil are other nonrenewable resources. Both coal and oil formed from the remains of plants buried in the earth's crust. The process took millions of years. Scientists believe that our oil supply will be used up in a few hundred years. The supply of coal will last longer.

Many people are becoming very concerned about an energy shortage. One solution is to use renewable energy sources. These might include energy from the sun, wind, running water, and burning garbage. Others believe that atomic energy is the solution.

A very large part of the energy crisis is the way we waste energy. Think of the gasoline that is wasted by cars that get poor gas mileage. Count the number of cars that pass on a highway with only one person in them. Many of our throw-away containers are made of plastic. Plastic is made from oil. Fuels are burned to provide electricity. How much electricity do you waste in a day? Tonight, check the number of lights, radios, T.V. sets, and other appliances being used at home. Are all of them used wisely? Is there any way you could conserve our natural resources?

Materials
(Will vary.)

A. The use and misuse of natural resources is always in the news. Collect articles, photos, and cartoons that are related to this problem from newspapers and magazines. You may add your own photos or cartoons.

B. With the rest of your class, make a bulletin board display. As a background, sketch a large, equal-arm balance. Label the bulletin board "The Environment–Hanging in the Balance." On one side, place articles that deal with conserving natural resources. On the other side, put the articles that deal with the misuse of these natural resources.

C. When finished, have a class discussion on what the display shows.

People are among the few organisms that can upset the balance of nature. We demand more and more of the resources our environment has to offer. Until very recently, we paid very little attention to the effects of our demands. We must learn to conserve and protect these resources. In many cases, once they are gone, they are gone forever.

1. Make a list of renewable and nonrenewable resources you used today.
2. List several ways we misuse our resources.
3. Suggest at least four things that you personally could do to conserve energy.
4. Wolves are classed as an endangered species. What does that mean? Why should we be concerned about their possible extinction?
5. Some scientists feel that humans should be classed as an endangered species. What does this mean?

REVIEW QUESTIONS

Choose the letter of the statement that best completes the statement, or answers the question.

1. Which of the following parts of our environment faces serious pollution problems? (a) air (b) fresh water (c) oceans (d) all of these parts.
2. The wise use of our natural resources is called (a) renewable (b) pollution (c) conservation (d) endangered.
3. A species that no longer exists on earth is called (a) extinct (b) endangered (c) threatened (d) dead.
4. Anything people take from their environment for their own use is called a (a) renewable resource (b) nonrenewable resource (c) energy source (d) natural resource.
5. A thick cloud of gases and particles that pollutes the air may be called (a) fog (b) smog (c) haze (d) daze.
6. If a piece of pH paper is dipped into a strong acid solution, the color change would indicate a number closest to (a) 14 (b) 9 (c) 5 (d) 0.
7. Which of the following is being considered as a way of using trash? (a) as a fuel (b) as building material (c) recycling glass and metal (d) all of these.

REVIEW EXERCISES

Give brief but complete answers to the following questions.

1. Why should people be concerned if a few animals or plants are in danger of becoming extinct?
2. What effects might cooling water from a power plant have on the life in a river?
3. Why might a city in a valley be more likely to have a smog problem than one on the open plains?
4. How could using the oceans as dumps damage that ecosystem?
5. Penguins in Antarctica have been found to have the pesticide DDT in their body tissues. Explain how it got there.
6. How could students and teachers in your school reduce the amount of paper they throw away?
7. Soil is continually being formed by natural processes. Why then is it considered a nonrenewable resource?
8. How does the expansion of urban and suburban areas contribute to the extinction of wildlife?

APPENDIX How To Use This Text

TABLE OF CONTENTS

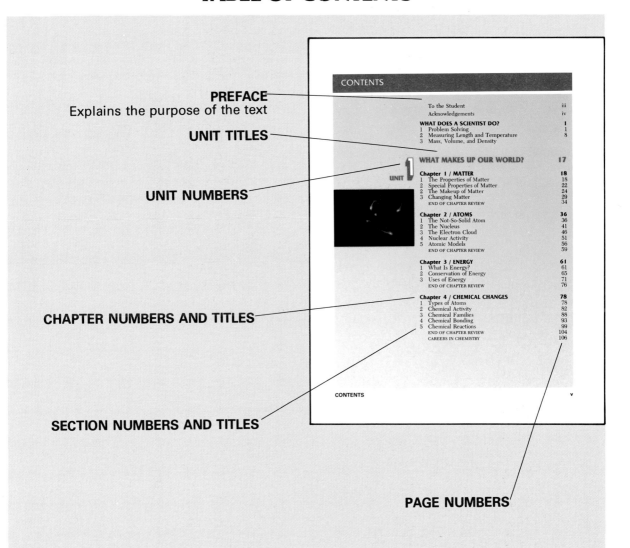

PREFACE — Explains the purpose of the text

UNIT TITLES

UNIT NUMBERS

CHAPTER NUMBERS AND TITLES

SECTION NUMBERS AND TITLES

PAGE NUMBERS

EXERCISE

Using this text, answer the following questions. Do not write in this book. (Spaces for answers can be found in section A of the *Holt General Science Resource Book*.)

1. Can the preface be found before or after Chapter 1?
2. What are the five major areas discussed in this text?
3. How many chapters are in this text?
4. On what page does Chapter 16 begin?
5. How many pages are there in Unit 3?

SECTION OPENER

PHOTOGRAPH OR ART
Each section opens with a photograph or piece of art that is tied into the opening paragraph and related to the contents of the section.

2 THE NUCLEUS

How small are atoms? That is very hard to answer. Hydrogen atoms are so small that it would take more than two billion to stretch across this page. This paper is only about one million atoms thick. Yet the atom is mostly empty space.

Imagine that you are in the top row of seats in a large stadium. With powerful binoculars you look at a baseball lying on the pitcher's mound. Suddenly you see an ant on the ball. If an atom of hydrogen were the same size as the stadium, you would be sitting in the path of the electron, looking at the ant-sized nucleus. When you finish this section, you will be able to:

• Define the terms *atomic number* and *atomic mass*.

• Describe the *neutron*.

• Draw the structure of the atom when given its *atomic number* and *mass*.

▲ Use a simple model to show the location of *subatomic particles* within an atom.

Chapter 2–2 The Nucleus   41

OBJECTIVES
The objectives state the important concepts to be learned in each section. The objectives marked by a ▲ relate to a skill to be learned in the activity.

EXERCISE

1. When you finish studying section 2 of Chapter 5, what will you be expected to be able to do?
2. What is the objective related to the activity at the end of section 1 of Chapter 12?
3. What is pictured in the photograph at the beginning of section 2 of Chapter 5?
4. How is the photograph at the beginning of section 2 of Chapter 5 related to the contents of that section?
5. What is the title of section 3 of Chapter 20?

TEXT PAGES

BOLDFACE TYPE
Boldface type is used to point out new terms being introduced.

MARGINAL DEFINITIONS
Given for each new term.

ILLUSTRATIONS
Help to make ideas easier to understand. Each illustration has a figure number that begins with the number of the chapter it is found in.

ITALIC TYPE
Words or phrases related to the understanding of new material are printed in *italic type*.

or soft coal. It takes about 10 meters of peat to make 1 meter of bituminous coal. Bituminous burns with a somewhat smoky but hotter flame than lignite. Most of the coal in the United States is bituminous.

In a few places, the formations of bituminous were squeezed and compressed even more. This happened mostly in areas where mountains were forming. The pressures and temperatures were very high. Under these conditions, bituminous becomes *anthracite* (an-thruh-site). Anthracite is a hard, black, and brittle form of coal. It burns with a blue smokeless flame and gives off much heat. Anthracite makes up only about 2% of the coal reserves in the United States. Most of this is located in the Appalachian Mountains.

The other two major fossil fuels are **petroleum** and **natural gas**. The formation of these two fuels is not yet fully understood. Both are thought to form from the decay of organisms that once lived in the sea. The remains of these animals and plants settled to the sea floor and were gradually covered by sediments. After millions of years, the remains were slowly changed by heat and pressure into a substance made mostly of hydrogen and carbon. Such substances are called **hydrocarbons**. *Petroleum*, which is often called oil, and *natural gas* are mixtures of many different *hydrocarbons*.

The tremendous pressure deep underground forces the oil and natural gas out of the rock material in which they usually form. These substances begin to move upward through spaces or fractures in the rocks. Often they meet a rock type or a formation they cannot pass through. In this way, they are trapped and form *oil pools* and *gas pockets*. Water, also trapped in the rocks, often forces the

11-2. *Chemical changes convert peat into lignite, then to bituminous, and finally to anthracite coal.*

Petroleum: A black, oily liquid formed from the remains of ancient sea life.

Natural gas: A colorless, odorless gas formed from the remains of ancient sea life.

Hydrocarbons: Substances made up only of carbon and hydrogen.

11-3. *Oil and natural gas are often found together in underground traps or pools.*

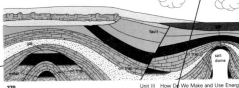

270   Unit III   How Do We Make and Use Energy?

EXERCISE

1. What is illustrated in Figure 21-14?
2. How is Figure 21-14 related to Figure 21-15?
3. What new words are defined on page 384?
4. Why is the term *electromagnetic spectrum* printed in italic type on page 196?
5. What three terms on page 336 are related to the understanding of sedimentary rock?

SUMMARY AND QUESTIONS

ACTIVITIES

TABLE 1

Object	Diameter (Earth = 1.0)
Sun	110.0
Mercury	0.4
Venus	1.0
Earth	1.0
Mars	0.5
Jupiter	11.2
Saturn	9.4
Uranus	4.0
Neptune	3.9
Pluto	0.2

TABLE 2

Planet	Av. Distance from Sun (Earth = 1.0)
Mercury	0.4 AU
Venus	0.7
Earth	1.0
Mars	1.5
Jupiter	5.2
Saturn	9.5
Uranus	19.2
Neptune	30.1
Pluto	39.5

C. Use a compass to draw circles on paper to represent each planet. Cut out the circles.

D. On your model, it would be impossible to use the same scale to show the size of the planets and their distance from one another. If you tried to do this, you would either have to make some circles so small you could hardly see them or you would have to place Earth 2,300 cm from the sun. For this reason, you will use a different scale for distance here.

E. Earth is about 150 million km from the sun. This distance is called an Astronomical (as-truh-**nom**-ih-kul) Unit (AU). Table 2 gives the distances of

the planets from the sun in AU's.

3. How far from the sun would Earth be on this scale?

4. How far from the sun would Jupiter be?

5. How long a piece of adding machine tape would you need to stretch from the sun to Pluto?

F. Assume the sun is at one end of this piece of tape. Measure off the distances from the sun to each planet using the information in Table 2. Attach your model of each planet in its proper place on the tape.

6. Note the large gap between Mars and Jupiter. What objects belong in this gap?

G. Label your model.

SUMMARY & QUESTIONS

Humans have learned more about our solar system in the past twenty years than in all recorded history. We have come to know the planets as other worlds on which many of Earth's processes are also at work. However, there are enough differences and puzzles to keep humans exploring and questioning for centuries to come.

1. Name five different types of objects that form the family of the sun.
2. Sketch and label a cross section of the sun.
3. List two major differences between the inner and outer planets in terms of their structures.
4. List five ways the inner planets are similar to each other.

424   Unit IV   How Is Our Planet Changing?

1-11. (top) Burning is a chemical change.

1-12. (bottom) Rusting is a chemical change. The dents are physical changes.

Chemical change: A change in matter in which new molecules are formed.

The materials that make up a mixture can usually be separated by a simple physical change. If you do not like the pepper and onions on the pizza, you can pick them off. It would be difficult to separate the cheese from the tomato sauce, but not impossible. Even the salt in the water can be removed by a physical change. If the solution is heated, the water can be boiled away, leaving the salt behind. These physical changes would not change the makeup of the molecules of each substance.

Changes in the second group are different. In these cases, the chemical makeups of the substances change. The molecules are different. They may break apart. New molecules are formed. New substances are produced. These are called **chemical changes**. Think of what happened to the wood in the fire. The ashes and soot are no longer wood. The molecules of wood have been changed into new substances. Some of these were given off as gases as the wood burned. Others were left as ash and soot.

Burning gasoline in the engine of a car is a *chemical change*. The gasoline molecules are broken down and new molecules are formed. Energy is released. Think of all the different materials we use that are not natural. Most were produced by chemical changes.

Some chemical changes need heat to occur. Others, such as rusting of metal, do not need heat. A piece of iron left outdoors will rust. The iron atoms join with oxygen atoms in the air to make a new compound.

Often both chemical and physical changes occur together. The fire we discussed earlier is one example. So is the peeling of paint on a house. Some of the changes are chemical, some are physical. The next time you see an old car, look for examples of both types of changes.

A. Obtain the materials listed in the margin.

B. Break one wood splint in half. Break these pieces into smaller pieces. Break the wood into the smallest pieces you can.
1. Are the smallest pieces you have still wood?

2. Is this a physical or a chemical change? Why?

C. Attach the large candle to the glass plate with a few drops of melted wax.
CAUTION: WEAR SAFETY GOGGLES AND TIE BACK LONG HAIR.

D. Place some pieces of the

Unit 1   What Makes Up Our World?

SUMMARY
Reviews the main ideas covered in the section.

QUESTIONS
Allow you to review and test what you have learned in the section.

ACTIVITY
Further develops concepts covered in the section. Each procedure is lettered and each question is numbered.

NOTES OF CAUTION
Warn of possible dangers that may result from carelessness.

EXERCISE

1. What are the main ideas covered in section 2 of Chapter 3?
2. How is the activity in section 3 of Chapter 3 related to the contents of that section?
3. What skill would you be learning if you did the activity in section 1 of Chapter 18?
4. How many procedures are there in the activity in section 3 of Chapter 9?
5. What precautions should you take when doing the activity in section 3 of Chapter 1?

GLOSSARY AND INDEX

GLOSSARY

abiotic: Refers to the nonliving materials and energy in an ecosystem.
absolute age: A term referring to age in approximate years as determined by radioactive dating.
absolute zero: The temperature (-273°C) at which particles in matter stop moving.
abyssal plain: A wide deposit of sediment that covers the deepest part of the ocean basin.
acceleration: The change in speed during a given interval of time, calculated by dividing the change in speed by the time it took for the change in speed to happen.
adaptation: A trait that enables an organism to survive better than similar organisms.
air mass: A large mass of air that has taken on the temperature and humidity of a part of the earth's surface.
algae: Protists that contain chlorophyll and make their own food.
alkali metals: A group of elements whose atoms all have one electron more than the stable number.
alternating current: An electric current that changes direction in an electric circuit.
ampere (amp): A measure of the number of electrons moving past a given point in an electrical circuit in 1 sec.
amplitude: The strength of a wave. One half the distance from the bottom of the trough to the top of the crest.
angiosperms: Vascular plants with seeds covered by a protective tissue.
anticyclone: A large mass of air spinning out of a high-pressure area.
area: The amount of surface within a given set of lines, measured in square metric units (m², cm², mm², etc.).
asexual reproduction: Reproduction that requires only one parent.
association neuron: A neuron that carries impulses from sensory neurons to motor neurons.

asteroids: Chunks of rock and metal that orbit the sun between Mars and Jupiter.
astronomer: A scientist who studies the stars, planets, and other heavenly bodies.
atmosphere: The layer of gases that surrounds the earth.
atom: The smallest particle of an element that has all the properties of that element.
atomic mass: The sum expressed in amu of the mass of protons and neutrons in an atom.
atomic mass unit (amu): A special unit used to express the mass of atomic particles and atoms.
atomic number: The number of protons in the nucleus of an atom.
atomic particles: The building blocks of atoms.
ATP (adenosine triphosphate): A compound that stores energy and makes it available to the cell.
autumnal equinox: The first day of fall. Daylight and darkness are the same length.
axis: The imaginary line through the center of the earth about which the earth rotates.

BTU (British thermal unit): The amount of energy needed to raise the temperature of 250 g of water 1°C.
bacteria: Very small protists with simple cell structure and no nucleus.
batholith: A large body of igneous rock formed underground.
binding energy: A powerful force that holds the nucleus of an atom together.
biomass fuel: Any plant or animal matter that is used as a source of energy.
biosphere: That area at or near Earth's surface where life is possible.
biotic: Refers to the living organisms in an ecosystem.
blood: Specialized tissue that transports materials to and from the body cells.
boiling point: The temperature (at ordinary air pressure) at which the particles of a liquid have enough energy to become a gas.

574   Glossary

---

process of, equation showing, 537
respiratory system, human, 538
response, 442
Reston (Virginia), 484
reusable materials, recycling of, 433, 558, 566
ribosomes, 450, 455
Richter Scale, 293
ridge, formation of, 308. See also mid-ocean ridges
rift, 306
ringworm, 458
ripple tank, 174, 177
rock candy, 134
rock collection, 328–329
rock crystals, 327, 330, 333, 334
rock cycle, 344, 344, 345
rock salt, 337, crystals of, 135, 136
rocks, faults in (see faults); folding of, 313, igneous, 329, 330–335, 341; joints in, 312; layers of, 347, 348–349, 350–352; metamorphic, 329, 330, 340–341, 342–345; moon, 408, 412–413; Precambrian, 360–361; sedimentary, 329, 330, 336–339, 341, 344; unconformity of, 350, 352
roller coaster, 69
rope, waves in a, 169, 171, 175
roundworms, 464
rubber, 264
rubidium, 91–92
Rumford's cannon boring experiment, 111–112
rural community, 488
rusting, 32
Ruth, Babe, 154

salamanders, 463, 486
salinity, 394; of sea water, 394, 397
saliva, 531–532
salivary glands, 531, 532
salt, table (see sodium chloride)
salt solutions, 30

San Andreas fault, 291
sand, 26, 336, 337, 347
sand dollars, 464
sand dunes, 318; layers in, 348
sandstone, 336, 338, 343
sandworms, 464
saturated air, 377
saturn, 418, 419, 420, 422–423; characteristics of (table). 419; moons of, 420
schist, 345
science fiction, 439
Scientific Law, 7
scientific model(s), 38
scientific problem solving, steps in, 8
scientific thinking, 1–5
scorpions, 463
screw, 163
sea anemones, 464
sea floor, exploration of, 394–395; new, creation of, 314, 332, 397; North Atlantic Ocean, 396; organisms of, 432; profile of, 395; sediments on, 397; spreading of, 306–307; trenches on, 396
sea urchins, 464
sea water, density of, 397–399; evaporation of, 394, 397; salinity of, 394, 397
seasons, causes of, 405–406
secretion, 422
sediment(s), 334–335, 336–340; chemical, 337, erosion and, 344; layers of, 336, 341; organic, 337; on sea floor, 397
sedimentary rocks, 329, 330, 336–339, 341, 344
seed(s), 504, 519; typical, structure of, 504
seed plants, 464–465. See also plants
segmented worms, 464
seismic stations, 302
seismograph, 292–293
selenium, symbol for, 49
sense organs, 543

sensory neuron(s), 545, 546
series circuit, 232–233, 234
sex chromosomes, 513–514
sex-influenced traits, 513
sex-linked traits, 513–514
sexual reproduction, 508–509
shadow stick, use of, 407–408
shale, 336, 338, 343
sheep, 487; bighorn, study of, 2–3, 4–8
shield volcanoes, 300
shiprock volcano, 332
"shooting stars," 420
shrews, 487
sickle-cell anemia, 512, 520
Sierra Nevadas, 333
sills (igneous rock), 333
silt, 336
silver, symbol for, 49
simple machines, 158–161
skeletal muscles, 527–528, 529
skeleton, 525; human, 525, 526
sky, color of, 194, 195, 198, 199
Skylab, 471–473
slate, 343, 345
sleet, 380
small intestine, 532, 533
smog, 553, 553
smooth muscles, 528
snails, 464
snakes, 463
snow, 380, 430, 434, 473
snowflakes, 134, 380
snowshoe hares, 493–494
snowstorm (February, 1978), 376, 381
social communities (human), 488–489
sodium, 394; atomic number of (table), 48; electron shells for (table), 48; symbol for, 48
sodium atoms, 94, 95
sodium chloride, 95; crystals of, 95, 134, 135; melting point of (table), 138
soil conservation, 566
solar cell(s), 283
solar eclipse, 411

596   Index

NEW VOCABULARY — All new vocabulary words are listed alphabetically.

DEFINITIONS — Given for each new word.

TOPICS LISTED ALPHABETICALLY

PAGE NUMBERS — Indicate where discussion of the topics can be found. Pages in **boldface** type indicate where illustrations can be found. Pages in *italic* type indicate where definitions can be found.

EXERCISE

1. What is the definition of the term *friction*?
2. On what page of the Glossary can you find a definition for the term *consumer*?
3. On what page of the text could you find a diagram of the air sacs of the lungs?
4. On what page of the text could you find a discussion of atomic particles?
5. On what page of the text could you find a definition of the term *chromosomes*?

GLOSSARY

abiotic: Refers to the nonliving materials and energy in an ecosystem.

absolute age: A term referring to age in approximate years as determined by radioactive dating.

absolute zero: The temperature ($-273°C$) at which particles in matter stop moving.

abyssal plain: A wide deposit of sediment that covers the deepest part of the ocean basin.

acceleration: The change in speed during a given interval of time, calculated by dividing the change in speed by the time it took for the change in speed to happen.

adaptation: A trait that enables an organism to survive better than similar organisms.

air mass: A large mass of air that has taken on the temperature and humidity of a part of the earth's surface.

algae: Protists that contain chlorophyll and make their own food.

alkali metals: A group of elements whose atoms all have one electron more than the stable number.

alternating current: An electric current that changes direction in an electric circuit.

ampere (amp): A measure of the number of electrons moving past a given point in an electrical circuit in 1 sec.

amplitude: The strength of a wave. One half the distance from the bottom of the trough to the top of the crest.

angiosperms: Vascular plants with seeds covered by a protective tissue.

anticyclone: A large mass of air spinning out of a high-pressure area.

area: The amount of surface within a given set of lines, measured in square metric units (m^2, cm^2, mm^2, etc.).

asexual reproduction: Reproduction that requires only one parent.

association neuron: A neuron that carries impulses from sensory neurons to motor neurons.

asteroids: Chunks of rock and metal that orbit the sun between Mars and Jupiter.

astronomer: A scientist who studies the stars, planets, and other heavenly bodies.

atmosphere: The layer of gases that surrounds the earth.

atom: The smallest particle of an element that has all the properties of that element.

atomic mass: The sum expressed in amu of the mass of protons and neutrons in an atom.

atomic mass unit (amu): A special unit used to express the mass of atomic particles and atoms.

atomic number: The number of protons in the nucleus of an atom.

atomic particles: The building blocks of atoms.

ATP (adenosine triphosphate): A compound that stores energy and makes it available to the cell.

autumnal equinox: The first day of fall. Daylight and darkness are the same length.

axis: The imaginary line through the center of the earth about which the earth rotates.

BTU (British thermal unit): The amount of energy needed to raise the temperature of 250 g of water $1°C$.

bacteria: Very small protists with simple cell structure and no nucleus.

batholith: A large body of igneous rock formed underground.

binding energy: A powerful force that holds the nucleus of an atom together.

biomass fuel: Any plant or animal matter that is used as a source of energy.

biosphere: That area at or near Earth's surface where life is possible.

biotic: Refers to the living organisms in an ecosystem.

blood: Specialized tissue that transports materials to and from the body cells.

boiling point: The temperature (at ordinary air pressure) at which the particles of a liquid have enough energy to become a gas.

bone: A hard, living tissue made up of bone cells and deposits of calcium and phosphorous compounds.

breeder reactor: A nuclear reactor that creates more nuclear fuel than it uses.

calorie: The amount of heat needed to raise the temperature of 1 gram of water by 1°C.

carnivore: A consumer that eats only animal tissue.

cartilage: A firm but flexible tissue that gives shape and support to many parts of the bodies of some animals.

cell: The smallest organized unit of living protoplasm.

cell division: The process by which one cell divides to form two new cells.

cell membrane: The outer boundary of the cell.

cell wall: The rigid protective layer that surrounds the plant cell.

Celsius (C): The name of the most commonly used temperature scale. The Celsius scale is always used in science.

centimeter (cm): One one-hundredth (0.01) of a meter.

chain reaction: A reaction in which some of the products can be used to further the reaction.

chemical activity: The way an atom reacts with other kinds of atoms.

chemical bond: A force that joins atoms together.

chemical change: A change in matter by which new molecules are formed.

chemical digestion: The process by which foods are changed into more simple chemical compounds.

chemical equation: A description of a chemical reaction using chemical formulas for the substances used and produced.

chemical family: A group of elements that are alike in their chemical behavior.

chemical reaction: A reaction in which a chemical change takes place.

chemical symbols: One or two letters used to represent an atom of a particular element.

chlorophyll: Green material found in green plants that use sunlight to make food.

chloroplasts: Oval-shaped structures containing chlorophyll.

chromatin: Material located in the nucleus of a cell that carries the genes.

chromosomes: Long rod-shaped bodies in the nucleus that control a cell's activities.

classify: To arrange organisms in groups according to similar characteristics.

coal: The remains of plants squeezed and changed into a solid burnable substance.

cold-blooded: An animal whose body temperature changes with the temperature of its environment.

comets: Icy bodies, usually found on the edge of the solar system.

community: All the plants and animals living in a particular area or ecosystem.

compound: A substance made by the joining of atoms of two or more elements.

compound machine: A machine containing more than one simple machine.

concave: A lens shape in which the edges are thicker than the center. (The center is *caved in.*)

condensation: The process by which a gas turns into a liquid.

conduction: The transfer of heat by direct contact.

conservation: The protection and wise use of natural resources.

consumer: An organism that depends on producers for its food needs.

continental rise: An apron of sediments deposited along the base of the continental slope.

continental shelf: The flattened top part of the continental slope formed by sediments from the continent.

continental slope: The sloping part of the sea floor that marks the boundary between the floor and the continents.

controlled experiment: Two experimental tests in which all factors are the same except the one being tested.

convection: The transfer of heat by movement of the heated part of a gas or liquid.

convex: A lens shape in which the edges are thinner than the center.

core: The two-layered center of the earth.

Coriolis effect: The apparent curving of the path of moving objects as a result of Earth's rotation.

covalent bond: A chemical bond formed when atoms share two or more electrons.
crust: The very thin top layer of the lithosphere.
crystal: A solid whose orderly arrangement of particles gives it a regular shape.
current: Water moving in a particular direction.
cyclone: An area of low pressure with winds circling into the center.
cytoplasm: The protoplasm surrounding the nucleus.

data: Information collected from observations.
decay bacteria: Tiny organisms that obtain their food by breaking down dead plants and animals.
decimeter (dm): A unit of length that is one-tenth (0.1) of a meter.
decomposer: An organism that obtains its food from wastes and dead organisms.
density: The amount of matter in a given unit of volume.
dew point: The temperature at which air becomes saturated with water vapor.
diaphragm: A sheet of muscle that separates the chest cavity from the stomach cavity.
dicot: A plant that produces seeds with two halves.
diffraction: The ability of waves to bend around an obstacle in their path.
digestion: The process that breaks down food so that it can be used by the cells.
direct current: An electric current that flows in one direction in an electric circuit.
dome mountains: Mountains produced when forces below the surface lift a part of the crust.
dominant: Refers to the stronger of a pair of traits.
dominant species: Organism present in the greatest numbers.
Doppler effect: An apparent change in frequency of waves that comes from the fact that the observer or the source of the waves is moving.
duct: A tube through which materials can flow from a gland.

eclipse: The Earth or Moon passing through the shadow of the other.

ecological niche: The role an organism plays in a community. Its occupation.
ecology: The study of the relationship between organisms and their environments.
ecosystem: A group of organisms and their physical environment.
egg: A reproductive cell from a female.
electric charge: The result of something being given an amount of electric energy.
electric circuit: A complete path that allows electrons to move from a place rich in electrons to a place poor in electrons.
electric current: The result of electrons moving from one place to another.
electric field: A region of space around an electrically charged object in which electric forces on other charged objects are noticeable.
electric force: The force that causes two like-charged objects to move apart or two unlike-charged objects to move toward each other.
electromagnet: A temporary magnet made when electric current flows through a coil of wire wrapped around a piece of iron.
electromagnetic induction: The production of electric current by motion in a magnetic field.
electromagnetic spectrum: The series of waves with properties similar to light.
electromagnetic waves: A form of energy able to move through empty space at very high speed.
electron: A very light, negatively charged atomic particle.
electron shells: The area around an atomic nucleus in which electrons move.
element: A simple form of matter that cannot be changed into any simpler form of matter by ordinary means.
embryo: A developing organism in its earliest stages of development.
endangered: Refers to an organism that is in danger of becoming extinct.
endocrine glands: Ductless glands that release hormones into the bloodstream.
energy: That property of something that makes it able to do work.
environment: Everything in its surroundings that affects the way an organism lives and acts.
environmental factors: The several conditions in an environment an organism needs to survive.

epicenter: The point on the Earth's surface directly over the focus of an earthquake.
equation: A simple way to show chemical changes.
era: A major division of the geologic time scale, marked by major changes in living things or the Earth's surface.
erosion: All processes that cause rock to be carried away.
evaporation: The process by which water changes from a liquid to a gas.
excretion: The process by which an organism gets rid of metabolic wastes.
experimental factor: The aspect that varies in an experiment to test a hypothesis.
extinct: Refers to an organism that no longer exists anywhere on the Earth.

fault: A crack in the Earth's surface along which there has been movement of one or both sides.
fault-block mountains: Mountains formed when large blocks of rock are tilted over.
feces: The solid wastes formed from undigested food materials in the large intestine.
feedback: An automatic "turn on" or "turn off" process that occurs when a certain level of a hormone is reached.
fertilization: The process during which a sperm cell and egg cell join.
fission: A nuclear reaction in which a large unstable nucleus splits into smaller nuclei.
focal length: The distance from the center of a lens to the focal point.
focal point: The point at which parallel light rays meet after being refracted.
focus: The point on a fault at which movement causes an earthquake.
folding: The bending of rocks under steady pressure without breaking.
food chain: The passing of energy in the form of food from one organism to another.
food web: All the feeding patterns in an ecosystem.
force: Any push or pull that causes something to move or change its speed or direction of motion.
fossil: Any indication of life that existed long ago.
fossil correlation: Determining the relative age of rocks by matching the fossils they contain.

fossil fuel: Mineral fuel found in the Earth's crust.
frequency: The number of complete waves that pass a given point each second.
friction: A force that opposes or slows down motion.
front: The boundary separating two air masses.
fruit: A ripened ovary that contains the seeds.
fuel: Any material used as a source of chemical energy.
fungi: Plantlike protists that cannot make their own food.
fuse: The part of an electric circuit that prevents too much current from flowing.
fusion: A nuclear reaction in which two small nuclei join to form one larger nucleus.

galaxy: A huge system of stars.
generation: A level in the succession of a family.
genes: Structures on chromosomes that determine inherited traits.
genetics: The science that studies the passing of traits from parent to offspring.
genus: A group of related species of organisms differing only in a few characteristics.
geothermal energy: Energy obtained from heat in the Earth's crust.
glacier: A large, slowly moving mass of ice on the land.
gland: An organ that makes and releases materials that have special functions in the body.
gram: A small unit of mass in the metric system.
gravity: The force that pulls an object toward the center of a planet.
gymnosperms: Vascular plants with seeds that are not covered.

habitat: Where an organism is usually found in a community. Its address.
halogens: A group of elements whose atoms all have one electron less than the stable number.
heat engine: A machine that changes heat energy into mechanical energy.
heat of fusion: The amount of heat required to change one gram of a solid to a liquid at the same temperature.

heat of vaporization: The amount of heat required to change 1 gram of a liquid to a gas at the same temperature.
herbivore: A consumer that eats only plant tissue.
heredity: The passing of traits from parent to offspring.
hertz: A unit used to measure the frequency of a wave. One hertz means that one complete wave passes a given point each second.
hormone: A chemical messenger that regulates and balances body functions. Hormones are carried in the blood.
humidity: The amount of water vapor in the air.
hurricane: A small but intense cyclonic storm formed over warm parts of the sea.
hybrid: Refers to an organism or cell in which the paired genes for a trait are different.
hydrocarbons: Substances made up only of carbon and hydrogen.
hydroelectric power: Energy obtained from water moving in rivers and streams.
hypothesis: A prediction or "educated guess" based on patterns in observations.

igneous rock: Rock formed when molten material cools.
image: The picture formed by a lens.
impulse: Information from a stimulus that travels along a nerve fiber.
inclined plane: A surface that slopes from one level to another.
incomplete dominance: A condition in which neither of the two genes that cause the appearance of a trait is dominant.
inert: A description of an atom that does not react readily with other atoms.
insulator: A material that does not allow energy to flow through it easily.
involuntary muscles: Muscles that cannot be controlled at will.
ion: An atom or molecule with an electric charge.
ionic bond: A kind of chemical bond formed when atoms transfer electrons from one to another.
ionosphere: A region of the atmosphere containing many electrically charged particles.

isostasy: The state of balance between different parts of the lightweight crust as it floats on the heavy mantle.
isotopes: Atoms of the same element whose nuclei contain the same number of protons but different numbers of neutrons.

joint: A crack in rock with no movement along either side; a place where two bones meet.

Kelvin temperature scale: A temperature scale on which 0°K is equal to absolute zero.
kidneys: Excretory organs that remove wastes from the blood.
kilogram (kg): One thousand (1,000) grams.
kilometer (km): One thousand (1,000) meters.
kilowatt-hour: The amount of energy supplied in one hour by one kilowatt of power. Used to measure amount of electric energy consumed.
kinetic energy: Energy that moving things have as a result of their motion.
Kinetic Theory of Matter: The scientific principle that says that all matter is made of particles whose motion determines whether the matter is solid, liquid, or gas.

laser: A beam of concentrated light energy.
lava: Magma that has flowed out on the surface.
Law of Conservation of Energy: A natural law that says that energy cannot be created or destroyed but can be changed from one form to another.
Law of Superposition: The idea that in undisturbed layers of rock the oldest layers are at the bottom and the youngest layers at the top.
lens: A piece of transparent material with curved surfaces that refracts light passing through it.
lever: A rigid bar that moves on a fixed point and can change the direction and size of the force applied.
life processes: Activities or processes carried on by all living organisms.
ligaments: Tough strands of elastic tissue that connect bones at movable joints.
light rays: Straight lines showing the path followed by light.

limiting factor: A condition in the environment that may cause a change in the size of a population.

lithosphere: The strong, solid outer layer of the Earth.

longitudinal wave: A wave in which the particles vibrate back and forth in the direction of the wave.

loudness: The amplitude or amount of energy contained in a sound wave.

lunar eclipse: The Moon passing through the Earth's shadow in space.

magma: Molten rock beneath the Earth's surface.

magnetic field: A region of space around a magnet in which magnetic forces are felt.

magnetic force: The force that causes two magnets to push or pull each other when their ends are brought together.

magnetic pole: The part of a magnet where the magnetic forces are strongest.

mantle: A very thick layer of the Earth reaching about 2,900 km beneath the lithosphere.

mass: A measure of the amount of matter contained in an object.

mating: A behavior in which organisms are together for the purpose of reproduction.

matter: Anything that takes up space and has mass.

mechanical advantage: The number of times a simple machine multiplies an effort force.

mechanical digestion: The process by which food is broken down into very small pieces.

melting point: The temperature at which a solid becomes a liquid.

mesosphere: The atmospheric layer above the stratosphere.

metamorphic rock: Rock that has changed by the action of heat or pressure without melting.

meteor: A bright streak of light in the sky.

meteorology: The branch of science that studies the atmosphere.

meteorites: Chunks of rock from space that collide with a planet or a moon.

meter (m): The basic unit of length in the metric system; slightly more than one yard in length.

mid-ocean ridge: A chain of underwater mountains found near the center of the oceans.

migration: The movement of animals from one place to another.

millimeter (mm): One one-thousandth (0.001) of a meter, or one-tenth (0.1) of a centimeter.

mirage: An illusion caused by the refraction of light in which distant objects are seen upside down or floating in the air.

mitochondria: Structures in the cytoplasm that release energy from food.

mixture: A substance made of two or more elements or compounds that have not combined with each other.

molecule: The smallest particle of a compound that has the properties of the compound.

monadnock: An isolated hill on a peneplain.

monocot: A plant that produces seeds with one part.

Moon phases: The changing appearance of the Moon as more of the sunlit side becomes visible from Earth.

motion: A change in position of an object when compared to a reference point.

motor neuron: A neuron that carries impulses from the brain or spinal cord to a muscle or gland.

music: Any tone produced by a regular pattern of vibrations.

mutation: A change in one or more genes that results in a new trait.

natural gas: A colorless, odorless gas formed from the remains of ancient sea life.

natural resource: Any natural substance of use to humans.

negative charge: The electric charge given to a hard rubber rod when rubbed with fur.

neuron: The basic unit of the nervous system; a complete nerve cell.

neutral: Describes an object having neither a positive nor negative charge.

neutron: An atomic particle with no charge found in the nucleus.

newton: The unit used to measure force in the metric system.

Glossary

nitrogen-fixing bacteria: Tiny organisms that combine nitrogen and oxygen from the air to make nitrogen compounds.

noble gases: The six elements whose atoms have completely filled electron shells.

noise: Sounds produced by irregular vibrations.

non-vascular: Plants without specialized tissue for transporting water, minerals, and food.

nuclear fusion: A nuclear reaction in which two small nuclei join to form one large nucleus.

nuclear power: Energy obtained by splitting atoms in a nuclear reactor.

nuclear reactor: A machine used to carry on a controlled nuclear chain reaction.

nucleolus: Small body found in the nucleus.

nucleoplasm: A type of protoplasm found inside the nucleus of a cell.

nucleus: The small central core of an atom.

observation: Anything that can be learned by using the senses—sight, hearing, taste, touch, smell.

ohm: A measure of the amount of resistance in an electric circuit.

omnivore: A consumer that eats both plant and animal tissue.

orbit: The curved path of one object around another object. The Earth is in orbit around the sun.

organ: Different tissues that work together to perform a function or functions.

organism: A complete living thing.

parallel circuit: An electric circuit with its various parts in separate branches.

peneplain: An almost completely flat surface produced by erosion.

periodic chart: An arrangement of all the elements in which chemical families are shown.

petrified: Refers to living matter that has been replaced by minerals and has "turned to stone."

petroleum: A black, oily liquid formed from the remains of ancient sea life.

physical change: A change in matter that does not change the individual molecules.

pitch: The level of a sound, either high or low. Pitch is related to frequency.

plate: A section of the Earth's lithosphere and crust.

plate tectonics: The theory that the earth's crust is made up of large moving plates.

plateau: A large region of elevated land.

polar front: The boundary where cold polar air meets warm air.

pollination: The process in which pollen is transferred from the anther to a stigma.

pollutants: Any materials such as gases, particles, and chemicals released into the environment.

population: A group of one kind of organism in a particular community.

population density: The number of organisms found in a certain area at a given time.

positive charge: The electric charge given to a glass rod when rubbed with silk cloth.

potential energy: Energy stored in an object as a result of a change in its position.

power: How fast work is done. Power = work ÷ time.

predator: An animal that hunts other animals for food.

prey: Animals that are hunted by other animals for food.

Principle of Uniform Process: The idea that the processes that act on the Earth today also acted on the Earth in the past.

prism: A specially shaped clear material. A prism divides white light into its separate colors.

producer: A green plant able to make its own food by photosynthesis.

protists: Kingdom of organisms that are neither plants nor animals.

proton: A positively charged atomic particle found in the center of an atom.

protoplasm: All living material found in a cell capable of carrying on all the life processes.

protozoans: Single-celled, animal-like protists.

pulley: A grooved, freely turning wheel over which runs a rope or chain.

pure: Refers to an organism or cell in which the paired genes for a trait are identical.

radiation: The transfer of heat through space by infrared rays.

radioactive atom: An atom whose nucleus is changing in order to become more stable.

recessive: Refers to the weaker of a pair of traits.

reflection: A process in which a wave is thrown back after striking a barrier that does not absorb the energy of the wave.

reflex: An automatic response to a stimulus in which the brain is not directly involved.

refraction: The bending of a wave caused by a change in its speed.

relative age: A time scale in which objects or events are older than or younger than others. Does not give exact age.

relative humidity: The amount of water vapor in the air compared with the amount of water vapor the air could hold at that temperature.

resistance: All conditions that limit the flow of electrons in an electric circuit.

respiration: The process in which oxygen combines with glucose to release useful energy for life activities.

RNA (ribonucleic acid): A large molecule that functions in the making of proteins.

ribosomes: Structures in the cytoplasm that make proteins from amino acids.

rift: A long, deep valleylike crack in the Earth's surface.

rock: A piece of the earth that is usually made up of two or more minerals mixed together.

rock cycle: The endless process by which rocks are formed, destroyed, and formed again in the Earth's crust.

salinity: The number of grams of dissolved salts in 1,000 g of seawater.

saliva: A liquid produced by the salivary glands. It contains special starch-digesting chemicals.

scientific law: A theory that has been tested many times and has always been found to be true.

scientific model A way of representing something that cannot be observed, based on properties that have been observed.

screw: An inclined plane wrapped around a cylinder.

sediment: Any substance that settles out of water.

sedimentary rock: A kind of rock made when a layer of sediment becomes solid.

seed: A ripened ovule that contains the embryo and a stored food supply and is protected by a seed coat.

seismograph: An instrument that records earthquake waves.

sensory neuron: A neuron that carries impulses from sense organs to the spinal cord or brain.

series circuit: A circuit in which all the parts are connected one after the other.

sex chromosomes: A pair of chromosomes that determine the sex of an individual.

sex-influenced trait: A trait that is dominant in one sex and recessive in the other.

sex-linked trait: A trait controlled by genes on the X chromosome.

sexual reproduction: Reproduction that requires two parents.

simple machine: A device that changes the size, direction, or speed of a force.

skeleton: The supporting framework of an animal's body.

smog: An irritating mixture of gases, particles, chemicals, and fog.

solar cell: A device that changes sunlight directly into electricity.

solar eclipse: The Moon passing between the Sun and the Earth, casting its shadow on the Earth.

solution: A mixture in which the molecules of one substance fit between the molecules of another. One of the substances is usually a liquid or a gas.

sound: A wave consisting of vibrations of the particles of the material through which the wave passes.

special properties: Characteristics that make one type of matter different from another.

species: A group of related organisms that have the same characteristics.

sperm: A reproductive cell from a male.

stable electron arrangement: An electron arrangement in which the outermost electron shell is filled.

stimulus: Any change in the environment that causes a response in an organism. *Stimuli* is plural for more than one stimulus.

stratosphere: The atmospheric layer above the troposphere.

subatomic particles: The tiniest particles of matter that make up atoms.

Glossary

summer solstice: The time at which a shadow stick casts the shortest shadow when the sun is highest in the sky.

system: A group of organs working together and forming a functional unit.

temperature: A measurement of the movement of particles in matter.

tendons: Tough nonelastic tissue that attaches some skeletal muscles to bones.

theory: A hypothesis that has withstood repeated testing.

thermosphere: The atmospheric layer above the mesosphere.

thrust fault: A slanting fault in which one slab of rock is pushed up over the other.

tide: A regular rise and fall of water level in the sea.

tissue: A group of cells that are similar in structure and function.

trait: A characteristic that is passed from parent to offspring.

transformer: The part of an electric circuit that changes the voltage.

transverse wave: A wave in which the molecules vibrate at right angles to the direction of the wave.

trench: A deep valley on the ocean floor. Found along the edges of ocean basins.

troposphere: The most dense layer of the atmosphere. It lies closest to the earth.

unconformity: A boundary in rocks that indicates a gap in the rock record. Usually due to erosion of some rocks.

uniformitarianism: The idea that the processes that are at work on Earth today have acted throughout Earth's history.

universe: All the galaxies and the space between them.

urine: Liquid waste removed from the blood by the kidneys.

vacuoles: Storage areas located in the cytoplasm.

valence: The number of electrons gained, lost, or shared by an atom when it forms a chemical bond.

valence electrons: Electrons in the outer shell of an atom.

vascular: Plants having specialized tissue for transporting water, minerals, and food.

vernal equinox: The first day of spring. Daylight and darkness are the same length.

virus: Particles that are not cells but can reproduce in the cells of living organisms.

visible spectrum: The band of colors produced when white light is divided into its separate colors.

volcano: A structure that may result when magma reaches the earth's surface.

volt: A measure of the amount of work done in moving electrons between two points in an electric circuit.

volume: The amount of space an object takes up.

voluntary muscles: Muscles that can be controlled at will.

warm-blooded: An animal that maintains a constant internal body temperature.

watt: The metric unit used to measure power. 1 watt = 1 newton-meter per second.

wave: A disturbance caused by energy moving from one place to another in a substance.

wavelength: The distance between two successive identical points on a wave.

weathering: The processes that break rock into small pieces.

wedge: Two inclined planes attached base to base.

weight: A measure of the pull of gravity on the mass of an object. On Earth the terms *mass* and *weight* are used as though they had the same meaning.

wheel and axle: A large wheel fixed to a small axle so that they turn together.

winter solstice: The time at which a shadow stick casts the longest shadow when the sun is highest in the sky.

work: The force applied to an object multiplied by the distance the object moves. $W = F \times d$

PHOTO CREDITS

HRW Photos by Russell Dian appear on the following pages: 1(br); 22(b); 73(tl,bl); 99; 115; 137; 141(t); 151; 168; 202(t,b); 221; 225; 236(b); 236–237(t); 240(b); 241(r,m); 244; 317; 320(tl); 377; 456; 458(bl); 470; 474; 488(bm); 497; 503–504.

Photos by Michael Brown appear on the following pages: 73(br); 113; 119(b); 120; 124(b); 128; 130–131; 134(t); 135–136; 146; 151(t); 189; 191; 204–205; 217; 222; 228; 231; 241(l); 242; 325(br); 330; 334; 378; 380(tl); 392; 399; 441.

Photos by Samuel Campbell appear on the following pages: 240(t); 475; 488(b1); 547; 557; 559; 560(t); 561.

HRW Photos by Ken Lax appear on the following pages: 1(tl,tr); 8(b); 12(b); 19; 23(t); 24(tr); 27; 30(t); 31(t,b); 155; 158(b); 159(b); 162(t,bl); 163(t,mr,b); 169; 257; 354; 531.

What Does a Scientist Do?: p.1(bl) Wide World; p.2 Reggie Tucker/Taurus; p. 3 Timothy Eagan/Woodfin Camp; p.8(t) Jerry Cooke/Photo Researchers; p.12(t) NASA; p.13 NASA.

Unit 1: p.17 Fritz Goro/Life Magazine
Chapter 1: p.18 NASA; p.20 Scott Ransom/Taurus; p.22(t) Jeffery Jay Foxx/Woodfin Camp; p.22(tm) ATT Western Electric: p.22(bm) William Hubbell/Woodfin Camp; p.23(tm) van Bucher/Photo Researchers; p.23(bm) Townsend P. Dickinson/Photo Researchers; p.23(b) Dr. E. R. Degginger, FPSA; p.24(l) Bettmann Archive; p.25 Bettmann Archive; p.30(t) Joern Gerdts/Photo Researchers; p.32(t) Alexander Ward/Photo Researchers; p.32(b) HRW Photo Rona Weissler-Tuccillo.
Chapter 2: p.36 Carl Frank/Photo Researchers; p.43 Brookhaven National Laboratory; p.46 HRW Photo Rona Weissler-Tuccillo; p.51 NASA; p.52 Bettmann Archive; p.54 Jerry Cooke/Photo Researchers; p.56 The Granger Collection.
Chapter 3: p.61 NASA; p.62 Daniel Brody/Stock Boston; p.63(l) AgPhoto by Webb, (r) EPA-Documerica; p.63(t) Albert Einstein College of Medicine, (b) Dan McCoy/Rainbow; p.65 The Granger Collection; p.66 Albert Schoenfield, *Swimming World;* p.68(t) Charles Harbutt/Magnum, (b) Raimondo Borea/Photo Researchers; p.71 National Parks Service; p.73(tr) HRW Photo by John King; p.74 Fred Ward/Black Star.
Chapter 4: p.78 Union Carbide Corporation Nuclear Division/Oak Ridge National Laboratory; p.82 The Granger Collection; p.88(l,m) South African Tourist Corporation, (r) Raimondo Borea/Editorial Photocolor Archive; p.89 New York Public Library; p.93 South African Tourist Corporation; p. 106 General Electric Co.; p. 107(t) Farrel Grehan, (b) General Electric Co.; p.108(t) Eric Leigh Simmons/The Image Bank, (b) Olin Corporation.

Unit 2: p.109 Gerald Brimacombe/The Image Bank.
Chapter 5: p.110 Swiss National Tourist Office; p.111 The Granger Collection; p.114 Bettmann Archive; p.117 Arthur Grace/Stock Boston; p.118 Mark Godfrey/Magnum; p.119(t) Howard Soucherek/Woodfin Camp; p.124(tl) Dresser Industries, (tr) UPI; p.134(b) American Museum of Natural History; p.138 Tom McCarthy/The Image Bank; p.140 Owens-Corning Fiberglass; p.141(b) University of Wisconsin at Madison, Institute of Environmental Studies.
Chapter 6: p.144 HRW Photo by Lois Ciesla-Safrani; p.149 New York Public Library; p.154(t) Marion H. Levy/Photo Researchers; p.158(t) Phil Degginger; p.152(br) Dr. E. R. Degginger; p.163(ml) Russ Kinne/Photo Researchers.
Chapter 7: p.167 Bill Wood/Bruce Coleman; p.173 Porterfield-Chickering/Photo Researchers; p. 177 NASA; p.178 HRW Photo William Hubbell.
Chapter 8: p.185 German Information Center; p.187 NASA; p.190 Craig Aurness/Woodfin Camp; p.192 Life Science Library/*Light & Vision;* p.194(t) Stephen J. Krasemann/Peter Arnold, Inc.; pp.194(b)–195 Courtesy of William Ramsey; p.196 New York Public Library; p.201 Courtesy of Russell Dian; p.202(m) Townsend B. Dickinson/Photo Researchers; p.203 Bausch and Lomb.

Unit 3: p.211 Joe Munro/Photo Researchers.
Chapter 9: p.212 HRW Photo by Lois Ciesla-Safrani; p.218 General Electric Co.; p.227 Albert Schoenfield, *Swimming World;* p.237(b) American Telephone and Telegraph Co.
Chapter 10: p.239 Eric Meola/Woodfin Camp; p.239 German Information Center; p.251(t) U.S. Department of Commerce, (b) Georg Gerster/Photo Researchers; p.253 Majestic, an American Standard Company; p.264 U.S. Department of Commerce; p.268 Denver Public Library Western Collection; p.269 Dr. E. R. Degginger, FPSA; p.270 U.S. Bureau of Mines; p.274(t) Bettmann Archives, (b) Alok Kavan/Photo Researchers; p.275(t) A.Z./Bruce Coleman, (b) Culver Pictures; p.276 NASA; p.277 New England Power Company.
Chapter 11: p.281 Jonathan Blair/Woodfin Camp; p.282 James Balog/Bruce Coleman; p.238 Jim Goodwin © 1977/Photo Researchers; p.284(t) Van Bucher/Photo Researchers, (b) Tom McHugh/Photo Researchers, (m) French Embassy Information Division; p.285 Tom McHugh/Photo Researchers.

Unit 4: p.289 Wide World.

Chapter 12: p.290 Wide World; p.291 U.S. Forest Service; p.293 Eric Kroll/Taurus Photos; p.298 Nicholas Devore/Bruce Coleman; p.300(t) W. Stoy/Bruce Coleman, (b) Jose Honorez/Bruce Coleman; p.311 J. Messerschmict/Bruce Coleman; p.313(t) Union Pacific Railroad, (b) U.S. Geological Survey; p. 314(t) Hans Wendler/The Image Bank, (bl) National Park Service, (br) Kenneth H. Forman; p.315 EROS; p.318(t) U.S. Corps of Army Engineering, (bl) Fred Maroon/Louis Mercier, (br) Malcolm Kirk/Peter Arnold, Inc.; p.319(tl) Al Green/Bruce Coleman, (tr) Thomas W. Putney/Delaware Stock Photo Library, (b) Courtesy of William Ramsey; p.320(tr) Union Pacific Railroad, (bl) Courtesy of William Ramsey, (br) John Shelton; p.321(tl,br) Ward's Natural Science Establishment, (tr,bl) National Park Service; p.324(t) Utah State Historical Society; pp.324(b) 325(t,bl), 326–327 HRW Photos by John Cubitto.

Chapter 13: p.328 Dan McCoy/Rainbow; p.329 John Running; p. 331(t) Hawaii Visitors Bureau, (b) American Museum of Natural History; p.322 Bruce Coleman; p.333(tl) M. Liacos, (tr) Dallas Peck/U.S. Geological Survey, (b) B. M. Shaub; p.335 Keith Grunner/Bruce Coleman; p.336(t) G. R. Roberts, (bl,bm) B. M. Shaub, (br) U.S. Geological Survey; p.337 B. M. Shaub; p.340 Dan McCoy/Rainbow; pp.340(b)–341,343(tl,bl) B. M. Shaub; p.343(tr) Dan McCoy/Rainbow, (br) Smithsonian Institute.

Chapter 14: p.347 Julian Baum; p.348(t) Tom McHugh/Photo Researchers; p.348(b) Tom McHugh/Photo Researchers-Natural History Museum Los Angeles County; p.349(t) Tom McHugh/Photo Researchers-Natural History Museum Los Angeles County, (b) Eric V. Grave/Photo Researchers; p.350(t) P. W. Grace/Photo Researchers, (b) Jan Lukas/Photo Researchers; p.351 Courtesy of University of Nebraska State Museum; p.353(t) HRW Photo Katherine Jenson, (ml) Jen and Des Bartlett/Photo Researchers, (mr) Tom McHugh/Photo Researchers, (r) R. T. Bird/Courtesy of American Museum of Natural History; p.355(t) Wide World, (b) Keith Gunnar/Bruce Coleman; p.356 HRW Photo Richard Weiss; p.357(t) EPA Documerica; p.360 Charles R. Belinky/Photo Researchers; p.362(t) Joe Tomala, Jr./Bruce Coleman, (b) Photo Researchers/Field Museum; p.363(l) Bucky and Avis Reeves/Photo Researchers, (tr) Tom McHugh/Photo Researchers; p.364(t) Tom McHugh/Photo Researchers, (b) Photo Researchers-Field Museum; p.365(t) Tom McHugh/Photo Researchers.

Chapter 15: p.369 Margaret Durrence/Photo Researchers; p.370 U.S. Department of Energy; p.376 A. Avis/Bruce Coleman; p.379(t) Thomas W. Putney/Delaware Stock Photo Library, (bl) Dan McCoy/Rainbow, (br) G. R. Roberts; p.380(tr) Dan McCoy/Rainbow, (bl) Thomas W. Putney/Delaware Stock Photo Library, (br) Robert H. Wright/Photo Researchers; p.381 UPI; p.385 NOAA; p.388 NASA; p.389 Frederic Lewis, Inc.

Chapter 16: p.393 Dan McCoy/Rainbow; p.394 Jay Lurie; p.395 Woodshole Oceanographic Institute; p.400 Russ Kinne/Photo Researchers; p.401 David Moore/Photo Researchers; p.402 Hale Observatories; pp.408–416 NASA; p.417 Naval Research Laboratory; pp.418–421,422(t,b) NASA; p.422(t,m) Courtesy of Charles Kohlhase and James Blinn of NASA/JPL; p.423(t) Dr. E. R. Degginger, FPSA.

Unit 5: HRW Photo

Chapter 17: p.428(l) NASA, (r) John Shaw/Bruce Coleman; p.430(l,m,r) NASA; p.431 John Shaw/Bruce Coleman; p.432(t) Woodshole Oceanographic Institute, (b) © Courtesy of Lucasfilm, Ltd.; p.434 Paul Kuhn/Bruce Coleman; p.438 Keith Gunnar/Bruce Coleman; p.440(t,b) NASA; p.441 Herbert Eisenbert/Photo Trends; p.442 Dr. E. R. Degginger; p.443(l) New York Public Library, (r) Bausch and Lomb; p.444 Photo Classics.

Chapter 18: p.447 © Elizabeth Crews/Jeroboam, Inc.; p.448 Grant Heilman; p.450(l) Joel Gordon, (r) Grant Heilman; pp.453–454 Runk/Schoenberger, from Grant Heilman; p.458(t) Courtesy of James McGuirk, (br) W. H. Hodge/Peter Arnold, Inc.; p.460 Carolina Biological Supply Co.; p.461(l) Manfred Kage/Peter Arnold, Inc., (r) National Medical Audio-Visual Center.

Chapter 19: p.471 NASA; p.473(t) Robert Salmonson, (b) Courtesy of James McGuirk; p.476 A. Cosmos Blank/Photo Researchers; p.477 Robert Salmonson; p.479 Dr. E. R. Deggenger; p.480 Len Lee Rue III; p.481 Grant Heilman; p.484 George Hall/Woodfin Camp; p.485 Annan/Photo Trends; p.486 J. M. Donald/Bruce Coleman; p.487 Fairchild/Peter Arnold, Inc.; p.488(t) Courtesy of James McGuirk, (br) George and Judy Manna/Photo Researchers; p.490 National Park Service; p.491 W. C. Frase/Bruce Coleman; p.492 Jane Burton/Bruce Coleman; p.495 Stouffer Prod. Ltd./Bruce Coleman; p.500(t) Martin Adler Levick/Black Star, (b) St. Vincent's Hospital.

Chapter 20: p.501 New York Zoological Society; p.502 Fisher Scientific Co.; p.506 Roy Ahaway/Photo Researchers; p.511 HRW Photo William Hubbell; p.512(l) Courtesy of Carolina Biological Supply Co., (r) Phillip A. Harrington/Peter Arnold, Inc.; p.514 Courtesy of Carolina Biological Supply Co.; p.516 Robert P. Carr/Bruce Coleman; p.518(tl) Jerome Wexler/Photo Researchers, (mr) Lynwood M. Chall/Photo Researchers, (bl) Jeff Foott/Bruce Coleman; p.519(tl) Norman Owen Tomalin/Bruce Coleman, (mr) Robert L. Dunne/Bruce Coleman, (bl) Jerome Wexler/Bruce Coleman; p.520 Michael Tweedie/Photo Researchers.

Chapter 21: p.524 Lois Greenfield/Bruce Coleman; p.528(tl,mr,bl) Courtesy of Carolina Biological Supply Co.; p.534(l) Eric V. Graves; p.536 Phil Degginger/Bruce Coleman; p.540 HRW Photo by Lois Ciesla-Safrani; p.542 CBS; p.544 Ward's Natural Science.

Chapter 22: p.522 Horst Schaefer/Photo Trends; p.533 Dr. E. R. Degginger; p.554(t) HRW Photo by Edith Reichmann, (b) Christopher R. Harris/Photo Trends; p.560(b) Dr. E. R. Degginger; p.562 HRW Photo by Lois Ciesla-Safrani; p.565(t) George Rodger/Magnum, (b) George Hall/Woodfin Camp; p.566(t) Soil Conservation Service, (bl) William E. Ferguson, (br) Mark Newton/Animals, Animals.

INDEX

(Note: Page numbers in **boldface** type include illustrations and those in *italic* type include definitions.)

Aaron, Henry (Hank), 154
abiotic parts of an ecosystem, *472*, *473*, **474**, **475**
absolute age (rocks), *357*–358
absolute zero, *132*
abyssal plain(s), *397*
accelerated motion, speed and, 144–148
acceleration, *147*
adaptation, environmental, 517–520
Age of Mammals, 365
Age of Reptiles, 364
air, density of, 373, 383, 384; gases in, 27, 80; as insulator, 217–218; moisture in, 376–381; particles of in sound waves, **180**; pollution of, 140, 489, 553–554; pressure of, 139, 373, 375; resistance of, 150, 218; saturated, 377; and winds, 373–375. *See also* atmosphere
air conditioning, **141**
air mass(es), 382–383, 389
air molecules, 371
air sacs in lungs, **538**
air thermometer, instructions for making, **130–131**
albino, *519*
albite, *327*
alchemists, **29**, 34
alcohol, 22
algae, *457*–458
alkali metals, 86, 87, 89, 90. *See also specific types*
Altar Stone, **401**, 402

alternating current (AC), *226*
alveoli, 538
amber, insect in, **354**
ameba, study of, 459, **460**
amperage, 228–230, 240
ampere, *228*
amphibians, 463
amplitude, wave, *170*–*171*
analogies, *66*
Andromeda, **402**
aneroid barometer, **373**
angiosperms, 465
animal cells, **448–450**, 455
Animal Kingdom, 457
animals, amphibians, 463; arthropods, 463; and carbon dioxide-oxygen cycle, 436; carnivores, *479*, 487; classification of, 462–464; cold-blooded, 463; competition among, 487, 517–518; as consumers, 478; endangered (table), 564; extinct (table), 564; habitats of, 486–487; herbivores, *479*, **479**, 487; invertebrates, 463–464; mammals, 463; many-legged, 463; migration of, *492*; omnivores, *479*, **480**, 487; predators, 481; prey, 481; vertebrates, 462–463; warm-blooded, 463; and water cycle, 435; wildlife, conservation of, **565**. *See also specific animals*
annelids, 464
Antarctica, coal found in, 337
anthracite coal, **270**
anticyclone(s), *385*
Appalachian Mountains, 313
appliances (*see* electric appliances)
aquarium, cycles in, **438**; as an ecosystem, 475

arachnids, characteristics of, 463
arc lights, 236
Archimedes, 159
Arctic Circle, 405
area, calculation of, 13
argon, 80; on moon, 412
argon atom, **79**
argon-40 atom, **57**
arthropods, 463
asexual reproduction, *502*, 506, 519
Asimov, Isaac, 439
aspirin, 27–28
association neuron(s), *545*, **546**
asteroid(s), *422*
astronauts, **18**, **429**, 471; on moon, **413–414**
astronomers, 402
athlete's foot, 458
atmosphere, of Earth, *370*–**371**, *372*–**373**; of Mars, **430–431**; of moon, 413; of Venus, **430–431**. *See also* air
atom(s), 27; argon, **79**; argon-40, **57**; carbon, **42–43**, 93, 98, 100; chemical activity of, 82–88; chemical bonding and (*see* chemical bonds); chlorine, **94**, 95; electron dot models of, 95–97; fluorine, **53**, **86**; helium, **79**, **86**, 93, 417; hydrogen (*see* hydrogen atoms); inert, *83*; iron, 34; lithium, **47**, **86**, 98; models of, 36–38, 40, 56–58; nitrogen, 44; neon, **79**, **86**; nucleus of (*see* atomic nucleus); oxygen, **33**, 39, **43**, 44, 98; radioactive, *269*; sodium, 94, 95; stable, 52; structure of, 41–45; sulfate group, 98; sulfur, 98; types of, 78–82, 94; uranium, 64, 275; valences of, 98

585

atomic bomb, **268,** 270, 271, 273–274
atomic (nuclear) energy, 64, 72, 74, 250, 268–278
atomic mass, *43–44*; arrangement of atoms by, **89,** 90
atomic mass unit, *42*
atomic models, 37–38
atomic nucleus, 39, 41–44; unstable, 269–270
atomic number, 42–43
atomic particles, 38–40. *See also* electrons; neutrons; protons
ATP, energy stored in, 536, **537,** 539
automobiles, air pollution and, 553, **554**
autumnal equinox, *403*
axis, of earth, *403–405*; of moon, *409*
azores, 397

bacteria, characteristics of, 460, **461**; decay, 437, *478–479*; diseases caused by, 460; nitrogen-fixing, *437–438*; uses for, 460
balanced equation, 100–103
ball-and-socket joint, **526**
balloon, hot air, **118,** 369
bar magnet, **214**
barium, symbol for, 49
batholith(s), 333, 342, 344
batteries, 219, **224, 232**; dry cell, 221–**222,** 226, 227, **228,** 234
bauxite, **324**
bears, 493
beryllium, 92; atomic data for, 48
bicycle pump, **128**
bighorn sheep, study of **2–3,** 4–8
bile, 533
bile duct, **532,** 533
bimetallic thermometer, **124**
binding energy, *52,* 269, **269**
biomass fuels, *283*
biosphere, *429–430*

biotic parts of an ecosystem, *472,* **472,** 474, **475**
birds, 463, 486, 487, 493
bituminous coal, 269–**270**
Black Hills (South Dakota), **314**
blackout, New York City (1977), **239**
bladder, urinary, **541**
blood, 533, **538,** 540, 541
blood cells, **538,** 540
blood types, inheritance of, 512–513 (table)
blood vessels, **538,** 540, 541
blue light, 198–199
bobcats, 487
Bohr, Niels, 57
boiling point, 23, 122, *139*
bone, *526*; *See also* human skeleton
boron, atomic number of (table), 48; electron shells for (table), 48; symbol for, 48
bracket fungi, **458**
breathing, 538
breathing capacity, measurement of, 538–539
breeder reactors, 277–278
bricks, lifting of, **157**
British Thermal Units (BTU's), *259*
bromine, symbol for, 49
bronchi, 538
bronchial tubes, **538**
bronchioles, **538**
bryophytes, 464
burning, as chemical change, **32**
butterflies, 478

Cajon Pass, 315–316
calcite, **327**
calories, **125**–126
cameras, **201**–209
canyon wall, **351**
carbohydrates, digestion of, 533
carbon, 30, 95; atomic number of (table), 44; electron shells for (table), 47; symbol for, 49
carbon atoms, 42–**43,** 93, 98, 100

carbon dioxide, 100, 375; importance of to life, 430, 432; and oxygen, exchange of, **538;** and respiration, 537, 539
carbon dioxide-oxygen cycle, 436–437
cardiac muscle, 527
careers, in electricity, **236–237;** in heating, **140–141;** in life science, **470, 500;** weather forecasting, **392**
carnivores, *479,* 487. *See also* food chain(s)
cartilage, *526*
casts, fossil, **355**
cattle, 485, 487
cell(s), *442;* animal, 448–450, **455;** blood, **538,** 540; cheek, examination of, 451–452; cork, **442, 443,** 453; egg, **502;** elodea, examination of, **454**–455; and factory, comparison of, 447–450; muscle, 527; nerve (*see* neurons); nucleus of, *448,* **453;** onion, examination of, 452, **453;** plant, 453–455; reproductive, **502;** respiration in, 450; sperm, **502;** structure and functions of, 448–450
cell body of neurons, 543
cell division, *513*
cell membrane, *448,* **448,** 450, **453**
cell theory, 443–444
cell wall, **453, 453**
Celsius temperature scale, 122, **123,** 126, 131, **132**
Cenozoic era, 362, **365** features of (table), 359
centimeter, *11*
centipedes, 463
chain reaction, nuclear, 53–54, 275
charcoal burning, **99,** 100
cheek cell, examination of, 451–452
chemical activity of atoms, 82–88

chemical bond(s), 93–94, 95–99; covalent, 95; ionic, 95
chemical change, 32
chemical digestion, 531, 533 See also digestion
chemical energy, 72, 74
chemical equation(s), 100–103, 536
chemical families, **85,** 88–92
chemical reactions, 99, 100–103
chemical sediments, 337
chemical symbols, **46**–47
chloride, 394
chlorine atom, **94,** 95
chlorophyll, 454, 457, 458
chloroplasts, **453,** 454
Christmas tree lights, **233**
chromatin, 449, **453,** 455
chromosomes, 502, **502**–503, 508–**509,** 512, 513–**514**
chromosphere, 418–419
cinder cones, **300**–301
circuit breakers, **241**
cirrus clouds, 378–**379,** 384
clams, 464
classes of living things, 457
classification, 456–468; of animals, 462–464; genes and, 456; levels of, 457; Linnaeus's system, 456; of plants, 464–467; of protists, 457–**458, 459**–460
clay, 336, 343; red, 397
cleavage of minerals, 326
climbing of stairs, **155**–156
cloud seeding, **392**
clouds, 378; cirrus, 378–**379,** 384; cumulus, 378, **379;** stratocumulus, 379; stratus, 378; **379,** 384
coal, 62, 74, 251, 252, 269–270; in Antarctica, 337; limited supply of, 74, 271; as sedimentary rock, 337
coal fields, **271**
coelenterates, 464
cold-blooded animals, 463
cold front (weather), **383**–384
color, light and, 194–200

color blindness, 514
color mutation, **520**
Colorado River, 321, 351
colors of minerals, **325**
Columbia Plateau, **314, 332**
comets, **423**
common cold, 461
communities, 484–490; changing physical factors in, 486; ecological niches in, 486–488; field, 486–487; forest, **485,** 486; habitats in, 486–488; human, 487–489; pond, 485; types of, 485
compass, 230, **231,** 232, 244, **245**
composite volcano, 301
compound(s), 27–**28,** 29; breakdown of, **30**–34; nitrogen, 436–437
compound machines, 161
concave lens, 208, **208**–209
concrete pavement, 334
condensation, 377, 436
conduction, heat transfer by, **116, 117,** 253
conductivity, 22
conductors, of electric current, 219, 221–222, 225–226; of heat, 116
conglomerate(s), **336,** 338
conifers, 465
Conservation of Energy, 65–71, 257, 259, 266, 566; Law of, 70, 250
Conservation of Matter, Law of, 100
conservation of natural resources, 564–566
consumer(s), 478, 479, **483.** See also food chain(s)
continental drift, **305**–306
continental rise, 395, **396**
continental shelf, 395, **396,** 397
continental slope(s), 395–**396,** 397
controlled experiment, 7
convection, heat transfer by, **117,** 253
convex lens, 203–204, 208–209

Copernicus, 403
copper mine, 566
coral, 464
core, of earth, 294–295; of sun, 417
Coriolis effect, 373, **374,** 398
cork cells, 442, **443,** 453
corona, of sun, 418
corundum, 327
Cousteau, Jacques Yves, 393
covalent bond, 95
craters, on Mars, **424,** 425; on moon, 412, **413**–414; volcanic, **307, 308**
crescent phase (moon), **410**
Crick, Francis, 512, 516
crocodiles, 463
crust, of earth, 294; of moon, 412
crustaceans, 463
crystal(s), 134–136; diamond, 93, 95, 134; graphite, **95;** ice, **134,** 378–380; mineral, **326,** 342; rock, **327,** 330, 333, 334; rock salt, 135, **136;** sodium chloride, 95, **134,** 135
cumulus clouds, 378, **379**
Curie, Marie, 269; and radioactivity, **52**
current(s), electric (see electric current); ocean, 397–398
cyclone(s), 384–385
cytoplasm, 448, 448, 449, **453**

Dalton, John, 26–27, 37
dandelions, adaptation of, **518**–519; leaves, 518
Darwin, Charles, 517–519
data, 8
daylight, length of, 402–403
deaf students, teaching of, **179**
Death Valley, **314**
decay bacteria, **437;** as decomposers, **478**–479
decimeter, 11
decomposer(s), 478, **478**–479 See also food chain(s)
deer, 486, 487, 490
Deimos, **424**

587

Index

density, *10*; of air, 373, 383, 384; population, *492*–*493*; of sea water, 397–399; of sun, 417; of water, 14
density currents, 397, 399
detergents, 264
dew, 379–**380**
dew point, *377*, 378
diamond(s), 93, 133, 134, 139, **327**
diamond crystal, 93, 95
diaphragm, *538*, **538**
dicots, 465
diffraction, *186*; of light waves, **186**, 189
diffraction grating, 194; use of, 198
digestion, 530, *531*–534, chemical, *531*, **532**, 533; mechanical, *531*, 535; organs of, **532**. *See also* human digestive system
digestive tract, 531
dikes (igneous rock), 333, 344
Dinosaur National Monument, **368**
dinosaurs, 355, **356**, **363**; footprints made by, **353**, 355
diphtheria, 460
direct current (DC), *226*
disease(s), air pollution and, 553; bacteria and, 460; common cold, 461; diphtheria, 460; fungi and, 458; lung cancer, 553; mumps, 461; pneumonia, 460; polio, 461; respiratory, 553; tuberculosis, 460; viruses and, 460–**461**
distance, measurement of, 145–**146**, 147, 150
doldrums, 374
dome mountains, **314**
dominant species, *491*–*492*
dominant trait, 507
Doppler effect, *175*
drafts, 262
drizzle, 380
dry cell, 221–**222**, 226, 227, **228**, 234

dry ice, 138
duct, *533*; bile, **532**, 533
ductility, 22
dunes, sand, **318**, **348**

ear, **180**
earth, air circulation on, **373**–**374**; air pressure on, **373**–**374**; atmosphere of, 370–373, **430**–**431**; biosphere of, 429–430; conditions necessary for life on, 428–433; core of, 294–295 (*see also* crustal plates); history of, 361–365 (*see also* geologic eras); environments on, 429–430, 431–432; interior of, 294–296; layers of, 291–**292**, 297; mantle of, *294*, 344; mass of (table), 419; moon's orbit around, 408–409, 415; orbit of **404**–406; orbital time of (table), 419; percent of surface covered by oceans, 393; as planet, 418–**419**; rotation of, 374, 403, **404**; rotation period (table), 419; as seen from space, **429**; solar radiation on, **372**–**373**; temperature range of, 431; wind belts of, **374**
earth science, revolution in, 290–291
earthquakes, 280, 292–296; measurement of intensity of, 292, **293**–294; turbidity currents, 397–398; waves produced by, 292–296
earthworms, 464, 486
echinoderms, 463–464
echo, **181**
echo sounding, **395**
eclipse(s), **411**
ecological niche(s), 486–488
ecology, *474*
ecosystem(s), 471–472, 473–476; abiotic parts of, *472*, 473, **474**, **475**; biotic parts of, *472*, **472**, 474, **475**; food chains in, 481–483 (*see also* food chains);

food webs in, 481; grasslands, **473**; pond, **472**, **474**, **475**; recycling in, 474; solar energy and, **473**; types of, 474. *See also* communities
Edison, Thomas A., 236
egg, *502*
egg cells, **502**
elasticity, 22
Einstein, Albert, 64
electric appliances, 240–243; cost of running, 260–261 (table); estimated annual energy consumption by, 258
electric charge(s), *213*–221; of atomic particles, 38–40; electrons and, 216, 220; like and unlike, 213, **214**, 215; negative, *213*, 221; positive, 79, *213*, 217
electric circuit(s), 221, *222*–227; amperage in, 228–230, 240; circuit breakers used in, **241**; complete, 225; fuses used in, **241**; parallel, **233**, 234, 240; parts of, 232–234; resistance in, 229–230, 240; series, 232–**233**, 234; voltage in, 228–230, 240
electric current, *217*; alternating (AC), *226*; conductors of, 219, 221–222, 225–226; direct (DC), *226*; electrons and, 217, 219, 220; lightning as, 217–**218**
electric energy, 72, 239–243, 250, 257
electric field, *216*
electric force(s), 212, *213*–221
electric generators, **246**–**247**, 250
electric home-heating system, 253
electric light bulb, 236, 240, **241**, 242
electric meter, **157**, 243
electrical ore finder, 248
electricity, 263; careers in, **236**–**237**; electric generators and production of, **246**–**247**, 250; hydroelectric power and

280; magnetism and, 244–250; measurement of, 227–234; nuclear energy and production of, 250, **276,** 277; solar cells and, 282: tidal power and generation of, 280; waste of, 566
electromagnet, **246**
electromagnetic energy, 72
electromagnetic induction, 245–250
electromagnetic levitation, 249
electromagnetic radiation, 117
electromagnetic spectrum, *187*–189, 196
electromagnetic train, **249**
electromagnetic waves, 186–*187*, 188
electron(s), 37
electron cloud, 47–**48,** 49–50
electron dot models, **95–97**
electron microscope, **450**
electron shell(s), *48*–50, 79–80, 83
electronic computer, **236**
electronic thermometer, **124**
electrons, 37–39; in conductor, 219, 221–222, 225–226; and electric charges, 216, 220; and electric current, 217, 219, 220; and lightning, 217–**218;** sharing of, 94–95, 97; stable arrangement, *80,* 86–87; transfer of, 95; valence, 98, 99
element(s), *26;* alkali metals, 86, 87, 89, 90; atomic numbers of first twelve (table), *48;* chemical families of, **85,** 88–92; halogens, 87, 89, 90; models of structure of first twelve, **49;** noble gases, 80, 81, 87, 89; periodic chart of, *91,* 92; periodic table of, **90**–92; symbols for, 48–49. *See also specific elements*
elephants, 353, **354**
ellipse, **404**
elodea cells, examination of, **454–455**

embryo, *504*
endangered animals (table), 564
endocrine glands, *548*
energy, *17;* alternative sources of, 281, **282, 283, 284, 285;** in ATP, 536, **537,** 539; binding, *269,* **269;** characteristics of, 61–65; chemical, 72, 74; conservation of, 65–71, 261, **262, 263,** 264, 566; consumption of in United States (table), **258–259;** conversion of one form into another, 63, 70, **72;** costs of, **252,** 257–258; electric, 72, 239–243, 250, 257; electromagnetic, 72; from food, 72, 74, 126; future sources of, 279–282; geothermal, 281; heat, 61, 63, 72, 74, 110–115; kinetic, 67–68, 70, 71, 168; Law of Conservation of, 70, 250; light, 61, 63; and mass, 64; and matter, 62–64; mechanical, 71–72; nuclear, 64, 72, 74, 250, 268–278; physical (*see* physical energy); population growth and consumption of, 265–268; potential, 67, 70, 227, 228, 234; radiant, 281, 417; release of, 535–539; shortage in sources of, 74, 262, 264–265; solar (*see* solar energy); and sound waves, 181; sources of, 268–272; transfer of, 63, **159, 161–164;** uses of, 71–75; waste of, 566; and water waves, **63,** 167–172; in wind, **63,** 71, 280; and work, 62, 63, 65, 157
energy crisis, 566
environment, adaptation to, *519*–521; of earth, *429,* **430, 431, 432;** pollution of (*see* pollution)
environmental factors, *429*–430
Environmental Quality Index, 562–564

environmental scientist, **411**
enzymes, 533
epicenter, *292*
equation(s), *100*–103, 536
equator, **373,** 374
equinoxes, *405;* autumnal, *403;* vernal, *403,* 406
eras in earth's history (*see* geologic eras)
erosion, *317*–322; by glacial action, **318;** gravity and, 317; on Mars, *425;* and metamorphic rock, 344; on moon, *412;* by ocean waves, 319; products of, 347; and rock layers, 350; by running water, 317–318, 335; and sand, **318;** sediments as result of, 344; of topsoil, 565–**566;** wind, **318**
esophagus, **532**
evaporation, *376,* **377;** of sea water, 394, 397; of water, 138–139, 376, 434, 435
evergreens, 465
excretion, 442, *540*–541
excretory system, human, **541**
exobiology, *432*
experiment, controlled, 7
experimental factor, 7
extinct animals (table), 564
extrusive igneous rock, 332, 344
eye, lens of, **207,** 208
eyeglasses, **202**

factory and cell, comparison of, 447–450
Fahrenheit temperature scale, 122, **123,** 126
falling object, speed of, **147**
family in classification, 457
Faraday, Michael, 248
fats, digestion of, 533
fault(s), *291,* relative age and, **350;** thrust, *313;* fault zone, **308**
fault-block mountains, 312–**313**
feces, 533

feedback, hormones and, 548–549
feldspar, **327**
Fermi, Enrico, 64
ferns, 464
fertilization, *503*
fertilizers, 438
field community, 486–487
filaments (muscle), 527, 529
fireplaces, 252–**253**
first quarter (moon), **410**, 411
fish, characteristics of, 463
fish and game laws, 565
fission (*see* nuclear fission)
flatworms, 464
floating ships, **312**
flowering plants, 465. *See also* plants
flowers, hand pollination of, **507**; natural pollination of, *503*–*504*; parts of, **503**, 505; sexual reproduction in, *503*–*504*; structure of, **503**
fluorine, atomic data for (table), 48
fluorine atom, **53**, 86
fluorite, **327**
focal length, *204*
focal point, *204*
focus of earthquake, *292*
fog, *379*, *380*
folded mountains, **313**
folding of rocks, 313
food, digestion of (*see* digestion); energy supplied by, 72, 74, 126; weights of, **12**
food chain(s), *477*, **478**–**481**; carnivores, *479*, 487; consumers, 478, 479, **483**; decomposers, 478, **478**–479; in an ecosystem, 481–483; herbivores, *479*, **479**, 487; levels in, **480**, 483–484; omnivores, *479*, **480**, 487; predator/prey relationship in, *481*, **481**
food-making process in plants, 435, **454**, **477**

food web(s), *481*, 487, 488
"fool's gold," 324
force(s), 148, *149*–154; applying to a book, **151**; balanced, 152; electric, 212, *213*–221; equal, 152, 153, 155; friction, *146*–147, 149, 153, 164; gravity (*see* gravity); magnetic, *214*–215; mass and, 150–153; opposite, 152, 153, 155; unbalanced, 152; work done by, 154–158, 161
force gauge, **156**, 157
forest community, **458**, 486
forest ranger, **470**
fossils, 347–352; casts, **350**; formation of, 349–350; molds, **350**; petrified, 349; preserved, **349**; reading record of, 350–352; trace, 351
fossil correlation, 357
fossil fuels, 268–**269**, 270, 271, 272; limited supplies of, 271–**272**
foxes, 493
fracture of minerals, 326, **327**
Franklin, Benjamin, 213
frequency, *171*; of electromagnetic waves, 187–188; of light waves, 187–188, 198–200; of sound waves, 180–181; of water waves, **171**–173, 175–**176**, 178
friction, *146*–147, 149, 153, 164
frogs, 463
fronts (weather), *383*, **383**–384, 389
frost, **380**
fruit, 465, **504**, *504*
fuel(s), 74, *251*; biomass, 281; fossil (*see* fossil fuels); heating, 251–255; synthetic, 281–**282**; uranium used as, 278
full moon, **410**, 411
Fundy, bay of, tides in, **400**
funji, 458; bracket, **458**; as decomposers, 478–479
furnace, 253

fuses, **241**
fusion, heat of, *137*; nuclear (*see* nuclear fusion)

galaxy, *402*
Galileo, 130
gall bladder, **532**, 533
gamma rays, 188
gas(es), 127–133; in air, 27, 80; liquids and, 138–139; in magma, 308, 331; on moon, 412; movement of particles in, **128**–129, 131–132; natural, 80, 251, 252, 264, 265; noble, 80, 81, 87, 89; pressure of, **128**–**129**, 308; in sun's layers, 417; temperature of, 129, 131–132. *See also specific gases*
gastric juice, 533
gear wheels, **163**, 164
gems, **326**
generating plant, **275**
generation, *506*
genes, *449*, *455*, *506*, *508*–*509*; and classification, 456; mutated, *519*
genetics, *507*, *See also* heredity
genus, 456, 457
geologic eras, 362–364
geologic time scale, 361–365
geothermal energy, **284**, *285*
geyser, **138**
giant waves, 308
giantism, 548
gibbous phase (moon), **410**, 411
giraffes, **518**, *519*
glaciers, erosion by, **318**
gland(s), 533; endocrine, *548*, **548**; salivary, 531, **532**; types and functions of (table), 549
glass, 137–138; volcanic, **331**
glucose, 536, 537, 539–541
gneiss, 343, 345
gold, 29, 34; discovery of in California (1848), 324; "fool's," 324; symbol for, 49
gold rush, **324**

Goodyear blimp, 82, 87
gram, *12*
Grand Canyon, 314, 321, 351, **362**; fossils found in, 356
Grand Teton Mountains, **313**
granite, **333**
graphite, 95; crystal of, **95**
grasslands ecosystem, **473**
gravel, 336
gravity, *19*, *66*, 150, **152**–153; and erosion, 317; of moon, 412; of sun, **152**–153, 416, 422
great plains, 314
green light, 200
green mold, **458**
green plants (*see* plants)
ground water, 434–435
growing season, **519**
growth as life process, 442
guitar string, vibration of, 179, **180**
gymnosperms, 465
gypsum, **327, 337, 338**

habitat(s), *486*–488
half-life, 358
hailstones, **380**
Halley's Comet, 423
halogens, 87, 89, 90. *See also specific types*
hardness of minerals, 326, **327**
hawks, 478, **487**
hearing, 180; importance of sense of, 179
heat, behavior of, 111; conductors of, 116; of fusion, *137*; temperature and (*see* temperature); transfer of (*see* heat transfer); of vaporization, *139*
heat energy, 61, 63, 72, 74, 110–115
heat engine, *114*
"heat fluid" theory, 111
heat transfer, 115–121; by conduction, **116**–117, 253; by convection, **117**, 253; by radiation, 117–119, 253
heating, careers in, **140**–141; home (*see* home heating systems)
helium, 46, 80, 82–**83**, 87, 412; atomic number of (table), 48; characteristics of, 83, 93; electron shells for (table), 48; on moon, 412; symbol for, 48
helium atoms, **79**, **86**, 93, 417
hematite, **324**
hemophilia, 519
herbivores, *479, 479*, 487. *See also* food chain(s)
heredity, *507*; DNA and (*see* DNA); genes and (*see* genes); Mendel's experiments and, 506–508; patterns of, 506–**507**, **508**–509. *See also* traits
hertz (Hz), *171*, 181, 187, 188
high tide, 400
highlands of moon, 413
highs (anticyclones), 385
Hindenburg explosion, **82**
hinge joint, **526**
hip joint, **527**
home heating systems, **117**, 251–255
home insulation, **140**, 253–**254**, 255
Hooke, Robert, 442, 453; microscope of, **443**
hormone(s), *548*–549; functions of (table), 549
horsepower, *155*–156
hot air balloon, **118**, 369
hot air heating system, forced, 252, **253**
hot water heating system, forced, 253, **254**
human body, systems in (*see specific systems*); water content of, 429
human communities, 487–489
human digestive system, 531, **532**, 533–**534**. *See also* digestion
human excretory system, **541**
human inheritance of traits, 511–514
human nervous system, 543, 545–546, 549. *See also* neurons
human population, changes in, 495–496; and energy consumption, 265–268; social problems caused by increase in, 489; world, estimated changes in, **496**
human respiratory system, **538**
human skeleton, **525, 526**
humidity, *376*; relative, *376*–377
hurricane(s), *388*, **388**
hybrid, 509
hydraulic press, 163, *164*
hydrocarbons, *270*–271
hydroelectric power, 275–276
hydrogen, 30, 33, 34, 82, **83**; atomic number of, 48; characteristics of, 83; electron shells for (table), 48; and solar energy, 417; symbol for, 48
hydrogen atoms, 42, 48, 96–97, 99–101; in core of sun, 417; electron dot model for, 95
hydrogen molecule, 95–97, 99
hydrogen peroxide, breakdown of, 31
hypothesis, 3–4

ice, 22–23; dry, 138
ice cap, Martian, **425**
ice crystals, **134**, 378–380
identical twins, 512, 518
igneous rock, **329**, *330*–335, 341; extrusive, 332, 344; intrusive, 332, **333**, 344
image (camera lens), *201*
impulse(s), nerve, *544*, 545, **546**
inclined plane, *161*, **163**
incomplete dominance, *512*–513
inert atom, 83
inferior mirage, **193**
infrared radiation, 117, 118
infrared waves, 187

inheritance, patterns of, 506–**507**, **508–509**
insects, 486, 487; characteristics of, 463
insulation, 116, **117, 262**; home, **140**, 253–**254**, 255
insulator(s), **116**, 219; air as, 217–218
intrusive igneous rock, 332, **333**, 344
invertebrates, 463–464; Paleozoic, 361
investigating, 5–8
involuntary muscles, *527*
iodine, symbol for, 49
ion(s), *79*, *371*; negative and positive, 95
ionic bond, 95
ionosphere, *372*
iron, 29, 34, 291–292; melting point of (table), 138; symbol for, 49
iron atoms, 34
iron pyrite, 324
Isle Royal National Park, **490–494**, 496
isostasy, *312*
isotopes, *44*

jellyfish, 464
jet stream, *371*–372
joints, bone, *526*, 527 rocks, *312*
junk car, **340**
Jupiter, *418*, **419–420**, 422; characteristics of (table), 419; moons of, **420**

kelvin temperature scale, **132**
kidneys, human, 540–541
kilocalorie, *126*
kilogram, *12*
kilometer, *11*, 145
kilowatt-hour, *259*
kilowatts, 157
kinetic energy, 67–68, 70, 71, 168
kinetic theory of matter, *128*, 132

King Kong, 65, 67
kingdoms of living things, 457
knee-jerk reflex, 547–**548**
Krakatoa, **308**
krypton, 80, 91–92

LaBrea tar pits, **354**
land section, profile of, 337–**338**, 339
land-use planning, 565
landfill, **558**
large intestine, **532**, 533
Lassen, Mount, **309**
last crescent (moon), **410**, 411
last quarter (moon), **410**, 411
lava, *300*. *See also* volcanoes
lava plateau, **314**, 331, **332**
Law of Conservation of Energy, 70, 250
Law of Conservation of Matter, 100
Law of Machines, 158
Law of Superposition, 349, 350
lead, melting point of (table), 138; symbol for, 49
Leeuwenhoek, Anton von, 443
lemmings, **492–493**
lens(es), camera, *202*, **202**–209; concave, *208*, **208–209**; convex, *203*–204, **208–209**; of eye, *207*, 208; eyeglass, 202; magnifying glass, 202–**203**; microscope, 202, 208; telescope, 202, 208
lever, *159*, **160**, 162
Lewis, Gilbert, 95
life, conditions necessary for, 428–433
life processes, 440–**442**, 444
life science, careers in, **470, 500**
ligaments, *526*
light, 185–210; blue, 198–199; and color, 194–200; green, 200, and green plants, 431–**432**; red, 196, 199–200; sunlight, 61, 63, **190, 195**, 196, 200; violet, 196, **197**; visible, 187–188; white, 195,

196, 198, 199. *See also* light rays; light waves
light energy, 61, 63
light rays, *190*; refraction (bending) of, 195, **196–197**, 201–209
light waves, 174–**175**, 185–194; diffraction of, **186**, 189; frequencies of, 187–188, 198–200; movement of, 189–194; reflection of, **186**, 189, **190**, 199–200; refraction of, **186**, 189, **192–193**; speed of, 186–187, 192
lighting, **263**
lightning, 217–**218**
lignite, 269–**270**
limestone, **337**, 343
limonite, **324**
Linnaeus, Carolus, classification system of, 456
liquids, 137; boiling point of, 122, *139*; evaporation of, 138–139; and gases, 138–139; movement of particles in, **128**, 138–139. *See also* water
lithium, 47; atomic number of (table), 48; electron shells for (table), 48; symbol for, 48; valence of, 98
lithium atoms, **47**, **86**, 98
lithium oxide, 98
lithosphere, *294*
liver, human, **532**, 533
liverworts, 464
lizards, 463
load (running water), 335
lodestone, **327**
longitudinal wave, *169*–**170**
low tide, 400
lows (cyclones), 384–**385**
lungs, air sacs of, **538**; cancer of, 553
luster of minerals, **326**
lynx, 493–494

M31 Galaxy (Andromeda), **402**
machines, compound, 161, **163**;

perpetual motion, **45**; simple, *158*–161

magma, *299*–300; and igneous rocks, *329*–334, 344

magnesium 93; atomic number of (table), 48; electron shells for (table), 48; symbol for, 48

magnetic field, *216*, **217**, 245–247, 250

magnetic force(s), *214*–215

magnetic pole(s), **214**–*215*

magnetism, 72; and electricity, 244–250

magnetite, **324**

magnets, 214–216; bar, **214**; electromagnet, **246**

magnifying glass, 202–**203**

malleability, 22

mammals, characteristics of, 463

mantle, of earth, *294*, 344; of moon, 412

marble, **343**, 345

Maria of moon, 412

marine biologist, **470**

mars, 418, **419**, 422; atmosphere of, **430**–431; characteristics of (table), 419; temperature of, 431

mass, *18*; atomic, *51*–55; energy and, 64; and force, 150–153

matter, *17*; energy and, 62–64; general properties of, 16–19, 22, 23, 127; kinetic theory of, *128*, 132; Law of Conservation of, 100; special properties of, 19–20. *See also* gas(es); liquids; solids

measurement, of area, 13; of distance, 145–**146**, 147, 150; of electricity, 227–234; of energy use, 259–261; metric system, 10–**11**, 12, 145, 157; of speed, 145–147, 150; of temperature (*see* thermometers); of volume, 13, 14

mechanical advantage, *158*

mechanical digestion, *531*–**532** 535 *See also* digestion

mechanical energy, 71–72

medicines, 264

melting point(s), 23, *137*–138

Mendel, Gregor, 506–508

Mendelev, Dmitri, 89–91

mercury, melting point of (table), 138; symbol for, 48

Mercury (planet), 418, **419**; characteristics of (table), 419

mesosphere, 372

Mesozoic era, 362, **364**, 365

metals, alkali, 86, 87, 89, 90; change in state of, **30**–31; as conductors of heat, 116; properties of, **19**. *See also specific types*

metamorphic, *341*

metamorphic rock, 329, 330, **340**–*341*, 342–345

meteorites, *414*, 423

meteoroids, *423*

meteorology, 378

meteors, *423*

meter, *11*

meters, electric, **260**; gas, **260**; reading of, **260**

metric system, 10–**11**, 12, 145, 157

microscopes, **203**, 208; electron, **450**

microwaves, 187

midgets, 548

Mid-Atlantic Ridge, **306**

mid-ocean ridges, **306**–307, 314, 397

migration, *492*

mildew, 458

millimeter, *11*

millipedes, 463

minerals, 324; cleavage of, 326; colors of, **325**; crystals of, **326**, 342; fracture of, 326, **327**; gem, **326**; hardness of, 326, **327**; luster of, **326**; streak of, **325**. *See also specific minerals*

mirages, 192–**193**

mirror, reflection of light from, **190**

Mississippi River, **318**, 335

mitochondria, *449*–450, **453**, 455, 536, **537**, 539

mixture(s), *31*–32

model(s), 38

mold(s), 458; green, **458**; penicillium, 458

molecule(s), 27. *See also specific molecules*

mollusks, characteristics of, 464

monadnock, **319**

money, metric units and (table), 12

monocots, 465

moon, 408–415; age of, 413; astronauts on, 413–**414**; atmosphere, 413; axis of, 409; craters on, 412, **413**–414; crust of, 412; dust on surface of, **412**; eclipse of, **411**; erosion and, 412; formation of, 409; gases on, 412; gravity force on, 412; highlands on, 413; history of, 414; mantle of, 412; maria of, 412; orbit of, 408–409, 415; phases of, **409**, **410**, 411; rays on, **413**, 414; rock samples from, **408**, 412–413; rotation of, 409; surface features of, **412**–413, 414; temperatures on, 412; and tides, 415; volcanoes on, 413

moonquakes, 414

moose, 490, **491**–495

mosses, 464

moth balls, 138

motion, *145*; accelerated, speed and, 144–148; First Law of, Newton's, 149; forces and (*see* force); Second Law of, Newton's, 151–152; Third Law of, Newton's, 152–153

motor neurons, 545, **546**

Mount St. Helens, eruption of, **301**–302

mountains, 290; dome, **314**; fault-block, 312–**313**; folded, **313**

Index

movable joint, 526
movement, *442*
mucus, 532–533
mud, 336, 337, 347
mumps, 461
muscle, cardiac, **528**; smooth, **528**; skeletal, **528**
muscular system, **527–528**, 529
mushrooms, 458
music, 181
mutation(s), *516–517*

natural community (human), 487–488
natural gas, 80, *270–271*
natural resources, *563*; conservation of, 564–566; nonrenewable, 564; renewable, 564, 566
natural selection, 516, *517–521*
nematodes, 464
neon, 79–80; atomic number of (table), 48; electron shells for (table), 48; on moon, 412; symbol for, 48
neon atom, **79**, **86**
neptune, 418, **419**, **421**
nerve fibers, 543, 544
nervous system, human, 543, 545–546, 549. *See also* neurons
neuron(s), 543, **544**, 549; association, 545, **546**; cell body of, 543; motor, 545, **546**; sensory, 545, **546**; typical, **544**
neutrons, 43, 269–**270**, 271; and atomic mass, 51, 53, 55; atomic mass of, 51
new moon, **410**
New York City blackout (1977), **239**
Newton, Isaac, **149**, 151–153, **196**
newtons, *155*
nickel, 291–292
nitrates, 436–437
nitrogen, in atmosphere (table), 370; atomic number of (table), 48; electron shells for (table),

48; importance of to life, 430, 432, 436; symbol for, 48
nitrogen compounds, 436–437
nitrogen cycle, 435, **436–438**
nitrogen-fixing bacteria, *437–438*
noble gases, 80, 81, 87, 89. *See also specific types*
noise, *181*
nonvascular plants, *464*
nuclear energy, 64, 72, 74, 250, 268–278; released by sun, 417
nuclear power; 275–276, 277
nuclear reactions, chain, 271–272; controlled, 275–278; fission, 53–54, 270–271, 274, 278; fusion, 272–**273**, 274
nuclear reactors, 53–54, 277–278
nucleolus, *448*, **453**
nucleoplasm, 448–449, **453**
nucleus, of atom (*see* atomic nucleus); cell, *448*, **453**
nursing aides, **500**

observation(s), 2–5
obsidian, **331**
ocean bottom (*see* sea floor)
ocean waves (*see* water waves)
oceanauts, 393
oceanographers, 394–395
oceans, beginning of, 360; bottom of (*see* sea floor); currents in, 397–398; depth of, method of finding, 305; percent of earth's surface covered by, 393; tides in, 399–400. *See also* sea water
octopus, 464
Oersted, Hans Christian, 248–249
ohm, *230*
Ohm's Law, 230
oil (*see* Petroleum)
oil spills, **554**
Olympus Mons, 418
omnivores, *479*, **480**, 487. *See also* food chain(s)
onion cells, examination of, 452, **453**
ooze, 397

optical properties, 23
orbit, of earth, **404–406**; of moon, 408–409, 415
orders of living things, 457
organic sediments, 337
organism, *429*; structures of, **525**
organs, 525
oxygen, 30, 33, 34; in atmosphere (table), 370, 430; atomic mass of, 51; atomic number of, 48; and carbon dioxide, exchange of, **538**; chemical test for, **32**; electron shells for (table), 48; importance of to life, 430, 432; properties of, 31–32; and respiration, 537, 539; symbol for, 48; valence of, 98
oxygen atoms, **33**, 39, 98, 100; protons in, **43**, 44
oxygen molecules, 27, 100
oxygen nucleus, **51**
oysters, 464
ozone, *371*

Paleozoic era, 362–**363**, 364
pancreas, **532**, 533
pancreatic juice, 533
pangaea, 364–365
paracutin, **309**
Paradichlorobenzene (PDB), use of, **330–331**
parallel circuit, **233**, 234, 240
paramecium, asexual reproduction in, **502**; study of, 459–**460**
paramedics, **500**
pea plant, inheritance patterns of, **507–508**
peat, 269–**270**
pendulum, **69–70**
peneplain, **319**
penicillium, 458
peppered moth, adaptation of, **520**
periodic chart, *91*, 92
periodic table, **90–92**

perpetual motion machine, **45**
petrified tree, **355**
petrify, 355
petroleum, 74, *270–271*, 566. See also fossil fuels
phobos, 424
photosphere, 417
photosynthesis, 473
phylum, 457
physical change, *30–32*
physical energy, 154–164; food and, 72, 74, 126; simple machines and, 158–161
pistons, **163**
pitch, sound and, *175*
pituitary hormone, 548, 549
planets, 418–419, **420–421**; characteristics of (table), 419; inner, 418; life on other, 439–440; outer, 418. See also *specific planets*
plant cells, **453–455**
plant kingdom, 457
plants, angiosperms, 465; bryophytes, 464; classification of, 464–467; dicots, 465; flowering, 465; food-making process in, 435, 454, **477**; gymnosperms, 465; light and, 431–**432**; monocots, 465; nonvascular, *464*; photosynthesis and, 473; as producers, 473, 486; seed, 464–465; solar energy and growth of, 431–432, **477**; tracheophytes, 464; vascular, *464*–465; and water cycle, 435–436
plasma, 540
plastic, **136–138**, 264, 566
plate tetonics, **307, 308, 309**
plateau(s), 313–*314*; lava, 314, 331, **332**
Pluto, 418, **419, 421**
pneumonia, 460
polar easterlies, 375
polar front, **384**, *384*
polio, 461
pollination, *503*–**504**; hand, **507**

pollutants, *553*
pollution, 552–557; air, 140, 489, 553–554; Environmental Quality Index and, 562–564; solid waste, **552,** 557–559, 561; water, 489, **554,** 560
pond community, 485
pond ecosystem, **472,** 474, **475**
population(s), *491*; changes in, 490–497; density of, *492–493*; dominant species in, *491*–492; human (*see* human population); limiting factors in, 493–495
population crash, 492, 496
population cycle, **494**
potential energy, *67*, *70*, 227, 228, 234
power, *156*
power plants, and air pollution, 553–554; nuclear, 277–278
Precambrian time(s), 361–**362**, 363
precipitation, types of, 380
predator, *481*
predator/prey cycle, 493
predator/prey relationship, *481*, **481**
Presidential Mountain Range, 319
prey, *481*
prism(s), *195*, 196, **197**, 200, **203**
producer(s), *478*, 486. See also food chains
protein molecules, 540
proteins, digestion of, 533; DNA and, 516
Protista Kingdom, 457
protists, 457–**458,** 459–**460**
protons, 39, 217, 220, **269**
protoplasm, *443*
protozoans, 458–460
pulleys, *160*–**161**, 162
pumice, 331
pumped storage, **277**
Punnett Square, **509,** 510
pure (genetically), *507*
pure water, 27

quartz, **325, 327**
quartzite, **343**

rabbits, 478, 486
radar, 187
radiant energy, 281, *417*
radiation, electromagnetic, 117; heat transfer by, 117–**119**, 253; infrared, 117, 118; mutations caused by, **519**; ultraviolet, 118, 371
radiators, **262**
radio waves, 72, 187, 371
radioactive wastes, disposal of, 277–**278**
radioactivity, 52, 269, 277
radium, 269
radon, 80
rain, 380, 430, 434, 473
rainbow, **194,** 196, 200
recessive trait, *507*
red clay, 397
red light, 196, 199–200
reference point, *145*, 147
reflection, *175*; of light waves, **186,** 189, **190,** 199–200; of sound waves, **181**; of water waves, 173, **175,** 176
reflex(es), *545*, 549; knee-jerk, demonstration of, 546–**547**
refraction, *174*; of light rays, 195, **196**–197, 201–209; of light waves, **186,** 189, **192–193**; of water waves, *174–175*
relative age, *356–357*
relative humidity, *376–377*
reproduction, *442*, 501–506; asexual, *502*, **502,** 506, 519; sexual (*see* sexual reproduction)
reproductive cells, 508–**509**
Reptiles, characteristics of, 463
resistance, air, *155*, 218; in electric circuits, *229–230*, 240
respiration, 442, *536*; carbon dioxide and, 537, 539; in cell, 450; oxygen and, 537, 539;

process of, equation showing, 537
respiratory system, human, **538**
response, *442*
Reston (Virginia), **484**
reusable materials, recycling of, 433, 558, 566
ribosomes, *450*, 455
Richter Scale, 293
ridge, formation of, **308**. See also mid-ocean ridges
rift, *306*
ringworm, 458
ripple tank, **174, 177**
rock candy, **134**
rock collection, **328**–329
rock crystals, **327**, 330, 333, 334
rock cycle, *344*, **344**, 345
rock salt, 337; crystals of, 135, **136**
rocks, faults in (*see* faults); folding of, 313; igneous, **329**, 330–335, 341; joints in, **312**; layers of, 347, **348**–349, 350–352; metamorphic, 329, 330, **340**–*341*, 342–345; moon, **408**, 412–413; Precambrian, 360–361; sedimentary, 329, 330, **336**–339, 341, 344; unconformity of, 350, 352
roller coaster, **69**
rope, waves in a, **169, 171, 175**
roundworms, 464
rubber, 264
rubidium, 91–92
Rumford's cannon boring experiment, **111**–112
rural community, **488**
rusting, **32**
Ruth, Babe, 154

salamanders, 463, **486**
salinity, *394*; of sea water, 394, 397
saliva, **531**–532
salivary glands, 531, **532**
salt, table (*see* sodium chloride)
salt solutions, 30

San Andreas fault, **291**
sand, 26, 336, 337, 347
sand dollars, 464
sand dunes, **318**; layers in, **348**
sandstone, **336, 338**, 343
sandworms, 464
saturated air, 377
saturn, 418, **419, 420**, 422–423; characteristics of (table), 419; moons of, **420**
schist, 345
science fiction, 439
Scientific Law, 7
scientific model(s), **38**
scientific problem solving, steps in, **8**
scientific thinking, 1–5
scorpions, 463
screw, **163**
sea anemones, 464
sea floor, exploration of, 394–395; new, creation of, 314, 332, 397; North Atlantic Ocean, **396**; organisms of, **432**; profile of, **395**; sediments on, 397; spreading of, **306**–307; trenches on, 396
sea urchins, 464
sea water, density of, 397–399; evaporation of, 394, 397; salinity of, 394, 397
seasons, causes of, 405–406
secretion, *422*
sediment(s), 334–335, 336–340; chemical, 337; erosion and, 344; layers of, **336**, 341; organic, 337; on sea floor, 397
sedimentary rocks, 329, 330, 336–339, 341, 344
seed(s), *504*, **519**; typical, structure of, **504**
seed plants, 464–465. *See also* plants
segmented worms, 464
seismic stations, 302
seismograph, 292–**293**
selenium, symbol for, 49
sense organs, 543

sensory neuron(s), *545*, **546**
series circuit, 232–**233**, 234
sex chromosomes, 513–**514**
sex-influenced traits, *513*
sex-linked traits, 513–514
sexual reproduction, 508–**509**
shadow stick, use of, **407**–408
shale, **336, 338**, 343
sheep, 487; bighorn, study of, 2–3, 4–8
shield volcanoes, **300**
shiprock volcano, **332**
"shooting stars," 420
shrews, 487
sickle-cell anemia, **512**, 520
Sierra Nevadas, **333**
sills (igneous rock), 333
silt, 336
silver, symbol for, 49
simple machines, *158*–161
skeletal muscles, 527–**528**, 529
skeleton, *525*; human, **525, 526**
sky, color of, **194**, 195, **198**, 199
Skylab, **471**–473
slate, **343**, 345
sleet, 380
small intestine, **532**, 533
smog, 553, **553**
smooth muscles, 528
snails, 464
snakes, 463
snow, 380, 430, 434, 473
snowflakes, **134**, 380
snowshoe hares, 493–494
snowstorm (February, 1978), 376, **381**
social communities (human), **488**–489
sodium, 394; atomic number of (table), 48; electron shells for (table), 48; symbol for, 48
sodium atoms, 94, 95
sodium chloride, 95; crystals of, **95, 134**, 135; melting point of (table), 138
soil conservation, 566
solar cell(s), *283*
solar eclipse, **411**

solar energy, 61–63, **74**, 283–**284**, **285**, 371; and change in seasons, 405–406; and ecosystems, **473**; and growth of plants, 431–432, **477**; hydrogen and production of, 417; importance of to life, 431–432; and water cycle, **436**
solar home heating, **281**
solar radiation, 372
solar system, structure of, 416–423. See also planets; sun
solid waste pollution, **552**, 557–559, 561
solids, 133–137; arrangement of particles in, 133–134; melting of, **137**; melting points of, 137–138; movement of particles in, **128**, 137–138
solstice, summer, 402–403, 405, 406; winter, *403*, 405, 406
solubility, 23
solution(s), 51
sound, *177*, loudness of, *179*–181; and pitch, 181; speed of, 180; vibration and, 179, **180**
Sound waves, 175, 177–181; creation of, **178**–179; frequency and pitch, 179. See also sound
source areas, **382**
spaceships **148**, 149
species, *456*, 457
spectrum, electromagnetic, *187*–189, 196; visible, *195*, **197**
speed, accelerated motion and, 144–148; calculation of, 145–147, 150; friction and, 146–147
sperm, *502*
sperm cells, **502**
spiders, 463
sponges, 464
squid, 464

stable electron arrangement, *80*, 86–87
starches, digestion of, **531**, 532
starfish, 428, 464
steam, 23, 27, 62, 111
steam engines, **62**, **114**, 158, 275
steam heating system, 252–**253**
stem, **518**
stimulus, *543*; response to, 543, 545–546, 549
stomach, human, **532**, 533
Stonehenge, **401**–403, 406
stones, 347
Stratocumulus clouds, 379
Stratosphere, *371*–372
Stratus clouds, 378, **379**, 384
streak of minerals, **325**
sub-atomic particle, *43*
submarine canyons, 395, 398
subscripts, *100*
suburban community, **488**
sugar, breakdown of, 31; melting point of (table), 138
sulfate group (atomic), 98
sulfur atom, 98
summer solstice, 402–*403*, 405, 406
sun, 416–422; apparent movement across sky, **403**; atmosphere of, 417–418; average distance from earth (table), 419; composition of, **417**; core of, 417; density of, 417; diameter of, 416; earth's orbit around, **404**–406; eclipse of, **411**; energy from (see solar energy); gaseous layers of, 417; gravity of, **152**–153, 416, 422; mass of, 416, **417**; temperature in core of, 417
sundial, shadow-stick, 407–408
sunlight, 61, 63, **190**, **195**, 196, 200
superior mirage, **193**
Superposition, Law of, 355–**356**
survival of the fittest, 518
symbols, chemical, 48
synchrotron, **39**

synthetic fuels (synfuels), 281–**282**
systems, 525

table salt (see sodium chloride)
talc, **327**
tapeworms, 464
taproot, **518**
telephone lineperson, **237**
telescopes, **202**, 208
television newsroom, **542**
television repair, careers in, 237
temperature(s), 121, **122**–126; absolute zero, *132*; in atmosphere (table), 370; boiling point, 122, *139*; calories and, 125–126; Celsius scale, 122, **123**, 126, 131, **132**; in core of earth, 292; Fahrenheit scale, 122, **123**, 126; of gases, 129, 131–132; Kelvin scale, **132**; measurement of (see thermometers); melting points, 137–138
tendons, **527**, **528**, 530
terrarium, cycles in, **438**
theory, 7
thermometers, **122**–123, 126; air, instructions for making, **130**–131; bimetallic, **124**; electronic, **124**
thermosphere, *372*
thermostats, **262**
Thompson, Benjamin, 111
Thrust fault, *313*, **313**
tidal power, 280, **284**
tide(s), 399–400; moon and, 415
tin, symbol for, 49
tissue, *525*
toads, 463
topaz, **327**
topsoil, erosion of, 565, **566**
tornadoes, 388–389
trace fossils, 355
trachea, 537–**538**
tracheophytes, 464
trade winds, 374

trait(s), *506;* dominant, *507;* heredity and, 506–508; recessive, *507*
transformers, electrical, **240**
transverse wave, *169,* **171**
trash (solid waste) pollution, **552,** 557–559, 561
trenches (sea floor), **308**
troposphere, *371*
tsunamis, 299
tuberculosis, 460
tuff, 331
tug of war, 151–152
turbidity currents, 397–398
turtles, 463
twins, identical, 512, 518
tycho (lunar crater), **413**

ultraviolet radiation, 118, 371
ultraviolet waves, 188
unbalanced forces, 152
unconformity of rock, 357
uniformitarianism, 355–356
universe, *402*
unstable atomic nucleus, 269–**270**
uranium, 277; limited amount of, 278; use of as fuel, 278
uranium atoms, 64, 275
uranium pellets, **277**
uranium-235, fission of, 53, **270,** 271
uranium-238, 270, 271
uranus, 418, **419, 421;** characteristics of (table), 419
urban community, **488**
ureter, **541**
urethra, **541**
urinary bladder, **541**

vacuoles, *449,* **453**
valence(s), *98*
valence electrons, 98, 99
valleys, **319**
vaporization, heat of, *139*
vascular plants, *464–465.* See *also* plants
vents, volcanic, **307**

venus, 418, **419;** atmosphere of **430**–431; characteristics of (table), 419; temperature of, 431
vernal equinox, *403,* 406
Verne, Jules, 439
vertebrates, 462–463; Paleozoic, 361, 362
vibration, sound and, 179, **180**
violet light, 196, **197**
viruses, 460–**461**
visible light, 187–188
visible spectrum, *195,* **197**
voice box, **538**
volcanic ash, 300
volcanic bombs, 300
volcanic dust, 300
volcanic eruptions, 298–302
volcanoes, 298–**299,** *300,* 301–302. See also *specific types.*
volt, *228*
voltage, 228–230, 240
volume, *10,* 18; measurement of, 13, 14; units of, 14
voluntary muscles, 527
Voyager spacecraft, 422–423

Wallace, Alfred, 517
warm-blooded animals, 463
warm front (weather), **383,** 384
wastes, metabolic, excretion of, 533, 540–541; radioactive, disposal of, 277–278; solid, pollution by, **552,** 557–559, 561
water, boiling point of, 139; breakdown of, **30,** 33, 34; conservation, 263; density of, 14; depletion of earth's supply, 436–438; on earth, 430–431; equation for formation of, 100; evaporation of, 138–139, 376, 434–435; ground, 434–435; in human body, 429; importance of to life, 429; particles of, 168–**169,** 180; pollution of, 489, **554,** 560; pure, 27; running,

erosion by, 317–318, 335; three forms of, **22**–23
water cycle, **435**–436
water molecules, 21–24, 27, 33, 112–114
water power, *280,* **280**
water vapor, 30, 375, 376
water waves, **167**–175, 399–400; behavior of, 173–178; crests of, **169,** 170, 171, 400; energy and, **63,** 168–171; erosion by, 319; frequency of, **171**–173, 175–**176,** 178; movement of particles in, 168–**169;** reflection of, 173, **175, 176;** refraction of, 173, **174, 177;** speed of, 173, 178; troughs of, **169,** 170, 400; tsunamis, 299
waterfall, **71**
Watson, James, 512, 516
watt(s), *156,* 242–243
Watt, James, and horsepower, 154–**155**
watt-hours, 243
wave(s), *168,* 169, **170;** earthquake 292–296; electromagnetic, 186–187, 188; infrared, 187; light (see light waves); microwaves, 187; ocean (see water waves); radio, 72, 187, 371; in a rope, **169, 171,** 175; sound (see sound waves); in a spring, 169, **170;** ultraviolet, 188; water (see water waves)
wavelength, *170,* **171**
weasels, 493
weather forecasting, 381; careers in, **392**
weather fronts, *383,* **383**–**384,** 389
weather map, interpretation of, 386, **387**
weathering, *317*
wedge, *161,* **163**
Wegener, Alfred, and continental drift, **305**–306
weight, *18,* 150, 155
Wells, H. G., 439

westerlies, 374, 375
wheel and axle, *160*, **163**
white light, 195, 196, 198, 199
wildlife, conservation of, **565**
Wilkins, Maurice, 512, 516
wind(s), earth's rotation and, 374; energy in, **63,** 71, 280; erosion by, **318;** movement of air and, 373–375; and ocean currents, 398; polar easterlies, 375; trade, 374; westerlies, 374, 375

wind belts, **374**
wind generators, **276**
windmills, **274**–275
winter solstice, *403*, 405, 406
winter storm (February, 1978), **376,** 381
Wisconsin Tokamak Fusion Reactor, **274**
wolves, 493–**495**
work, *62*; energy and, 62, 63, 65, 157; forces and, *155*–156; formula for, 155

world population, estimated changes in, **496**
worms, segmented, 464. *See also specific types*

Xrays, 72, 185, 188, 189
xenon, 80

yeast, 458

zincite, **324**

Index 599